Lecture Notes in Computer Science   9885

Commenced Publication in 1973
Founding and Former Series Editors:
Gerhard Goos, Juris Hartmanis, and Jan van Leeuwen

More information about this series at http://www.springer.com/series/7409

Jeff Z. Pan · Diego Calvanese
Thomas Eiter · Ian Horrocks
Michael Kifer · Fangzhen Lin
Yuting Zhao (Eds.)

Reasoning Web

Logical Foundation of Knowledge Graph Construction and Query Answering

12th International Summer School 2016
Aberdeen, UK, September 5–9, 2016
Tutorial Lectures

 Springer

Editors
Jeff Z. Pan
University of Aberdeen
Aberdeen
UK

Michael Kifer
Stony Brook University
Stony Brook, NY
USA

Diego Calvanese
Free University of Bozen-Bolzano
Bolzano
Italy

Fangzhen Lin
University of Science and Technology
Hong Kong
SAR China

Thomas Eiter
University of Technology Vienna
Vienna
Austria

Yuting Zhao
University of Aberdeen
Aberdeen
UK

Ian Horrocks
University of Oxford
Oxford
UK

ISSN 0302-9743   ISSN 1611-3349   (electronic)
Lecture Notes in Computer Science
ISBN 978-3-319-49492-0   ISBN 978-3-319-49493-7   (eBook)
DOI 10.1007/978-3-319-49493-7

Library of Congress Control Number: 2017931056

LNCS Sublibrary: SL3 – Information Systems and Applications, incl. Internet/Web, and HCI

Printed on acid-free paper

This Springer imprint is published by Springer Nature
The registered company is Springer International Publishing AG
The registered company address is: Gewerbestrasse 11, 6330 Cham, Switzerland

Preface

This volume collects some lecture notes from the 12th Reasoning Web Summer School (RW 2016), which was held during September 5–9, 2016, in Aberdeen, UK.

The Reasoning Web series of annual summer schools has become the prime educational event in the field of reasoning techniques on the Web, attracting both young and established researchers, since its first initiation in 2005 by the European Network of Excellence REWERSE. As with previous editions, this year's summer school was co-located with the 10th International Conference on Web Reasoning and Rule Systems (RR 2016). For many years, the lecture notes of the Reasoning Web Summer School have been the top choices for new PhD students working on Semantic Web and KR to understand the state of the art in the field. Some even regard the RW lecture notes as a yearly updated version of handbooks for description logics (and related work). The 2016 edition of the school was organized by the University of Aberdeen, with support from the TeamAberdeen.

In 2016, the theme of the school was "Logical Foundation of Knowledge Graph Construction and Query Answering." The notion of knowledge graph has become popular since Google started to use it to improve its search engine in 2012. The idea of knowledge graph comes from an early knowledge representation mechanism called "semantic networks." RDF (resource description framework) is a modern version of semantic networks standardized by W3C, adding many of the limitations of classic semantic networks. OWL (Web Ontology Language) is a W3C standard for defining ontologies as rich schema of RDF graphs. OWL chooses description logics as underpinnings, and provides two levels of implementability:

- OWL 2 DL is the most expressive yet decidable language in the OWL (version 2) family.
- The three profile languages OWL2-QL, OWL2-EL, and OWL2-RL are tractable sub-languages of OWL2-DL.

One key feature of OWL is that it allows for faithful approximate reasoning services, by approximate OWL2-DL ontologies to those in its tractable sub-languages. This opens new doors for the possibility of handling large ontologies efficiently. A well-known example of such faithful approximate reasoners is the TrOWL reasoner, which even outperforms some well-known sound-and-complete reasoners in time-constrained competitions designed for sound-and-complete ontology reasoners.

The aim of the lecture notes is to provide a logical foundation for constructing and querying knowledge graphs. Our journey starts from the introduction of knowledge graph as well as its history, and the construction of knowledge graphs by considering both explicit and implicit author intentions (Chapter 1), where explicit intentions can be useful to generate reasoning-based authoring tests, so as to control the quality and costs of knowledge graph construction, while implicit intentions can help indicate, e.g., how reasoners can be optimized so as to better support knowledge graph authors. We continue to discuss an important notion of inseparability, in Chapter 2, for constructing, revising, and reusing ontologies in a safe manner. Some of the notions of inseparability are also useful for ontology testing.

Now, given good-quality knowledge graphs, what the key aspects do we need to consider for querying them? We first explain, in Chapter 3, how to combine navigational queries, which are popular among graph databases, with basic pattern-matching queries, for SPARQL and beyond. Secondly, in Chapter 4, we introduce an infrastructure for allowing researchers to run experiments with linked data sets by querying, accessing, analyzing, and manipulating hundreds of thousands of linked data sets available online, so that researchers can worry less about some engineering issues related to data collection, quality, accessibility, scalability, availability, and findability. However, it is not always possible to clean up your knowledge graphs before querying them. Thus, in Chapter 5, we look into the problem of inconsistency-tolerant query answering over DL knowledge bases, explaining computational properties and reasoning techniques for various options of inconsistency-tolerant semantics. In addition, in Chapter 6, we provide a comprehensive survey on representation and reasoning with fuzzy RDF and OWL knowledge bases. In Chapter 7, we look into the knowledge graph construction and querying aspects together, by combining machine learning and reasoning in deployed applications, including some smart city applications. These chapters have been written as accompanying material for the students of the summer school. We hope they will be useful for readers who want to know more about the logical foundations of constructing and querying knowledge graphs.

Presentation slides of all the chapters of this book can be found at http://www.abdn. ac.uk/events/rr-2016/rw-summer-school-2016/programme/.

We thank everyone who made Aberdeen's summer school and these lecture notes possible. First and foremost, we thank the presenters of these lectures and their co-authors. We would like to thank our sponsors: Accenture Centre for Innovation, National Science Foundation (NSF), and EU FP7/Marie Curie IAPP project K-Drive. Furthermore, we would like to thank the local organization team at the University of Aberdeen: Martin Kollingbaum, Wamberto Vasconcelos, Diana Zee, and Nicola Pearce. Last but not least, we would like to thank the team of RR 2016, Umberto Straccia, Magdalena Ortiz, Stefan Schlobach, Rafael Peñaloza, Adila Krisnadhi, and Giorgos Stamou, for the great collaboration in putting together all the details of the two events.

September 2016

Jeff Z. Pan
Diego Calvanese
Thomas Eiter
Ian Horrocks
Michael Kifer
Fangzhen Lin
Yuting Zhao

About the Editors

Jeff Z. Pan is a Reader (Professor) at University of Aberdeen. He is the Chief Scientist of the EC Marie Curie K-Drive project and the Chief Editor of the book of 'Exploiting Linked Data and Knowledge Graphs in Large Organisations'. He is known for his work on knowledge construction, reasoning and exploitation.

Diego Calvanese is a Full Professor at Free University of Bozen-Bolzano. He is one of the editors of the Description Logic Handbook and a member of the Editorial Board of JAIR and Big Data Research. He has been nominated EurAI Fellow in 2015.

Thomas Eiter is a Full Professor at TU Wien. He is an expert on Knowledge Representation and Non-monotonic Logic Programming. He is a EurAI Fellow, a Corresponding Member of the Austrian Academy of Sciences (ÖAW) and a Member of the Academia Europea.

Ian Horrocks is a Full Professor at Oxford. He chaired the W3C OWL WG and is an Editor-in-Chief of the Journal of Web Semantics. He won the BCS Roger Needham award and is a Fellow of the Royal Society, a Fellow of EurAI and a Fellow of the Academia Europea. He is the Scientific Coordinator of SIRIUS.

Michael Kifer is a Full Professor at Stony Brook University. He won the ACM SIGMOD Test of Time Award twice for the work on object-oriented query languages and the work on F-Logic. He won 20-year Test of Time Award from the Association for Logic Programming for the work on Transaction Logic.

Fangzhen Lin is a Full Professor at Hong Kong University of Science and Technology. He was awarded the Croucher Senior Research Fellowship (2006) and won many best paper awards in top Knowledge Representation and Artificial Intelligence conferences including KR06, KR08, IJCAI97, and AIPS2000. He was an Associate Editor of AI Journal and JAIR.

Yuting Zhao is a Senior Researcher at the Center for Advanced Studies at IBM Italy. Before that, he was a Research Fellow in Professor Jeff Z. Pan's group at University of Aberdeen. He co-edited two books with Professor Pan on Semantic Web and Software Engineering, including 'Ontology Driven Software Development'.

Organization

Organizing Chair

Jeff Z. Pan   University of Aberdeen, UK

Scientific Advisory Committee

Diego Calvanese   Free University of Bozen-Bolzano, Italy
Thomas Eiter   Vienna University of Technology, Austria
Ian Horrocks   Oxford University, UK
Michael Kifer   Stony Brook University, USA
Fangzhen Lin   Hong Kong University of Science and Technology,
                                 Hong Kong, SAR China

Local Organizing Committee

Wamberto Vasconcelos   University of Aberdeen, UK
Martin Kollingbaum   University of Aberdeen, UK
Diana Zee   University of Aberdeen, UK
Nicola Pearce   University of Aberdeen, UK

Additional Reviewer

Mantas, Simkus

Collaborative Organizations

University of Aberdeen
TeamAberdeen
Visit Scotland
Visit Aberdcenshire
ORACLE
Expert System
IBM
Free University of Bozen-Bulzano
Institute of Information Science and Technologies (ISTI)
Wright State University
National Technical University of Athens

TU Wien
VU University Amsterdam
Rule ML

Sponsors

Accenture Centre for Innovation

National Science Foundation (NSF)

EU FP7/Marie Curie IAPP project K-Drive

Contents

List of Contributors

Wouter Beek Department of Computer Science, VU University Amsterdam, Amsterdam, The Netherlands

Meghyn Bienvenu LIRMM - CNRS, Inria, and Université de Montpellier, Montpellier, France

Fernando Bobillo Department of Computer Science and Systems Engineering, Universidad de Zaragoza, Zaragoza, Spain

Elena Botoeva Faculty of Computer Science, Free University of Bozen-Bolzano, Bolzano, Italy

Camille Bourgaux LRI - CNRS and Université Paris-Sud, Orsay Cedex, France

Filip Ilievski Department of Computer Science, VU University Amsterdam, Amsterdam, The Netherlands

Caroline Jay School of Computer Science, University of Manchester, Manchester, UK

Boris Konev Department of Computer Science, University of Liverpool, Liverpool, UK

Freddy Lecue Accenture Technology Labs, Dublin, Ireland;
Inria, Rocquencourt, France

Carsten Lutz Faculty of Informatics, University of Bremen, Bremen, Germany

Nico Matentzoglu School of Computer Science, University of Manchester, Manchester, UK

Jeff Z. Pan Department of Computing Science, University of Aberdeen, Aberdeen, UK

Juan L. Reutter Escuela de Ingeniería, Pontificia Universidad Católica de Chile, Santiago, Chile; Center for Semantic Web Research, Santiago, Chile

Laurens Rietveld Department of Computer Science, VU University Amsterdam, Amsterdam, The Netherlands

Vladislav Ryzhikov Faculty of Computer Science, Free University of Bozen-Bolzano, Bolzano, Italy

Stefan Schlobach Department of Computer Science, VU University Amsterdam, Amsterdam, The Netherlands

Umberto Straccia ISTI - CNR, Area Della Ricerca di Pisa, Pisa, Italy

Markel Vigo School of Computer Science, University of Manchester, Manchester, UK

Domagoj Vrgoč Escuela de Ingeniería, Pontificia Universidad Católica de Chile, Santiago, Chile; Center for Semantic Web Research, Santiago, Chile

Frank Wolter Department of Computer Science, University of Liverpool, Liverpool, UK

Michael Zakharyaschev Department of Computer Science and Information Systems, Birkbeck University of London, London, UK

Yuting Zhao Department of Computing Science, University of Aberdeen, Aberdeen, UK

Understanding Author Intentions: Test Driven Knowledge Graph Construction

Jeff Z. Pan[1](✉), Nico Matentzoglu[2], Caroline Jay[2], Markel Vigo[2], and Yuting Zhao[1]

[1] Department of Computing Science, University of Aberdeen, Aberdeen, UK
{jeff.z.pan,yuting.zhao}@abdn.ac.uk
[2] School of Computer Science, University of Manchester, Manchester, UK
matentzn@cs.man.ac.uk, {caroline.jay,markel.vigo}@manchester.ac.uk

Abstract. This chapter presents some state of the arts techniques on understanding authors' intentions during the knowledge graph construction process. In addition, we provide the reader with an overview of the book, as well as a brief introduction of the history and the concept of Knowledge Graph.

We will introduce the notions of explicit author intention and implicit author intention, discuss some approaches for understanding each type of author intentions and show how such understanding can be used in reasoning-based test-driven knowledge graph construction and can help design guidelines for bulk editing, efficient reasoning and increased situational awareness. We will discuss extensively the implications of test driven knowledge graph construction to ontology reasoning.

1 Introduction

Knowledge Graph has become popular in knowledge representation and knowledge management applications widely across search engine, biomedical, media and industrial domains [1] in recent years. In 2012 Google popularised the term *Knowledge Graph* (KG) with a blog post titled 'Introducing the Knowledge Graph: things, not strings',[1] while simultaneously applying the approach in their core business, fundamentally in the web search area. Google's Knowledge Graph is an entity-based knowledge base used for improving search engine's search results. It was added to Google's search engine in 2012 and displays results in a 'Knowledge Panel' (see, e.g., Fig. 1), providing structured and detailed information about the search entity in addition to a list of links to related sites. The idea is that users would make use of such information to resolve their query without having to navigate to other sites and collect the information themselves.

Inspired by the successful story of Google, knowledge graphs are gaining momentum in the World Wide Web arena. Recent years have witnessed increasing industrial take-ups by other internet giants, including Facebook's Open

---

[1] https://googleblog.blogspot.co.uk/2012/05/introducing-knowledge-graph-things-no
t.html.

© Springer International Publishing AG 2017
J.Z. Pan et al. (Eds.): Reasoning Web 2016, LNCS 9885, pp. 1–26, 2017.
DOI: 10.1007/978-3-319-49493-7_1

Fig. 1. Knowledge panel in Google's search results

Graph and Microsoft's Satori. Furthermore, there have been industrial research efforts, e.g. Knowledge Vault, community-driven events (Knowledge Graph Tutorial in WWW2015 and JIST2015[2]), academia-industry collaborations, and the establishment of start-ups specialised in the area, such as Diffbot and Syapse. All these initiatives, both in academic and industrial environments, have further developed and extended the initial Knowledge Graph concept popularised by Google. Additional features, new insights, and various applications have been introduced and, as a consequence, the notion of knowledge graphs has grown into a much broader term that encapsulates a whole line of community efforts on its own right, new methods and technologies.

Knowledge Graph is *not* really a brand new concept. The basic ideas were proposed in the knowledge representation formalism called Semantic Networks. There is a modern W3C standard for semantic networks, called RDF (Resource Description Framework),[3] which is used together with the W3C standard Web Ontology Language OWL[4], with Description Logics [2] as its underpinning, for defining schema for RDF graphs. OWL ontologies have been widely accepted in many application domains. For example, the well known SNOMED CT (Systematised Nomenclature of Medicine-Clinical Terms) [3] ontology has been regarded the most comprehensive, multilingual clinical healthcare terminology in the world. We will explain briefly about the history of Knowledge Graph in Sect. 2.

---

[2] http://www.slideshare.net/jeffpan_sw/linked-data-and-knowledge-graphs-constructing-and-understanding-knowledge-graphs.

[3] https://www.w3.org/TR/2004/REC-rdf-primer-20040210/.

[4] https://www.w3.org/TR/owl2-overview/.

Knowledge graphs can be seem as RDF/OWL graphs in the Big Data era. For example, the Google Knowledge Graph contains (in 2012) over 500 million objects and 3.5 billion facts about and relationships among these objects. While many NoSQL solution providers are still arguing about whether schema-less or schema-rich solutions are more useful, knowledge graphs exploit the 20 years of research from the Semantic Web community and offer schema-less, simple schema and rich schema solutions all under the same RDF/OWL framework. We will discuss more about what Knowledge Graph is and its connections to RDF, OWL and big data in Sect. 3.

In this book, we will focus on Knowledge Graph construction and query answering. Knowledge Graph construction is highly relevant to ontology authoring. Despite the usefulness and wide acceptance of ontologies, ontology authoring remains a challenging task. Existing studies on ontology authoring, such as OWL Pizzas [4] and those carried out in the NeOn project [5], suggest that ontology formalisms are in general not easily understandable by ontology authors and logical consequences can be quite difficult to resolve, since, in many situations, ontology authors are usually domain experts but not necessarily proficient in logic. Knowledge Graph query answering can be used not only in deployment time for exploiting knowledge graphs, but also in design time as a useful way to test the constructed knowledge graphs to check their qualities.

The key research question that we address in this chapter is how to understand the intentions of authors of knowledge graphs. *Author intention* is a mental state that represents a commitment to carrying out knowledge graph construction activities. Intention is closely related to but different from purpose: *purpose* of knowledge graph construction is the reason of authoring an ontology. In a sense, while both are related to the scope of the to be constructed ontology, purpose is more high level, and intention can be more fine-grained, subtle and even dynamics. For a non-trivial knowledge graphs, the actual purpose of the knowledge graphs might not be easily established up front, one might need to properly understand the author intentions, before we could, at a later stage, sum up the purpose accordingly.

We distinguish two types of intentions. The first type is *explicit intention*, referring to intentions that are indicated by explicit requirement statements from ontology authors. The second type is *implicit intention*, referring to intentions that are implicitly inferred by authoring and construction activities of authors of knowledge graphs. Accordingly, we will introduce, in Sect. 4, some techniques for understanding explicit intentions via the so called competency questions. We will then address, in Sect. 5, how to understand implicit intentions through user behaviour patterns. Understanding author intention is only one aspect, we will also need to help authors of knowledge graph to understand the consequences of their changes to their target knowledge graph, so see if the changes match their intention. Thus in Sect. 6, we discuss how to design tools to help authors of knowledge graphs to better understand consequences of their authoring activities.

2   A Brief History of Knowledge Graph

As a graphic way to present knowledge about entities and the relationships among entities, Knowledge Graph inherit from many classic Knowledge Representation formalisms, in particular such as *Semantic Networks*. In 2004, researchers in *Knowledge Representation and Reasoning* (KR) addressed some well known issues for Semantic Networks when standardising the modern version of Semantic Network, or RDF (Resource Description Framework) graph, which uses *Web Ontology Language* (OWL) to define its schema. OWL offers different levels of expressive power to modellers to define light-weight schemas or heavy schema for their knowledge graphs.

In this section, we will briefly introduce a few knowledge representation mechanisms that are relevant to Knowledge Graph, including Semantic Network, Frame, KL-ONE, RDF and OWL.

Semantic Networks

Semantic Networks [6–8] was firstly proposed by Quillian [9] to analysis the meaning of word concepts and the organization of human semantic memory. In Semantic Networks, knowledge is represented in the forms of graphs, in which nodes represent entities, concepts or situations, while arcs represent the relations between them, such as *is-a* (e.g., "a seagull is a kind of bird") and *part-of* (e.g., "wings are part of a bild"). After Quillian, many variants of semantic networks were proposed. Although Semantic Networks is flexible and is easy to trace associations in the (knowledge) graphs (see e.g., Fig. 2), there are no formal syntax and semantics for Quillian's semantic network, which makes semantic networks hard to be scalable and to be understandable by machines and programs. Furthermore, semantic networks do not allow users to define the meaning of labels on nodes and arcs, resulting in the constructed semantic networks to be ambiguous and hard to be contextualised.

Frame

Frame was proposed by Minsky [10] in order to 'represent knowledge as collections of separate, simple fragments.' Each of such record-like fragment is called slots. A frame consists of some class slots and entity slots (see Fig. 3). Each slot can be seen as a shallow knowledge graph about some concept or entity.

Frame addressed some limitations of Semantic Network. Firstly, it distinguishes language level relationships, such as (subclass-of, instance-of), from user defined relationships, such as part-of. Secondly, Frame not only allows to use atomic classes but also class expressions by using slot constraints, such as intersections, unions and negations. Furthermore, it was given a formal semantics based on first-order predicate logic [11]. However, there has not been a standard Frame language until the related OWL standards were set up in 2004 (version 1) and 2009 (version 2).

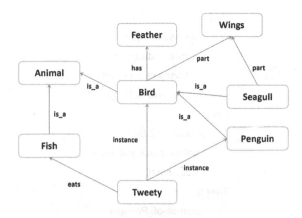

Fig. 2. An example of Semantic Networks.

Another feature of Frame is the support of default knowledge. For example, when talking about penguin, people might often think of the kind of black and white cute animal on snow-land. Thus black and white can be set as the default colour for the Penguin class. The default value can be overwritten in entity slots; in this example, we find "Tweety" as a baby penguin has grey colour.

KL-ONE

KL-ONE [12] is probably one of the most well known Knowledge Representation and Frame systems. It is widely recognised as being the first to use a deductive classifier, which is an automatic reason for computing subsumption relations among named classes (Frames in KL-ONE are called concepts), in a Frame system. This kind of reasoning service is called Classification.

Another important notion introduced in KL-ONE is "defined concepts", which are different from primitive concepts that are not fully defined. KL-ONE provides constructors for the users to build complex "defined concepts". It is important to highlight that class hierarchies from the Classification service are rather different from the class hierarchies in classic Frame systems, since the ones in classic Frame system are asserted class hierarchies, while the ones from Classification service are inferred class hierarchies, based on definitions of classes. Later, these kind of KR languages that KL-ONE once supported are called Description Logics [2], which are chosen to be the underpinnings of the W3C OWL standard. Most chapters in this book are relevant to Description Logics (DLs). Please refer to Chap. ? for a detailed and formal introduction of Description Logics.

RDF, OWL and Linked Data

Resource Description Framework (RDF) is the modern standard for Semantic Networks. It is a Recommendation (standard) from the *World Wide Web Consortium* (W3C), for describing entities, often referred to as resources in RDF.

Bird

      subclass-of: Animal

      warm_blooded: yes

      member-slot: has-part

        value-class: Wing

Penguin

      subclass-of: Bird

      * colour: black and white

      * size: small

Tweety:

      instance-of: Penguin

      colour: grey

Fig. 3. An example of Frame.

An RDF resource can be anything we can identify, such as a person, a shop or a product. Like Frame, RDF addresses some of the issues related to classic Semantic Networks, in terms of the lack of formal syntax and semantics.

An RDF statement is of the form 'subject property object.' We could visualise an RDF statement as two nodes (annotated with the subject and object) connected with an arc (annotated with the property). Accordingly, a set of RDF statement is called an RDF graph. In addition to entity level statements, RDF also provides a simple schema language called RDF Schema (RDFS), including four language level properties rdfs:subClassOf, rdfs:subPropertyOf, rdfs:domain and rdfs:range. For example, the *subclass-of* relation can be represented by the *rdfs:subClassOf* property in RDF, the semantics of which is clearly defined in the RDF specifications [13]. Having said that, RDF only provides limited expressive power (compared to OWL). For example, only 4 statements in the Frame mentioned in Fig. 3 can be represented in RDF as follows (OWL is needed for representing local attributes and properties, which will be shown later on in this section):

- Bird rdfs:subClassOf Animal.
- Penguin rdfs:subClassOf Bird.
- Tweety rdf:type Penguin.
- Tweety colour grey.

It should be pointed out that RDF does not address all the limitations of Semantic Networks, *e.g.* RDF does not allow users to have class expressions that Frame supports. This is, however, addressed by OWL, a Description Logic based W3C standard for defining vocabularies for RDF graphs. OWL provides a much wider selections of class constructors to build class expressions (called concept descriptions in Description Logics) than those that are allowed in many Frame systems, including the support of datatypes [14].

Since OWL also supports RDF syntax, an OWL ontology can be viewed as an RDF graph too. The OWL family consists of an expressive language OWL 2 DL, as well as three tractable profile (sub-languages) OWL 2 EL, OWL 2 QL and OWL2 RL. Therefore, RDF/OWL graphs can be schema-less (with vocabulary only), with light-weight schema (with PTIME-complete or even less computational complexity in RDFS,[5] OWL 2 EL, OWL 2 QL or OWL 2 RL), or with a rich schema (with OWL 2 DL).

A Description Logic knowledge base, also called an ontology, consists of a TBox (a set of schema axioms) and an ABox (a set of data axioms, or assertions). For example, most statements in the Frame mentioned in Fig. 3 (except the two statements representing default knowledge[6] about colour and size) can be represented in Description Logic.[7] Axioms 1–4 are schema axioms and axioms 5–6 are assertions.

1. Bird $\sqsubseteq$ Animal
2. Bird $\sqsubseteq$ $\exists$ warm_blooded.{yes}
3. Bird $\sqsubseteq$ $\exists$ has-part.Wing
4. Penguin $\sqsubseteq$ Bird
5. Penguin (Tweety)
6. Colour (Tweety, grey)

There are some standard reasoning services for RDF/OWL graphs, including:

- *Ontology Consistency Checking*: to check if the given RDF/OWL graph is consistent.
- *Classification*: to compute the subsumption relation between named classes.
- *Class Satisfiability Checking*: to check if a given class expression is satisfiable (meaning it is possible for such a class to have instances).
- *Class Subsumption Checking*: to check if there is a subsumption relation between two given class expressions.
- *Realisation/Materialisation*: to compute all the inferred instances of all the named classes and named properties.

Based on RDF and OWL, *Linked Data* is a common framework to publish and share data across different applications and domains, where RDF provides a graph-based data model to describe entities, while OWL offers a standard way of defining vocabularies for data annotations. In the Linked Data paradigm, RDF graphs can be linked together by means of mappings, including schema-level mapping (*rdfs:subClassOf*) and entity-level mappings (*owl:sameAs*). Indeed, with billions of logical statements, the Linked Open Data (LOD) is one of the biggest knowledge graphs ever built. Just like SQL for relational databases, SPARQL is the standard query language for RDF/OWL graphs, supporting both conjunctive queries and navigational queries. In Chap. 7, both kinds of queries will be explained and studied in details.

---

[5] Without the use of blank notes (https://www.w3.org/TR/rdf11-concepts/#section-blank-nodes).

[6] Although there are extensions of OWL for representing default knowledge, which is not included in the current version of OWL.

[7] cf. Chap. 2 for an introduction of the DL syntax.

3   Knowledge Graphs: Ontologies in the Big Data Era

As discussed in the previous section, although the term 'Knowledge Graph' became well known in 2012 when Google started to use knowledge graph in their search engine, the idea of Knowledge Graph has been developed since Semantic Networks. Over the years, Knowledge Graph becomes a technological integration of using graphic structure to present knowledge, applying logics (e.g. Description Logic) to regular the semantics, and empowering the knowledge management with efficient search algorithms and logical reasonings.

In general, a knowledge graph can be seen as an ontology with a entity centric view, consisting of a set of interconnected typed entities and their attributes, as well as an some schema axioms for defining the vocabulary (terminologies) used in the knowledge graph. The basic unit of a knowledge graph is (the representation of) a singular entity, such as a football match you are watching, a city you will visit soon or anything you would like to describe. Each entity might have various attributes; e.g., the attributes of a person include name, birthdate, and family information. Furthermore, entities are connected to each other by relations; e.g., some Person entities like some Product entities, which are on sale in some Shop entities.

Given that both linked data and knowledge graphs can be seen as RDF/OWL graphs,[8] one might wonder the differences between linked data and knowledge graphs. They are indeed very similar. Linked data emphasis more on the linkage aspect, hence requiring the use of owl:sameAs, while knowledge graphs emphasis more on the entity aspects, including the expectations that each entity has a type (so that one know about the context) and that entities can have local attributes (such as warm_blooded) and local properties (such as has-part).

Compared to ontologies and standard RDF/OWL graphs, the distinctive features of Knowledge Graph lie in its special combination of knowledge representation structures, data management processes, and search algorithms. Inspired by the success story of Google and the other enterprise applications [1], knowledge graphs are gaining momentum in the world's leading information companies and more, such as Facebook, Twitter, BestBuy, Guardian and Skyscanner. Within the first week that Facebook announced their Open Graph API, more than 50,000 web sites signed up for it.

Indeed, knowledge graphs can be seem as RDF/OWL graphs in the Big Data era. Indeed, knowledge graphs are often not small knowledge bases. In additional to the Google Knowledge Graph, there are well known knowledge graphs, such as YAGO [15], DBpedia [16] and NELL [17], containing millions of nodes and billions of edges. This means that special care is required to deal with such huge knowledge bases.

- *Knowledge Graph Reasoning*: Tractable reasoning services is needed to deal with large-scale knowledge graphs. The OWL family provides a few tractable

---

[8] In fact, knowledge graphs can be written in some proprietary syntax, as long as such syntax can be mapped to RDF and OWL.

profiles, including OWL 2 QL [18], OWL 2 EL [19] and OWL 2 RL. RDF becomes RDF-DL if we (1) disallow the use of blank nodes in the property position of any RDF triples, and (2) disallow the use of RDF vocabulary in the subject any object positions in an RDF triples. RDF-DL is a sub-language of OWL 2 QL. RDFS-FA [20] can be seen as a metamodeling extension of RDF-DL. Furthermore, strategies such as approximate reasoning[9] [21–26] and divide and conquer [27–35] have been shown to be helpful.

- *Knowledge Graph Completion*: The problem here is to take an existing knowledge graph as input and learn new facts (such as new types of entities, or new relationships between entities) about the knowledge graph. One could use some information extraction methods to collect candidate facts for enriching a knowledge graph. For example, such methods might return a fact claiming that Theresa May was born in Egypt, and suppose her birthplace was not yet included in the knowledge graph. Some Statistical Relational Learning model [36–38] might use related facts about May (such as her profession being a UK Priminister) to infer that this new fact is unlikely to be true and should be discarded. There are also works on learning schema [39–41] of knowledge graphs, so as to help bridge the gap between queries and data. In Sect. 4, we will introduce test driven approaches to checking the quality of candidate facts and axioms of knowledge graphs. Furthermore, in order to represent vague facts [42–44] or even uncertain facts [45], Chap. 6 will introduce the notion of fuzzy ontologies/knowledge graphs [46].
- *Knowledge Graph Processing*: There are other processes about knowledge graph other than reasoning and completion. In Chap. 4, an experimental platform will be presented for researchers and practitioners to design and carry out experiments on linked data and knowledge graphs on aspects including data collection, quality, accessibility, scalability, availability and findability. Furthermore, Chap. 7 will address how to use reasoning and learning together [47,48] to handle large scale knowledge discovery problems in the context of transportation in cities of Bologna, Dublin, Miami, Rio and spend optimisation in finance. In this context, stream reasoning services[10] [49–52] are ofter consider, in particular in the scenarios [47,48] of smart cities and mobile computing [53,56,57].

4   Understanding Explicit Intentions Through Authoring Tests and Competency Questions

In the previous sections, we have introduced Knowledge Graph and its history. In this section, we will investigate explicit author intentions, which are often indicated by authoring tests and requirements. While authoring tests (cf. Sect. 4.1)

---

[9] http://www.slideshare.net/jeffpan_sw/the-rise-of-approximate-ontology-reasoning-is-it-mainstream-yet.

[10] http://www.slideshare.net/jeffpan_sw/the-maze-of-deletion-in-ontology-stream-reasoning.

are more formal and thus directly relevant to reasoning and querying services, requirement documents (cf. Sect. 4.2) are often less formal and thus easier for authors to generate.

4.1 Test Driven Knowledge Graph Construction

The idea of Test Driven Knowledge Graph Construction (TDKGC) comes from the notion of Test Driven Software Development (TDSD), where tests are specified before the target software is implemented, and the improvement of the software is to pass tests, only. Similarly, in TDKGC, authoring tests are specified before the target knowledge graph is constructed, and the quality of the knowledge graph is guaranteed by making sure such authoring tests are passed. Many believe that writing authoring tests before the knowledge graph does not take any more efforts than writing them after the knowledge graph. TDKGC actually forces authors to think about requirements before constructing their knowledge graphs. On the other hand, writing authoring tests first will help authors to detect and remove errors sooner.

An authoring test is represented in the form of a query and answer pair $T = \langle q, a \rangle$, where q is a test query and a is the expected answer of q. In what follows, we discuss a few important notions related to test queries.

Queries for Inseparable Knowledge Graphs. The notion of inseparability of two ontologies indicates that inseparable ontologies can safely be replaced by each other for the tests we consider [54,55,58–60]. More precisely, given the current version of a knowledge Graph $\mathbf{K}_c$ and a sequence of revised versions of the knowledge graph $\mathbf{K}_{r_1}, ... \mathbf{K}_{r_n}$, these knowledge graphs are not separable w.r.t. a test query q if and only if they give the same answer for q. In other words, we want to make sure that, if the current version $\mathbf{K}_c$ passes the test query q, no matter how we further revise the knowledge graph to address further test queries, the revised knowledge graph should always pass the test query q. There are a few typical kinds of test queries in this contexts: subsumption queries, conjunctive queries and navigational queries. For example, when we consider the Pizza Ontology, a test subsumption query can be that 'AmericanHot is (always) a sub-class of SpicyPizza'.

In Chap. 2, more details will be provided for inseparable ontologies/ knowledge graphs. It should be noted that there can be different notions of inseparability. For example, in terms of subsumption, there are (1) subsumption inseparability and (2) concept hierarchy inseparability, where subsumption inseparability means that a test subsumption query should hold for all $\mathbf{K}_{r_1}, ... \mathbf{K}_{r_n}$, while concept hierarchy inseparability means that the concept hierarchy of all the $\mathbf{K}_{r_1}, ... \mathbf{K}_{r_n}$ should be the same as that of $\mathbf{K}_c$.

Navigational Queries. A knowledge graph is not only an ontology (knowledge base), it is also a graph. It is natural to consider navigational queries, in addition the usual subsumption queries and conjunctive queries [61–63]. The most common way to add navigation into graph queries is to start with a conjunctive query language and augment it with navigational primitives based on regular expressions. In the SPARQL query language, navigational queries are known as property paths. For example, the following query asks for all individuals in the friend-of-a-friend social network of Alice:

$$q(x, y) :\text{-} \text{foaf:know}(Alice, x), \text{foaf:know}^+(x, y)$$

In Chap. 3, more details of navigational queries will be presented. It should be noted that, compared to subsumption queries, it often takes more efforts to prepare the expected answers of navigational queries and conjunctive queries.
Inconsistency Tolerant Querying. Sometimes we might have to handle inconsistent knowledge graphs. In most cases, the top priority is to resolve the inconsistencies in knowledge graphs. On the other hand, there are situations where we have to live with inconsistent knowledge graphs, particularly when we do not have control about some imported sub-graphs from other sources, and disconnecting such sub-graphs might cause more problems. In such situations, probably the wise thing to do is to see how to get meaningful answers from inconsistent knowledge graphs; this is called *inconsistency tolerant querying* [64–66].

In Chap. 5, further investigations of inconsistency tolerant querying will be presented. Arguably one of the most natural, inconsistency-tolerant semantics is to consider query answers shared by all the maximal consistent sub-graphs of an inconsistent knowledge graph. It should be noted that we have many choices here, in terms of choosing the inconsistency tolerant semantics, some of which are inspired by earlier work on consistent query answering in relational databases.

Authoring tests are widely used in ontology authoring systems to check the quality of target ontologies. Well known systems, such as OWL Unit Test Framework in Protégé, Tawny-OWL [67] and OntoStudio[11], allow users to define authoring tests and run them in the authoring environment. Some systems, such as the Rabbit interface [68] and the Simplified English prototype [69], offer syntactic checking, including incorrect words or disallowed input. Other systems, such as Protégé[12] and OntoTrack [70], use reasoners to perform semantic checking such as inconsistency checking. Some systems, such as Roo [71], inform users the potential authoring consequences. It is believed that justification engine [70] are useful to explain why certain feedbacks are provided.

However, it should be noted that, generic tests, such as consistency and input validity, do not capture the requirements that are specific to the ontologies in question. On the other hand, customised author-defined tests allow the expression of such requirements but require further knowledge and skills of ontology technologies.

---

[11] http://www.semafora-systems.com/en/products/ontostudio/.
[12] http://protege.stanford.edu.

4.2   Competency Question Driven Knowledge Graph Construction

While the idea of Test Driven Knowledge Graph Construction is promising, there are three key challenges: (1) For novice authors, designing a test suite for a knowledge graph is hardly easier than designing the knowledge graph itself; (2) If one does not start with authoring tests, where do the test queries come from? (3) How to produce correct expected answer for test queries?

In Test Driven Software Development, tests are generated from requirements. Thus, this suggest a possible solution of (2) — in Test Driven Knowledge Graph Construction, authoring tests can be from requirements as well. As requirement documents tend to be more informal than authoring tests, they help to address the above challenge (1). In what follows, we will introduce a solution based on a special kind of requirement, competency questions. In doing so, we will introduce some key notions that are relevant to competency questions. We will comment on the challenge (3) at the end of this section.

What Forms of Requirements Should We Consider? We could start with some requirement formats that people have been using for knowledge representation and engineering. One good candidate would be competency questions. A competency question (CQ) [72] is a natural language sentence that expresses a pattern for a type of question people want to be able to answer with a knowledge graph/knowledge base. For example, in a knowledge graph about mobile apps, in order to answer a question "Which process implements a given algorithm?", the knowledge graph should contain concepts *Process* and *Algorithm*, and their instances should be able to have a relation called *Implements*.

In ontology engineering, competency questions (CQs) are already used a means of evaluating ontologies by posing queries that use ontology terms, relations, and attributes [72]. The NeOn methodology [73], for example, describes methods of producing competency questions from ontology author. Fernandes et al. [74] proposed a visual solution based on a goal-based methodology for capturing competency questions. There are existing implementations for transforming competency questions into SPARQL queries [76] and into DL queries [77]. In short, competency questions are considered to be a good starting point for constructing test queries.

How are Such Requirements Formulated? Now, one of the key questions is whether there exist some patterns of constructing competency questions. This is an important question. If we understand how real-world competency questions are formulated, we could extract features from competency questions for generating test queries. We could also provide an interface, following such patterns, for users to produce and edit their competency questions.

Ren et al. [75] studied two corpus of competency questions and concluded that there are clear patterns among these competency questions. They further tested such patterns against competency questions from existing literatures and confirmed that all but one competency questions are compatible with the identified patterns. Table 1 summarises the high level patterns from the study of Ren et al., where some core features, such as Predicate Arity (PA), Relation Type (RT), Modifier (M) and Domain-independent Element (DE), and elements, such as classes, properties and individuals, can be used to identify CQ patterns.

Example 1. Consider the CQ "Which pizza has some cheese topping?" There are three elements here: *pizza* as a class element [CE1], *has* as an object property element [OPE] and *cheese topping* as a second class element [CE2]. Since 'has' is a binary predicate, the above CQ belongs to pattern 1.

How to Generate Test Queries from Such Requirements? Now that we understand that competency questions have patterns, we can extract features and elements from them and construct test queries. Instead of just translating competency questions directly into SPARQL queries [76] or DL queries [77], it might be more interesting to look into the answerability problem competency questions.

This is relevant to the notion of presupposition, which refers to a special condition that must be met for a linguistic expression to be meaningful [78]. For example, the question "Did you meet Kevin when you walked your dog last Friday?" presupposes that the addressee has a dog and walked the dog last Friday. Although the question was about whether the addressee met Kevin, the question become meaningless if the two conditions are not satisfied.

Example 2. Consider the CQ "Which processes implement an algorithm?" There are three elements here: *Process* as a class element [CE1], *implement* as an object property element [OPE] and *Algorithm* as a second class element [CE2]. Since 'implement' is a binary predicate, the above CQ belongs to pattern 1.

In order to meaningfully answer the above CQ, it is necessary for the ontology to satisfy the following presuppositions:

(a) Firstly, [CE1], [OPE] and [CE2] should occur in the knowledge graph; i.e. Classes *Process*, *Algorithm* and property *implements* should occur in the ontology;
(b) The ontology allows the possibility of *Process*es implementing *Algorithm*s;
(c) The ontology allows the possibility of *Process*es *not* implementing *Algorithm*s.

Table 2 summarises the findings of Ren et al. [75] in terms of the kinds test queries that can be generated from competency questions, using the notion of presupposition. In Example 2, the test (a) is a occurrence test, and tests (b) and (c) are Relation Satisfiability tests.

Let us conclude this section by pointing out a few key points about Table 2, in terms of the implications on reasoning:

• Three are two types of reasoning tasks involved in the generated test queries: class satisfiability checking and class subsumption checking. Both of them are boolean queries. As discussed in the beginning of Sect. 4.2 of this Chapter, one of the challenges (3) is how to produce the correct answers for test queries.

Table 1. Archetypes of Competency questions (from [75]): PA = Predicate Arity, RT = Relation Type, M = Modifier, DE = Domain-independent Element; obj. = object property relation, data. = datatype property relation, num. = numeric modifier, quan. = quantitative modifier, tem. = temporal element, spa. = spatial element; CE = class expression, OPE = object property expression, DP = datatype property, I = individual, NM = numeric modifier, PE = property expression, QM = quantity modifier)

ID	Pattern	Example	PA	RT	M	DE
1	Which [CE1] [OPE] [CE2]?	Which pizzas contain pork?	2	obj.		
2	How much does [CE] [DP]?	How much does Margherita Pizza weigh?	2	data.		
3	What type of [CE] is [I]?	What type of software (API, Desktop application etc.) is it?	1			
4	Is the [CE1] [CE2]?	Is the software open source development?	2			
5	What [CE] has the [NM] [DP]?	What pizza has the lowest price?	2	data.	num.	
6	What is the [NM] [CE1] to [OPE] [CE2]?	What is the best/fastest/most robust software to read/edit this data?	3	both	num.	
7	Where do I [OPE] [CE]?	Where do I get updates?	2	obj.		spa.
8	Which are [CE]?	Which are gluten free bases?	1			
9	When did/was [CE] [PE]?	When was the 1.0 version released?	2	data.		tem.
10	What [CE1] do I need to [OPE] [CE2]?	What hardware do I need to run this software?	3	obj.		
11	Which [CE1] [OPE] [QM] [CE2]?	Which pizza has the most toppings?	2	obj.	quan.	
12	Do [CE1] have [QM] values of [DP]?	Do pizzas have different values of size?	2	data.	quan.	

Comparing to non-boolean queries, one advantage of boolean queries is that it is easier to get the expected correct answers. Indeed, since the idea is that all presuppositions should be satisfied, all the class satisfiability and subsumption tests are expected to be true.

- In terms of expressive power, we need the $\mathcal{ALCQ}$. It should be noted that, there is a nice *pay-as-you-go* property for EL ontologies, since (1) all the cardinality related presuppositions are satisfied, and (2) class satisfiability of $(CE1 \sqcap \neg \exists P.E2)$[13] can be reduced to non-subsumption checking of $CE1 \sqsubseteq \exists P.E2$.

---

[13] Full negation is beyond EL.

Table 2. Authoring Tests (from [75]): ($\sqcap$ means conjunction, $\neg$ means negation, $\exists P.E$ means having P relation to some E, $= nP.E$ ($\geq nP.E, \leq nP.E$) means having P relation(s) to exactly (at least, at most) n E(s), $\forall P.E$ means having P relation (if any) to only E, $\top$ means everything)

AT	Parameter	Realisation
Occurrence	[E]	E in ontology vocabulary
Class Satisfiability	[CE]	CE is satisfiable
Relation Satisfiability	[CE1]	$CE1 \sqcap \exists PE2$ is satisfiable, $CE1 \sqcap \neg \exists PE2$ is satisfiable
	[P]	
	[E2]	
Meta-Instance	[E1]	$E1$ has type $E2$
	[E2]	
Cardinality satisfiability	[CE1]	$CE1 \sqcap = nP.E2$ is satisfiable, $CE1 \sqcap \neg = nP.E2$ is satisfiable
	[n]	
	[P]	
	[E2]	
Multiple cardinality (on concrete quantity modifier)	[CE1]	$\forall n \geq 0$, $CE1 \sqcap \neg = nP.E2$ is satisfiable
	[P]	
	[E2]	
Comparative cardinality (on quantity modifier)	[CE1]	$\exists n \geq 0$, $CE1 \sqcap \neg \leq n\ P.E2$ and $CE1 \sqcap \geq n+1\ P.E2$ are satisfiable, $\exists m \geq 0$, $CE1 \sqcap \neg \leq m\ P.E2$ and $CE2 \sqcap \geq m\ P.E2$ are satisfiable
	[P]	
	[E2]	
Multiple value (on superlative numeric modifier)	[CE1]	$\forall D \subseteq range(P)$, $CE1 \sqcap \neg \exists PD$ is satisfiable
	[P]	
Range	[P]	$\top \sqsubseteq \forall P.E$
	[E]	

5 Understanding Implicit Intentions Through User Behaviour Patterns

In the previous section, we focus on understand author intentions by addressing user requirements and turning them into authoring tests. In this session, we will look into studying user behaviours in order to understand their implicit intentions. As indicated by the previous section, reasoning is central to several knowledge graph construction activities. It is not only key to checking the consistency of the knowledge graph or for classification purposes but also for other knowledge graph construction activities that pivot on reasoning. We have some new understanding [79–81] of reasoning activities in the context of a broader series of studies about how knowledge graph are constructed.

5.1   An Interview Study: What Ontologists Report

In an interview study with 15 ontologists we asked them about the challenges they typically face when constructing knowledge graph and the workarounds they employ to overcome them [79,80]. The reported problems can be classified into two groups: the user interface of authoring tools not being able to handle the complexity of knowledge graphs and current technology (reasoners and debuggers) not being able to deal with such complexity in an efficient manner.

Problems of the User Interface
We found that running the reasoner could have undesirable consequences, which were especially negative on ontologies that were large or individuals were unfamiliar with. *Contextual awareness*, which is often lost after invoking the reasoner, refers to the quality that describes how well the current environment is understood and consequently how in control the user is. This means that the user interface is not capable of explicitly highlighting the consequences of running the reasoner and sets such a burden on the user.

Limitations of the Current State of the Technology
The size of knowledge graphs and the complexity of axioms cause problems when running the reasoner. This is exemplified by a quote of Participant 15: *"I don't like to work with ontologies (knowledge graphs) that are so big or complicated that I can't reason reasonably quickly."* In order to address such problems, users take work-arounds to bypass existing limitations:

- Externalising reasoning by employing continuous integration.
- Complex entities are taken out of knowledge graphs and reasoning is done separately. For instance one participant reported that they kept the disjoints in a separate file.
- Using languages with low expressivity, such as RDF and OBO, as these tend to be more human readable, seen as less error prone and faster to reason.

Reasoning as the Means to Realise Other KG Construction Activities
For debugging purposes, the authors reported they run the reasoner frequently not to accumulate problems over time: *"Every axiom change, every class addition, every refactor I always run the reasoner."* (Participant 8) Users try to fix the knowledge graph by making changes in a trial and error fashion, without exhibiting any particular strategy: *"It's little bit trial and error, we don't really have a methodology."* (Participant 9).

5.2   An Observation Study: What Ontologists Actually Do

In a laboratory study with 15 participants, we observed how ontologists construct knowledge graphs by analysing their behavioural traces [81]. Participants were given a set of construction tasks that were carried out in an instrumented version of Protégé, ie. Protégé4US [82], that logs the events of the user interface. An eye-tracker was also employed in order to know how participants allocated their attention during the KG construction activities.

We discovered a number of activities that occur before and after the reasoner is run. For instance we found that, 40% of the time, users save the current knowledge graph before the reasoner is invoked. Similarly the reasoner is run 17% of the time after converting a class into a defined one and looking at the class hierarchy. Sometimes reasoning engines show unstable behaviour that causes the authoring environment to freeze. The observed activity of saving the current knowledge graph before running the reasoner might therefore be a strategy to prevent information loss.

Two activities occur after running the reasoner: 41% of the time participants observe the consequences of reasoning on the asserted hierarchy and the description area of Protégé (located at the bottom right part in the default setup). Then they select one or more classes at the hierarchy and check their descriptions again. Since the description area shows the inferred features of classes, this behaviour may indicate that participants are checking whether the reasoner had consequences for the features of classes, in addition to classifying classes. To check classification, participants expand the inferred class hierarchy and make analogous selections on inferred entities, which occurs 30% of the time after the reasoner has been run. In general, we found that navigation of the inferred class hierarchy tends to be of an exploratory nature, and occurs to check the state of the hierarchy. By contrast, navigation of the asserted hierarchy is more directed—users know what they are looking for—and focuses on finding the specific location of a class to which a modification is planned.

Frequent reasoning may be due to a particular authoring style, or an indicator of a 'trial and error' strategy for debugging as indicated by the interview study. This is because the (perceived) complexity of OWL leads some authors to invoke the reasoner at every modification they make, or when complex modifications are believed to potentially cause an error. Some authors adopt the frequent classification strategy in order to avoid the propagation of problems. Considering the high demands of reasoning, tools and reasoners should allow authors to reason over subsets of ontologies and axioms to speed up the process. Ideally, authoring tools should also provide background reasoning capabilities, as do the compilers of software development IDEs. *Incremental* reasoning, which is already supported by some OWL reasoners [23,49,50,52,83], seems like a promising approach to providing such background reasoning capabilities.

5.3   Implications for Design

We propose a set of recommendations that are informed by the insights we discovered in the above-mentioned studies. The implementation of these recommendations will enable tools to have a better support of the mentioned activities and, consequently, ontology authors will alleviate their cognitive burden alleviated.

- Increase situational awareness by **giving feedback about the consequences of actions carried out**, especially when axioms are modified, commands such as *undoing* are invoked and when the reasoner is run. **Providing a history of modifications** to which users can revert may alleviate this problem.

- **Anticipate reasoner invocation**: As saving the ontology is a good predictor of running the reasoner, this could be used to anticipate the invocation of the reasoner in the background to save time.
- **Automatic detection of authoring problems**: The 'trial and error' authoring strategy, which is operationalised as the concatenation of editing and reasoning activities, is used for debugging and indicates authoring difficulties [79]. Detecting this strategy automatically opens up new avenues to problem pre-emption.
- **Make changes to the inferred hierarchy explicit**: In order to check the (sometimes unexpected) consequences of running a reasoner, users engage in a detailed exploration of the inferred class hierarchy, which is time consuming and may be prohibitively difficult in large ontologies. This occurs because the changes made to the ontology as a result of reasoning are not made explicit by Protégé, and therefore they have to be actively sought. Authoring tools should give explicit feedback about the consequences of reasoning, without obliging users to perform a manual exploration of the class hierarchy.

6   Designing Tools for Improving Contextual Awareness

We have established that contextual awareness is reduced after performing a modelling action and that better tool support is needed to support regaining awareness (Sect. 5). Despite existing tool support, the problem of raising contextual awareness after a change remains to be an area of active research. In the following, we will discuss the key design considerations for tools that support knowledge graph authors at understanding the consequences of their actions better.

A *change* of knowledge graph in the context of this section is a kind of knowledge graph construction action that can be intuitively defined as follows: Given a knowledge graph $\mathcal{O}$, a previous version of the knowledge graph $\mathcal{O}'$, α an OWL 2 axiom, $\mathfrak{A}$ the set of all axioms α with $\alpha \in \mathcal{O}$ and $\alpha \notin \mathcal{O}'$ (additions), $\mathfrak{R}$ the set of all axioms α with $\alpha \notin \mathcal{O}$ and $\alpha \in \mathcal{O}'$ (removals), we call $(\mathfrak{R},\mathfrak{A})$ the change from $\mathcal{O}'$ to $\mathcal{O}$. We do not want to further qualify α: it could be anything from an annotation assertion to a "string that is OWL API parseable" (not even being OWL 2 according to the specification). By *evaluation* we refer to the comparison of the consequences of a change with the authors intentions. A consequence of a change can be functional, i.e. affecting system behaviour or non-functional, i.e. quality attributes of the knowledge graph. A functional consequence for example can be the introduction of a bug, such as an unsatisfiable class, or an undesired entailment. The expectation of a particular consequence can be made explicit for example in the form of a competency question (Sect. 4). Non-functional consequences include profile or expressivity elevations, for example from OWL 2 EL to OWL 2 DL or $\mathcal{EL}$ to $\mathcal{ALC}$, or increases in classification time. During the evaluation phase the authors compare their intention, for example the subsumption of a number of classes under a new defined class, with the actual consequences. If a certain degree of confidence is reached that the intentions are met, the process of change evaluation terminates. If an undesired consequence is detected, the authors attempts to fix it. This fix than can be treated as a new change that will be evaluated in the same way. Note that intentions are user dependent and a user can be aware of their intentions to varying degrees: an intention can be conscious (e.g. generating a definition that subsumes all vegetarian pizzas), subconscious (e.g. classes should be satisfiable) or unconscious (e.g. a previously unknown subsumption).

In the following, we will summarise the five key problems we need to consider when designing a tool that supports knowledge graph authors at raising their contextual awareness after a change. As can be seen in Fig. 4, the first step, designing a unit, is independent of the remaining four, which stand in a cyclic relationship that will be explained in the following.

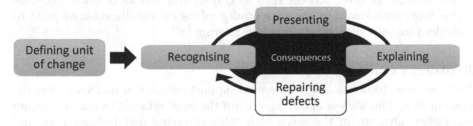

Fig. 4. 5 key problems that need to be addressed when developing tool support for raising contextual awareness after a modelling action.

Define Unit of Change

The first step is to define the unit of change we are interested in. This differs from use case to use case. Let's consider the problem of large ontologies that get published periodically. The maintainers of knowledge graphs are interested in making sure that any new version is still consistent with respect to a large given ABox. Therefore, their unit of change is the deployment of a new version. In the case of the Inference Inspector[14] case, we were interested in designing a tool that reflected changes compared to the last reasoner run. The unit of change was therefore defined to be the set of all changes that occurred between two runs of the reasoner. The authors of ROO defined a change as an atomic modelling action, i.e. the addition of an axiom [71]. The choice of a unit of change can have implications on many factors such as GUI responsiveness (cost of reasoning) or intelligibility of consequence (there could be too many if the unit is too wide).

Recognising Consequences

During the first main step, we decide which kinds of consequences we want to help the user recognise. The literature in this area is centered around recognising

---

[14] https://github.com/matentzn/inference-inspector.

logical defects, such as ontology inconsistency or unsatisfiable classes. Other approaches show logical weakening and strengthening of implications [84], lost and gained entailments with respect to key entailment sets, introduction of tautologies and stray signature [71] and non-functional criteria such as profile violations [85]. There is a wide range of methods available for recognising consequences. Diagnosis-based approaches are used to identify unsatisfiable classes. Diff-based approaches are used to compare versions of knowledge graphs. Typical problems with those approaches involve correctness and scalability. Examples of actively maintained tools for this task are ECCO[15], ORE[16], the Inference Inspector, and Protégé.

Presenting Consequences

Next, we define how we want to present the consequences. The literature in this area (visualisation of knowledge graph) typically deals with simplifying graphical representations of OWL axioms such as graphs, crop circles or other diagrammatic representations. For the presentation of logical entailments, we need to consider problems such as ordering and filtering [86].

Explaining Consequences

Next, we need to define how we want to support the user in understanding the consequence. This step is in our experience the most difficult to realise, despite significant advances in the areas such tableau-tracing and justifications, and remains largely a work in progress. Not only are current approaches to computing explanations (justifications or otherwise) often not scalable—there is also a lot of work ahead for the community in making them more easily understandable for authors of knowledge graphs that are not experts in logic (which also comes back to the presentation problem). Another problem in this area is the sheer number of possible explanations for any given consequence, and how to present the most relevant ones first.

Repairing Defective Consequences

Ideally, the author is satisfied at this point and moves on to perform the next authoring action. If she identified a defective consequence, and understands its origin, perhaps with the help of a good explanation (previous step), she might be able to decide on a repair strategy, such as removing an axiom or changing a definition all by herself. However, the sheer number of possible repairs makes this step often quite hard. Techniques for suggesting (or fully automated) repairs are often based on heuristics.

---

[15] https://github.com/rsgoncalves/ecco.
[16] http://aksw.org/Projects/ORE.html.

7   Discussions and Outlook

In this chapter, firstly we have introduced the concept of Knowledge Graph and its brief history. In the second part of this chapter we have focused on some state of the arts approaches to understanding intentions of ontology authors.

We attempt to distinguish explicit intentions from the implicit ones. Author intentions of the former type are usually indicated by explicit requirement statements such as competency question; they are thus more helpful in terms of identifying *what* the scope of the to be constructed ontology is. Authoring tests can be constructed, based on requirement statements, such as competency questions, so as to check the quality of the ontology by checking if it is compatible with the explicit intentions. Authoring intentions of the latter type, however, are usually implicitly inferred by authoring activities of ontology authors; they are thus more helpful in terms of showing *how* ontology authors plan to construct their ontology. Knowledge about the implicit intentions could help us build tools to deliver better user experience for ontology authoring and to make ontology authoring more effective.

To the best of our knowledge, this is the first attempt to address the problem of understanding intentions of ontology authors. There are many open questions and we just name a couple here. For example, there might be other requirement statements (than competency questions) that one should consider for explicit intentions. Also, it would be very interesting to investigate how to integrate explicit and implicit intentions and the benefits of such integrations.

Acknowledgments. This research has been partially funded by the EPSRC *WhatIf* project (EP/J014176/1) and the EU Marie Curie IAPP K-Drive project (286348). In particular, we would like to thank our colleagues Yuan Ren, Artemis Parvizi, Chris Mellish and Kees van Deemter from the University of Aberdeen and Robert Stevens from the University of Manchester for their joint work on ontology authoring.

References

1. Pan, J.Z., Vetere, G., Gomez-Perez, J.M., Wu, H.: Exploiting Linked Data and Knowledge Graphs for Large Organisations. Springer, Heidelberg (2016)
2. Baader, F., Calvanese, D., McGuinness, D.L., Nardi, D., Patel-Schneider, P.F.: The Description Logic Handbook: Theory, Implementation, and Applications. Cambridge University Press, Cambridge (2003). ISBN: 0-521-78176-0
3. Stearns, M.Q., Price, C., Spackman, K.A., Wang, A.Y.: SNOMED clinical terms: overview of the development process and project status. In: Proceedings of the AMIA Symposium, p. 662. American Medical Informatics Association (2001)
4. Rector, A., Drummond, N., Horridge, M., Rogers, J., Knublauch, H., Stevens, R., Wang, H., Wroe, C.: OWL pizzas: practical experience of teaching OWL-DL: common errors & common patterns. In: Motta, E., Shadbolt, N.R., Stutt, A., Gibbins, N. (eds.) EKAW 2004. LNCS, vol. 3257, pp. 63–81. Springer, Heidelberg (2004)
5. Dzbor, M., Motta, E., Gomez, J.M., Buil, C., Dellschaft, K., Görlitz, O., Lewen, H.: D4.1.1 analysis of user needs, behaviours & requirements wrt user interfaces for ontology engineering. Technical report, August 2006

6. Brachman, R.J., Levesque, H.J. (eds.): Readings in Knowledge Representation. Morgan Kaufmann Publishers Inc., San Francisco (1985). ISBN: 093461301X

7. Sowa, J.F.: Semantic networks. In: Encyclopedia of Artificial Intelligence. Wiley, New York (1987)

8. Russell, S.J., Norvig, P.: Artificial Intelligence: A Modern Approach, 3rd edn. Prentice Hall, Upper Saddle River (2010). ISBN: 978-0-13-604259-4

9. Quillian, M.R.: Word concepts: a theory and simulation of some basic semantic capabilities. Behav. Sci. **12**(5), 410–430 (1967)

10. Minsky, M.: A framework for representing knowledge. In: MIT-AI Laboratory Memo 306 (1974). Reprinted in the Winston, P. (ed.) Psychology of Computer Vision. McGraw-Hill (1975)

11. Hayes, P.J.: The logic of frames. In: Metzing, D. (ed.) Frame Conceptions and Text Understanding, pp. 46–61. Walter de Gruyter and Co. (1979)

12. Brachman, R.J., Schmolze, J.G.: An overview of the KL-ONE knowledge representation system. Cogn. Sci. **9**(2), 171 (1985)

13. Hayes, P.J., Patel-Schneider, P.F.: RDF 1.1 semantics. W3C Recommendation, February 2014

14. Pan, J., Horrocks, I.: OWL-Eu: adding customised datatypes into OWL. J. Web Semant. **4**(1), 29–39 (2006)

15. Suchanek, F., Kasneci, G., Weikum, G.: Yago: a core of semantic knowledge. In: Proceedings of the WWW (2007)

16. Bizer, C., Lehmann, J., Kobilarov, G., Auer, S., Becker, C., Cyganiak, R., Hellmann, S.: Dbpedia-a crystallization point for the web of data. J. Web Semant. **7**(3), 154–165 (2009)

17. Carlson, A., Betteridge, J., Kisiel, B., Settles, B., Hruschka Jr., E., Mitchell, T.: Toward an architecture for never-ending language learning. In: Proceedings of the AAAI (2010)

18. Calvanese, D., De Giacomo, G., Lembo, D., Lenzerini, M., Rosati, R.: Tractable reasoning and efficient query answering in description logics: the DL-Lite family. J. Autom. Reason. **39**(3), 385–429 (2007)

19. Baader, F., Brandt, S., Lutz, C.: Pushing the EL envelope further. In: Clark, K., Patel-Schneider, P.F. (eds.) Proceedings of the OWLED 2008 DC Workshop on OWL: Experiences and Directions (2008)

20. Pan, J.Z., Horrocks, I.: RDFS(FA) and RDF MT: two semantics for RDFS. In: Fensel, D., Sycara, K., Mylopoulos, J. (eds.) ISWC 2003. LNCS, vol. 2870, pp. 30–46. Springer, Heidelberg (2003). doi:10.1007/978-3-540-39718-2_3

21. Pan, J.Z., Thomas, E.: Approximating OWL-DL ontologies. In: The Proceedings of the 22nd National Conference on Artificial Intelligence (AAAI 2007), pp. 1434–1439 (2007)

22. Pan, J.Z., Thomas, E., Zhao, Y.: Completeness guaranteed approximation for OWL DL query answering. In: Proceedings of the DL (2009)

23. Ren, Y., Pan, J.Z., Zhao, Y.: Towards scalable reasoning on ontology streams via syntactic approximation. In: The Proceedings of IWOD2010 (2010)

24. Console, M., Mora, J., Rosati, R., Santarelli, V., Savo, D.F.: Effective computation of maximal sound approximations of description logic ontologies. In: Mika, P., et al. (eds.) ISWC 2014. LNCS, vol. 8797, pp. 164–179. Springer, Heidelberg (2014). doi:10.1007/978-3-319-11915-1_11

25. Zhou, Y., Nenov, Y., Grau, B., Horrocks, I.: Pay-as-you-go OWL query answering using a triple store. In: Proceedings of the AAAI (2014)

26. Pan, J.Z., Ren, Y., Zhao, Y.: Tractable approximate deduction for OWL. Artif. Intell. **235**, 95–155

27. Hogan, A., Pan, J.Z., Polleres, A., Decker, S.: SAOR: template rule optimisations for distributed reasoning over 1 billion linked data triples. In: Patel-Schneider, P.F., Pan, Y., Hitzler, P., Mika, P., Zhang, L., Pan, J.Z., Horrocks, I., Glimm, B. (eds.) ISWC 2010. LNCS, vol. 6496, pp. 337–353. Springer, Heidelberg (2010). doi:10.1007/978-3-642-17746-0_22

28. Urbani, J., Kotoulas, S., Maassen, J., Harmelen, F., Bal, H.: OWL reasoning with WebPIE: calculating the closure of 100 billion triples. In: Aroyo, L., Antoniou, G., Hyvönen, E., Teije, A., Stuckenschmidt, H., Cabral, L., Tudorache, T. (eds.) ESWC 2010. LNCS, vol. 6088, pp. 213–227. Springer, Heidelberg (2010). doi:10.1007/978-3-642-13486-9_15

29. Ren, Y., Pan, J.Z., Lee, K.: Parallel ABox reasoning of EL ontologies. In: Proceedings of the First Joint International Conference of Semantic Technology (JIST 2011) (2011)

30. Du, J., Guilin Qi, Y.-D.S., Pan, J.Z.: A decomposition-based approach to OWL DL ontology diagnosis. In: Proceedings of the 23rd IEEE International Conference on Tools with Artificial Intelligence (ICTAI 2011) (2011)

31. Urbani, J., Harmelen, F., Schlobach, S., Bal, H.: QueryPIE: backward reasoning for OWL horst over very large knowledge bases. In: Aroyo, L., Welty, C., Alani, H., Taylor, J., Bernstein, A., Kagal, L., Noy, N., Blomqvist, E. (eds.) ISWC 2011. LNCS, vol. 7031, pp. 730–745. Springer, Heidelberg (2011). doi:10.1007/978-3-642-25073-6_46

32. Ren, Y., Pan, J.Z., Lee, K.: Optimising parallel ABox reasoning of EL ontologies. In: Proceedings of the DL (2012)

33. Heino, N., Pan, J.Z.: RDFS reasoning on massively parallel hardware. In: Cudré-Mauroux, P., et al. (eds.) ISWC 2012. LNCS, vol. 7649, pp. 133–148. Springer, Heidelberg (2012). doi:10.1007/978-3-642-35176-1_9

34. Fokoue, A., Meneguzzi, F., Sensoy, M., Pan, J.Z.: Querying linked ontological data through distributed summarization. In: Proceedings of the AAAI (2012)

35. Kazakov, Y., Krtzsch, M., Simank, F.: The incredible ELK. J. Autom. Reason. **53**(1), 1–61 (2014)

36. Getoor, L.: Introduction to Statistical Relational Learning. MIT Press, Cambridge (2007)

37. De Raedt, L.: Logical and Relational Learning. Springer Science and Business Media, Heidelberg (2008)

38. Nickel, M., Murphy, K., Tresp, V., Gabrilovich, E.: A review of relational machine learning for knowledge graphs. Proc. IEEE **104**(1), 11–33 (2016)

39. Lehmann, J., Hitzler, P.: A refinement operator based learning algorithm for the $\mathcal{ALC}$ description logic. In: Blockeel, H., Ramon, J., Shavlik, J., Tadepalli, P. (eds.) ILP 2007. LNCS (LNAI), vol. 4894, pp. 147–160. Springer, Heidelberg (2008). doi:10.1007/978-3-540-78469-2_17

40. Vlker, J., Niepert, M

41. Pan, J.Z., Zhao, Y., Xu, Y., Quan, Z., Zhu, M., Gao, Z.: TBox learning from incomplete data by inference in BelNet+ Knowl. Based Syst. **75**, 30–40 (2015)

42. Alexopoulos, P., Villazon-Terrazas, B., Pan, J.Z.: Towards vagueness-aware semantic data. In: Proceedings of the URSW (2013)

43. Alexopoulos, P., Peroni, S., Villazon-Terrazas, B., Pan, J.Z.: Annotating ontologies with descriptions of vagueness. In: Proceedings of the ESWC (2014)

44. Jekjantuk, N., Pan, J.Z., Alexopoulos, P.: Towards a meta-reasoning framework for reasoning about vagueness in OWL ontologies. In: Proceedings of the ICSC (2016)

45. Sensoy, M., Fokoue, A., Pan, J.Z., Norman, T., Tang, Y., Oren, N., Sycara, K.: Reasoning about uncertain information and conflict resolution through trust revision. In: Proceedings of the AAMAS (2013)
46. Stoilos, G., Stamou, G., Pan, J.Z.: Fuzzy extensions of OWL: logical properties and reduction to fuzzy description logics. Int. J. Approx. Reason. **51**(6), 656–679 (2010)
47. Lécué, F., Pan, J.Z.: Predicting knowledge in an ontology stream. In: Proceedings of the 23rd International Joint Conference on Artificial Intelligence, IJCAI 2013, Beijing, China, 3–9 August 2013 (2013). http://www.aaai.org/ocs/index.php/IJCAI/IJCAI13/paper/view/6608
48. Lecue, F., Pan, J.Z.: Consistent knowledge discovery from evolving ontologies. In: Proceedings of the AAAI (2015)
49. Ren, Y., Pan, J.Z.: Optimising ontology stream reasoning with truth maintenance system. In: Proceedings of the ACM Conference on Information and Knowledge Management (CIKM 2011) (2011)
50. Kazakov, Y., Klinov, P.: Incremental reasoning in OWL EL without bookkeeping. In: Alani, H., et al. (eds.) ISWC 2013. LNCS, vol. 8218, pp. 232–247. Springer, Heidelberg (2013). doi:10.1007/978-3-642-41335-3_15
51. Urbani, J., Margara, A., Jacobs, C., Harmelen, F., Bal, H.: DynamiTE: parallel materialization of dynamic RDF data. In: Alani, H., et al. (eds.) ISWC 2013. LNCS, vol. 8218, pp. 657–672. Springer, Heidelberg (2013). doi:10.1007/978-3-642-41335-3_41
52. Ren, Y., Pan, J.Z., Guclu, I., Kollingbaum, M.: A combined approach to incremental reasoning for EL ontologies. In: Ortiz, M., Schlobach, S. (eds.) RR 2016. LNCS, vol. 9898, pp. 167–183. Springer, Heidelberg (2016). doi:10.1007/978-3-319-45276-0_13
53. Nguyen, H.H., Beel, D., Webster, G., Mellish, C., Pan, J.Z., Wallace, C.: CURIOS mobile: linked data exploitation for tourist mobile apps in rural areas. In: Supnithi, T., Yamaguchi, T., Pan, J.Z., Wuwongse, V., Buranarach, M. (eds.) JIST 2014. LNCS, vol. 8943, pp. 129–145. Springer, Heidelberg (2015). doi:10.1007/978-3-319-15615-6_10
54. Botoeva, E., Kontchakov, R., Ryzhikov, V., Wolter, F., Zakharyaschev, M.: Games for query inseparability of description logic knowledge bases. Artif. Intell. **234**, 78–119 (2016). doi:10.1016/j.artint.2016.01.010. http://www.sciencedirect.com/science/article/pii/S0004370216300017
55. Botoeva, E., Lutz, C., Ryzhikov, V., Wolter, F., Zakharyaschev, M.: Query-based entailment and inseparability for ALC ontologies. In: Proceedings of the 25th International Joint Conference on Artificial Intelligence (IJCAI 2016), pp. 1001–1007 (2016)
56. Nguyen, H., Valincius, E., Pan, J.Z.: A power consumption benchmark framework for ontology reasoning on android devices. In: Proceedings of the 4th OWL Reasoner Evaluation Workshop (ORE) (2015)
57. Guclu, I., Li, Y.-F., Pan, J.Z., Kollingbaum, M.J.: Predicting energy consumption of ontology reasoning over mobile devices. In: Groth, P., Simperl, E., Gray, A., Sabou, M., Krötzsch, M., Lecue, F., Flöck, F., Gil, Y. (eds.) ISWC 2016. LNCS, vol. 9981, pp. 289–304. Springer, Heidelberg (2016). doi:10.1007/978-3-319-46523-4_18
58. Konev, B., Lutz, C., Walther, D., Wolter, F.: Model-theoretic inseparability and modularity of description logic ontologies. Artif. Intell. **203**, 66–103 (2013)
59. Konev, B., Lutz, C., Wolter, F., Zakharyaschev, M.: Conservative rewritability of description logic TBoxes. In: Proceedings of the 25th International Joint Conference on Artificial Intelligence (IJCAI 2016) (2016)

60. Lutz, C., Wolter, F.: Deciding inseparability and conservative extensions in the description logic EL. J. Symbolic Comput. **45**(2), 194–228 (2010)
61. Kostylev, E.V., Reutter, J.L., Vrgoč, D.: Containment of data graph queries. In: ICDT, pp. 131–142 (2014)
62. Kostylev, E.V., Reutter, J.L., Vrgoč, D.: Static analysis of navigational XPath over graph databases. Inf. Process. Lett. **116**(7), 467–474 (2016)
63. Libkin, L., Martens, W., Vrgoč, D.: Querying graphs with data. J. ACM **63**(2), 14 (2016)
64. Baader, F., Bienvenu, M., Lutz, C., Wolter, F.: Query and predicate emptiness in ontology-based data access. J. Artif. Intell. Res. (JAIR) **56**, 1–59 (2016)
65. Bienvenu, M., Bourgaux, C., Goasdoué, F.: Explaining inconsistency-tolerant query answering over description logic knowledge bases. In: Proceedings of the 30th AAAI Conference on Artificial Intelligence (AAAI) (2016)
66. Bienvenu, M., Bourgaux, C., Goasdoué, F.: Query-driven repairing of inconsistent DL-Lite knowledge bases. In: Proceedings of the 25th International Joint Conference on Artificial Intelligence (IJCAI) (2016)
67. Lord, P.: The semantic web takes wing: programming ontologies with Tawny-OWL. In: OWLED 2013 (2013). http://www.russet.org.uk/blog/2366
68. Denaux, R., Dimitrova, V., Cohn, A.G., Dolbear, C., Hart, G.: Rabbit to OWL: ontology authoring with a CNL-based tool. In: Fuchs, N.E. (ed.) CNL 2009. LNCS (LNAI), vol. 5972, pp. 246–264. Springer, Heidelberg (2010). doi:10.1007/978-3-642-14418-9_15
69. Power, R.: OWL simplified english: a finite-state language for ontology editing. In: Kuhn, T., Fuchs, N.E. (eds.) CNL 2012. LNCS (LNAI), vol. 7427, pp. 44–60. Springer, Heidelberg (2012). doi:10.1007/978-3-642-32612-7_4
70. Liebig, T., Noppens, O.: OntoTrack: a semantic approach for ontology authoring. Web Semant. Sci. Serv. Agents World Wide Web **3**(2), 116–131 (2005)
71. Denaux, R., Thakker, D., Dimitrova, V., Cohn, A.G.: Interactive semantic feedback for intuitive ontology authoring. In: FOIS, pp. 160–173 (2012)
72. Uschold, M., Gruninger, M., et al.: Ontologies: principles, methods and applications. Knowl. Eng. Rev. **11**(2), 93–136 (1996)
73. Suárez-Figueroa, M.C., Gómez-Pérez, A., Motta, E., Gangemi, A.: Ontology Engineering in a Networked World. Springer, Heidelberg (2012)
74. Fernandes, P.C.B., Guizzardi, R.S., Guizzardi, G.: Using goal modeling to capture competency questions in ontology-based systems. J. Inf. Data Manag. **2**(3), 527 (2011)
75. Ren, Y., Parvizi, A., Mellish, C., Pan, J.Z., Deemter, K., Stevens, R.: Towards competency question-driven ontology authoring. In: Presutti, V., d'Amato, C., Gandon, F., d'Aquin, M., Staab, S., Tordai, A. (eds.) ESWC 2014. LNCS, vol. 8465, pp. 752–767. Springer, Heidelberg (2014). doi:10.1007/978-3-319-07443-6_50
76. Zemmouchi-Ghomari, L., Ghomari, A.R.: Translating natural language competency questions into SPARQL queries: a case study. In: WEB 2013, pp. 81–86 (2013)
77. Malheiros, Y., Freitas, F.: A method to develop description logic ontologies iteratively based on competency questions: an implementation. In: ONTOBRAS, pp. 142–153 (2013)
78. Beaver, D.: Presupposition. In: van Benthem, J., ter Meulen, A. (eds.) The Handbook of Logic and Language, pp. 939–1008. Elsevier (1997)
79. Vigo, M., Jay, C., Stevens, R.: Design insights for the next wave ontology authoring tools. In: Proceedings of the SIGCHI Conference on Human Factors in Computing Systems, CHI 2014, pp. 1555–1558 (2014). ISBN: 978-1-4503-2473-1. doi:10.1145/2556288.2557284

80. Vigo, M., Bail, S., Jay, C., Stevens, R.: Overcoming the pitfalls of ontology authoring: strategies and implications for tool design. Int. J. Hum.-Comput. Stud. **72**(12), 835–845 (2014). ISSN: 1071–5819. doi:10.1016/j.ijhcs.2014.07.005, http://www.scie ncedirect.com/science/article/pii/S1071581914001013

81. Vigo, M., Jay, C., Stevens, R.: Constructing conceptual knowledge artefacts: activity patterns in the ontology authoring process. In: Proceedings of the 33rd Annual ACM Conference on Human Factors in Computing Systems, CHI 2015, pp. 3385–3394 (2015). ISBN: 978-1-4503-3145-6. doi:10.1145/2702123.2702495

82. Vigo, M., Jay, C., Stevens, R.: Protégé4US: harvesting ontology authoring data with Protégé. In: Presutti, V., Blomqvist, E., Troncy, R., Sack, H., Papadakis, I., Tordai, A. (eds.) ESWC 2014. LNCS, vol. 8798, pp. 86–99. Springer, Heidelberg (2014). doi:10.1007/978-3-319-11955-7_8

83. Grau, B.C., Halaschek-Wiener, C., Kazakov, Y., Suntisrivaraporn, B.: Incremental classification of description logics ontologies. J. Autom. Reason. **44**(4), 337–369 (2010)

84. Gonalves, R.S.: Impact analysis in description logic ontologies. Ph.D. thesis, University of Manchester (2014)

85. Matentzoglu, N., Vigo, M., Jay, C., Stevens, R.: Making entailment set changes explicit improves the understanding of consequences of ontology authoring actions. In: Blomqvist, E., Ciancarini, P., Poggi, F., Vitali, F. (eds.) EKAW 2016. LNCS (LNAI), vol. 10024, pp. 432–446. Springer, Heidelberg (2016). doi:10.1007/ 978-3-319-49004-5_28

86. Parvizi, A., Mellish, C., van Deemter, K., Ren, Y., Pan, J.Z.: Selecting ontology entailments for presentation to users. In: Proceedings of the International Conference on Knowledge Engineering and Ontology Development, KEOD 2014, Rome, Italy, 21–24 October 2014, pp. 382–387 (2014). doi:10.5220/0005136203820387

Inseparability and Conservative Extensions of Description Logic Ontologies: A Survey

Elena Botoeva[1], Boris Konev[2], Carsten Lutz[3], Vladislav Ryzhikov[1], Frank Wolter[2(✉)], and Michael Zakharyaschev[4]

[1] Faculty of Computer Science, Free University of Bozen-Bolzano, Bolzano, Italy
{botoeva,ryzhikov}@inf.unibz.it
[2] Department of Computer Science, University of Liverpool, Liverpool, UK
{konev,wolter}@liverpool.ac.uk
[3] Faculty of Informatics, University of Bremen, Bremen, Germany
clu@informatik.uni-bremen.de
[4] Department of Computer Science and Information Systems,
Birkbeck University of London, London, UK
michael@dcs.bbk.ac.uk

Abstract. The question whether an ontology can safely be replaced by another, possibly simpler, one is fundamental for many ontology engineering and maintenance tasks. It underpins, for example, ontology versioning, ontology modularization, forgetting, and knowledge exchange. What 'safe replacement' means depends on the intended application of the ontology. If, for example, it is used to query data, then the answers to any relevant ontology-mediated query should be the same over any relevant data set; if, in contrast, the ontology is used for conceptual reasoning, then the entailed subsumptions between concept expressions should coincide. This gives rise to different notions of ontology *inseparability* such as query inseparability and concept inseparability, which generalize corresponding notions of conservative extensions. In this chapter, we survey results on various notions of inseparability in the context of description logic ontologies, discussing their applications, useful model-theoretic characterizations, algorithms for determining whether two ontologies are inseparable (and, sometimes, for computing the difference between them if they are not), and the computational complexity of this problem.

1 Introduction

Description logic (DL) ontologies provide a common vocabulary for a domain of interest together with a formal modeling of the semantics of the vocabulary items (concept names and role names). In modern information systems, they are employed to capture domain knowledge and to promote interoperability. Ontologies can become large and complex as witnessed, for example, by the widely used healthcare ontology SNOMED CT, which contains more than 300,000 concept names, and the National Cancer Institute (NCI) Thesaurus ontology,

© Springer International Publishing AG 2017
J.Z. Pan et al. (Eds.): Reasoning Web 2016, LNCS 9885, pp. 27–89, 2017.
DOI: 10.1007/978-3-319-49493-7_2

which contains more than 100,000 concept names. Engineering, maintaining and deploying such ontologies is challenging and labour intensive; it crucially relies on extensive tool support for tasks such as ontology versioning, ontology modularization, ontology summarization, and forgetting parts of an ontology. At the core of many of these tasks lie notions of *inseparability* of two ontologies, indicating that inseparable ontologies can safely be replaced by each other for the task at hand. The aim of this article is to survey the current research on inseparability of DL ontologies. We present and discuss different types of inseparability, their applications and interrelation, model-theoretic characterizations, as well as results on the decidability and computational complexity of inseparability.

The exact formalization of when an ontology 'can safely be replaced by another one' (that is, of inseparability) depends on the task for which the ontology is to be used. As we are generally going to abstract away from the syntactic presentation of an ontology, a natural first candidate for the notion of inseparability between two ontologies is their logical equivalence. However, this can be an unnecessarily strong requirement for most applications since also ontologies that are not logically equivalent can be replaced by each other without adverse effects. This is due to two main reasons. First, applications of ontologies often make use of only a fraction of the vocabulary items. As an example, consider SNOMED CT, which contains a vocabulary for a multitude of domains related to health case, including clinical findings, symptoms, diagnoses, procedures, body structures, organisms, pharmaceuticals, and devices. In a concrete application such as storing electronic patient records, only a small part of this vocabulary is going to be used. Thus, two ontologies should be separable only if they differ with respect to the *relevant* vocabulary items. Consequently, all our inseparability notions will be parameterized by a signature (set of concept and role names) Σ; when we want to emphasize Σ, we speak of Σ-inseparability. Second, even for the relevant vocabulary items, many applications do not rely on all details of the semantics provided by the ontology. For example, if an ontology is employed for conjunctive query answering over data sets that use vocabulary items from the ontology, then only the existential positive aspects of the semantics are relevant since the queries are positive existential, too.

A fundamental decision to be taken when defining an inseparability notion is whether the definition should be model-theoretic or in terms of logical consequences. Under the first approach, two ontologies are inseparable when the reducts of their models to the signature Σ coincide. We call the resulting inseparability notion *model inseparability*. Under the second approach, two ontologies are inseparable when they have the same logical consequences in the signature Σ. This actually gives rise to potentially many notions of inseparability since we can vary the logical language in which the logical consequences are formulated. Choosing the same language as the one used for formulating the ontologies results in what we call *concept inseparability*, which is appropriate when the ontologies are used for conceptual reasoning. Choosing a logical language that is based on database-style queries results in notions of *query inseparability*, which are appropriate for querying applications. Model inseparability implies all the resulting consequence-based notions of inseparability, but the converse does not hold for all standard DLs. The notion of query inseparability suggests some additional aspects. In particular, this type of inseparability is important both for ontologies that contain data as an integral part (*knowledge bases or KBs*, in DL parlance) and for those that do not (*TBoxes*, in DL parlance) and are maintained independently of the data. In the latter case, two TBoxes should be regarded as inseparable if they give the same answers to any relevant query *for any possible data*. One might then even want to work with two signatures: one for the data and one for the query. It turns out that, for both KBs and TBoxes, one obtains notions of inseparability that behave very differently from concept inseparability.

Inseparability generalizes conservative extensions, as known from classical logic. In fact, conservative extensions can also be defined in a model-theoretic and in a consequence-based way, and they correspond to the special case of inseparability where one ontology is syntactically contained in the other and the signature is the set of vocabulary items in the smaller ontology. Note that none of these two additional assumptions is appropriate for many applications of inseparability, such as ontology versioning. Instead of directly working with inseparability, we will often consider corresponding notions of entailment which, intuitively, is inseparability 'in one direction'; for example, two ontologies are concept Σ-inseparable if and only if they concept Σ-entail each other. Thus, one could say that an ontology concept (or model) entails another ontology if it is *sound* to replace the former by the latter in applications for which concept inseparability is the 'right' inseparability notion. Algorithms and complexity upper bounds are most general when established for entailment instead of inseparability, as they carry over to inseparability and conservative extensions. Similarly, lower bounds for conservative extensions imply lower bounds for inseparability and for entailment.

In this survey, we provide an in-depth discussion of four inseparability relations, as indicated above. For TBoxes, we look at Σ-concept inseparability (do two TBoxes entail the same concept inclusions over Σ?), Σ-model inseparability (do the Σ-reducts of the models of two TBoxes coincide?), and (Σ_1, Σ_2)-$\mathcal{Q}$-inseparability (do the answers given by two TBoxes coincide for all Σ_1-ABoxes and all Σ_2-queries from the class $\mathcal{Q}$ of queries?). Here, we usually take $\mathcal{Q}$ to be the class of conjunctive queries (CQs) and unions thereof (UCQs), but some smaller classes of queries are considered as well. For KBs, we consider Σ-$\mathcal{Q}$-inseparability (do the answers to Σ-queries in $\mathcal{Q}$ given by two KBs coincide?). When discussing proof techniques in detail, we focus on the standard expressive DL $\mathcal{ALC}$ and tractable DLs from the $\mathcal{EL}$ and *DL-Lite* families. We shall, however, also mention results for extensions of $\mathcal{ALC}$ and other DLs such as Horn-$\mathcal{ALC}$.

The structure of this survey is as follows. In Sect. 2, we introduce description logics. In Sect. 3, we introduce an abstract notion of inseparability and discuss applications of inseparability in ontology versioning, refinement, re-use, modularity, the design of ontology mappings, knowledge base exchange, and forgetting. In Sect. 4, we discuss concept inseparability. We focus on the description logics $\mathcal{EL}$ and $\mathcal{ALC}$ and give model-theoretic characterizations of Σ-concept inseparability which are then used to devise automata-based approaches to deciding concept inseparability. We also present polynomial time algorithms for acyclic $\mathcal{EL}$ TBoxes. In Sect. 5, we discuss model inseparability. We show that it is undecidable even in simple cases, but that by restricting the signature Σ to concept names, it often becomes decidable. We also consider model inseparability from the empty TBox, which is important for modularization and locality-based approximations of model inseparability. In Sect. 6, we discuss query inseparability between KBs. We develop model-theoretic criteria for query inseparability and use them to obtain algorithms for deciding query inseparability between KBs and their complexity. We consider description logics from the $\mathcal{EL}$ and $DL\text{-}Lite$ families, as well as $\mathcal{ALC}$ and its Horn fragment. In Sect. 7, we consider query inseparability between TBoxes and analyse in how far the techniques developed for KBs can be generalized to TBoxes. We again consider a wide range of DLs. Finally, in Sect. 8 we discuss further inseparability relations, approximation algorithms and the computation of representatives of classes of inseparable TBoxes.

2   Description Logic

In description logic, knowledge is represented using concepts and roles that are inductively defined starting from a set $\mathsf{N_C}$ of *concept names* and a set $\mathsf{N_R}$ of *role names*, and using a set of concept and role constructors [1]. Different sets of concept and role constructors give rise to different DLs.

We start by introducing the description logic $\mathcal{ALC}$. The concept constructors available in $\mathcal{ALC}$ are shown in Table 1, where r is a role name and C and D are concepts. A concept built from these constructors is called an $\mathcal{ALC}\text{-}concept$. $\mathcal{ALC}$ does not have any role constructors. An $\mathcal{ALC}$ *TBox* is a finite set of $\mathcal{ALC}$ *concept inclusions* (CIs) of the form $C \sqsubseteq D$ and $\mathcal{ALC}$ *concept equivalences* (CEs) of the form $C \equiv D$. (A CE $C \equiv D$ can be regarded as an abbreviation for the two CIs $C \sqsubseteq D$ and $D \sqsubseteq C$.) The *size* $|\mathcal{T}|$ of a TBox $\mathcal{T}$ is the number of occurrences of symbols in $\mathcal{T}$.

Table 1. Syntax and semantics of $\mathcal{ALC}$.

Name	Syntax	Semantics
Top concept	$\top$	$\Delta^{\mathcal{I}}$
Bottom concept	$\bot$	$\varnothing$
Negation	$\neg C$	$\Delta^{\mathcal{I}} \setminus C^{\mathcal{I}}$
Conjunction	$C \sqcap D$	$C^{\mathcal{I}} \cap D^{\mathcal{I}}$
Disjunction	$C \sqcup D$	$C^{\mathcal{I}} \cup D^{\mathcal{I}}$
Existential restriction	$\exists r.C$	$\{d \in \Delta^{\mathcal{I}} \mid \exists e \in C^{\mathcal{I}}\,(d,e) \in r^{\mathcal{I}}\}$
Universal restriction	$\forall r.C$	$\{d \in \Delta^{\mathcal{I}} \mid \forall e \in \Delta^{\mathcal{I}}\,((d,e) \in r^{\mathcal{I}} \rightarrow e \in C^{\mathcal{I}})\}$

DL semantics is given by *interpretations* $\mathcal{I} = (\Delta^{\mathcal{I}}, \cdot^{\mathcal{I}})$ in which the *domain* $\Delta^{\mathcal{I}}$ is a non-empty set and the *interpretation function* $\cdot^{\mathcal{I}}$ maps each concept name $A \in \mathsf{N_C}$ to a subset $A^{\mathcal{I}}$ of $\Delta^{\mathcal{I}}$, and each role name $r \in \mathsf{N_R}$ to a binary relation $r^{\mathcal{I}}$ on $\Delta^{\mathcal{I}}$. The extension of $\cdot^{\mathcal{I}}$ to arbitrary concepts is defined inductively as shown in the third column of Table 1. We say that an interpretation $\mathcal{I}$ *satisfies* a CI $C \sqsubseteq D$ if $C^{\mathcal{I}} \subseteq D^{\mathcal{I}}$, and that $\mathcal{I}$ is a *model* of a TBox $\mathcal{T}$ if it satisfies all the CIs in $\mathcal{T}$. A TBox is *consistent* (or *satisfiable*) if it has a model. A concept C is *satisfiable w.r.t.* $\mathcal{T}$ if there exists a model $\mathcal{I}$ of $\mathcal{T}$ such that $C^{\mathcal{I}} \neq \varnothing$. A concept C is *subsumed by a concept* D *w.r.t.* $\mathcal{T}$ ($\mathcal{T} \models C \sqsubseteq D$, in symbols) if every model $\mathcal{I}$ of $\mathcal{T}$ satisfies the CI $C \sqsubseteq D$. For TBoxes $\mathcal{T}_1, \mathcal{T}_2$, we write $\mathcal{T}_1 \models \mathcal{T}_2$ and say that $\mathcal{T}_1$ *entails* $\mathcal{T}_2$ if $\mathcal{T}_1 \models \alpha$ for all $\alpha \in \mathcal{T}_2$. TBoxes $\mathcal{T}_1$ and $\mathcal{T}_2$ are *logically equivalent* if they have the same models. This is the case if and only if $\mathcal{T}_1$ entails $\mathcal{T}_2$ and vice versa.

A *signature* Σ is a finite set of concept and role names. The *signature* $\mathsf{sig}(C)$ of a concept C is the set of concept and role names that occur in C, and likewise for TBoxes $\mathcal{T}$, CIs $C \sqsubseteq D$, assertions $r(a, b)$ and $A(a)$, ABoxes $\mathcal{A}$, KBs $\mathcal{K}$, UCQs q. Note that the universal role is not regarded as a role name, and so does not belong in any signature. Similarly, individual names are not in any signature and, in particular, not in the signature of an assertion, ABox, or KB. We are often interested in concepts, TBoxes, KBs, and ABoxes that are formulated using a specific signature. Therefore, we talk of a Σ-TBox $\mathcal{T}$ if $\mathsf{sig}(\mathcal{T}) \subseteq \Sigma$, and likewise for Σ-concepts, etc.

There are several extensions of $\mathcal{ALC}$ relevant for this chapter, which fall into three categories: extensions with (*i*) additional concept constructors, (*ii*) additional role constructors, and (*iii*) additional types of statements in TBoxes. These extensions are detailed in Table 2, where $\#\mathcal{X}$ denotes the size of a set $\mathcal{X}$ and double horizontal lines delineate different types of extensions. The last column gives an identifier for each extension, which is simply appended to the name $\mathcal{ALC}$ for constructing extensions of $\mathcal{ALC}$. For example, $\mathcal{ALC}$ extended with number restrictions, inverse roles, and the universal role is denoted by $\mathcal{ALCQI}^u$.

Table 2. Additional constructors: syntax and semantics.

Name	Syntax	Semantics	Identifier
Number restrictions	$(\leqslant n\, r\, C)$	$\{d \mid \#\{e \mid (d,e) \in r^{\mathcal{I}} \wedge e \in C^{\mathcal{I}}\} \leq n\}$	$\mathcal{Q}$
	$(\geqslant n\, r\, C)$	$\{d \mid \#\{e \mid (d,e) \in r^{\mathcal{I}} \wedge e \in C^{\mathcal{I}}\} \geq n\}$	
Inverse role	r^-	$\{(d,e) \mid (e,d) \in r^{\mathcal{I}}\}$	$\mathcal{I}$
Universal role	u	$\Delta^{\mathcal{I}} \times \Delta^{\mathcal{I}}$	$\cdot^u$
Role inclusions (RIs)	$r \sqsubseteq s$	$r^{\mathcal{I}} \subseteq s^{\mathcal{I}}$	$\mathcal{H}$

We next define a number of syntactic fragments of $\mathcal{ALC}$ and its extensions, which often have dramatically lower computational complexity. The fragment of $\mathcal{ALC}$ obtained by disallowing the constructors $\bot$, $\neg$, $\sqcup$ and $\forall$ is known as $\mathcal{EL}$.

Thus, $\mathcal{EL}$ concepts are constructed using $\top$, $\sqcap$ and $\exists$ only [2]. We also consider extensions of $\mathcal{EL}$ with the constructors in Table 2. For example, $\mathcal{ELI}^u$ denotes the extension of $\mathcal{EL}$ with inverse roles and the universal role. The fragments of $\mathcal{ALCI}$ and $\mathcal{ALCHI}$, in which CIs are of the form

$$B_1 \sqsubseteq B_2 \qquad \text{and} \qquad B_1 \sqcap B_2 \sqsubseteq \bot,$$

and the B_i are concept names, $\top$, $\bot$ or $\exists r.\top$, are denoted by $DL\text{-}Lite_{core}$ and $DL\text{-}Lite_{core}^{\mathcal{H}}$ (or $DL\text{-}Lite_{\mathcal{R}}$), respectively [3,4].

Example 1. The CI $\forall \mathsf{childOf}^-.\mathsf{Tall} \sqsubseteq \mathsf{Tall}$ (saying that everyone with only tall parents is also tall) is in $\mathcal{ALCI}$ but not in $\mathcal{ALC}$, $\mathcal{EL}$ or $DL\text{-}Lite_{core}^{\mathcal{H}}$. The RI $\mathsf{childOf}^- \sqsubseteq \mathsf{parentOf}$ is in both $\mathcal{ALCHI}$ and $DL\text{-}Lite_{core}^{\mathcal{H}}$.

$\mathcal{EL}$ and the $DL\text{-}Lite$ logics introduced above are examples of Horn DLs, that is, fragments of DLs in the $\mathcal{ALC}$ family that restrict the syntax in such a way that conjunctive query answering (see below) becomes tractable in data complexity. A few additional Horn DLs have become important in recent years. Following [5,6], we say that a concept C occurs positively in C itself and, if C occurs positively (negatively) in C', then

- C occurs positively (respectively, negatively) in $C' \sqcup D$, $C' \sqcap D$, $\exists r.C'$, $\forall r.C'$, $D \sqsubseteq C'$, and
- C occurs negatively (respectively, positively) in $\neg C'$ and $C' \sqsubseteq D$.

Now, we call a TBox $\mathcal{T}$ *Horn* if no concept of the form $C \sqcup D$ occurs positively in $\mathcal{T}$, and no concept of the form $\neg C$ or $\forall r.C$ occurs negatively in $\mathcal{T}$. For any DL $\mathcal{L}$ from the $\mathcal{ALC}$ family introduced above (e.g., $\mathcal{ALCHI}$), the DL *Horn-$\mathcal{L}$* only allows for Horn TBoxes in $\mathcal{L}$. Note that $\forall \mathsf{childOf}^-.\mathsf{Tall}$ occurs negatively in the CI α from Example 1, and so the TBox $\mathcal{T} = \{\alpha\}$ is not Horn.

TBoxes $\mathcal{T}$ used in practice often turn out to be *acyclic* in the following sense:

- all CEs in $\mathcal{T}$ are of the form $A \equiv C$ (*concept definitions*) and all CIs in $\mathcal{T}$ are of the form $A \sqsubseteq C$ (*primitive concept inclusions*), where A is a concept name;
- no concept name occurs more than once on the left-hand side of a statement in $\mathcal{T}$;
- $\mathcal{T}$ contains no cyclic definitions, as detailed below.

Let $\mathcal{T}$ be a TBox that contains only concept definitions and primitive concept inclusions. The relation $\prec_{\mathcal{T}} \subseteq \mathsf{N_C} \times \mathsf{sig}(\mathcal{T})$ is defined by setting $A \prec_{\mathcal{T}} X$ if there exists a TBox statement $A \bowtie C$ such that X occurs in C, where $\bowtie$ ranges over $\{\sqsubseteq, \equiv\}$. A concept name A *depends* on a symbol $X \in \mathsf{N_C} \cup \mathsf{N_R}$ if $A \prec_{\mathcal{T}}^+ X$, where $\cdot^+$ denotes transitive closure. We use $\mathsf{depend}_{\mathcal{T}}(A)$ to denote the set of all symbols X such that A depends on X. We can now make precise what it means for $\mathcal{T}$ to *contain no cyclic definitions*: $A \notin \mathsf{depend}_{\mathcal{T}}(A)$, for all $A \in \mathsf{N_C}$. Note that the TBox $\mathcal{T} = \{\alpha\}$ with α from Example 1 is cyclic.

In DL, data is represented in the form of ABoxes. To introduce ABoxes, we fix a set $\mathsf{N_I}$ of *individual names*, which correspond to constants in first-order logic. An *assertion* is an expression of the form $A(a)$ or $r(a, b)$, where A is a concept name, r a role name, and a, b individual names. An *ABox* $\mathcal{A}$ is just a finite set of assertions. We call the pair $\mathcal{K} = (\mathcal{T}, \mathcal{A})$ of a TBox $\mathcal{T}$ in a DL $\mathcal{L}$ and an ABox $\mathcal{A}$ an $\mathcal{L}$ *knowledge base* (KB, for short). By $\mathsf{ind}(\mathcal{A})$ and $\mathsf{ind}(\mathcal{K})$, we denote the set of individual names in $\mathcal{A}$ and $\mathcal{K}$, respectively.

To interpret ABoxes $\mathcal{A}$, we consider interpretations $\mathcal{I}$ that map all individual names $a \in \mathsf{ind}(\mathcal{A})$ to elements $a^{\mathcal{I}} \in \Delta^{\mathcal{I}}$ in such a way that $a^{\mathcal{I}} \neq b^{\mathcal{I}}$ if $a \neq b$ (thus, we adopt the *unique name assumption*). We say that $\mathcal{I}$ satisfies an assertion $A(a)$ if $a^{\mathcal{I}} \in C^{\mathcal{I}}$, and $r(a, b)$ if $(a^{\mathcal{I}}, b^{\mathcal{I}}) \in r^{\mathcal{I}}$. It is a *model* of an ABox $\mathcal{A}$ if it satisfies all assertions in $\mathcal{A}$, and of a KB $\mathcal{K} = (\mathcal{T}, \mathcal{A})$ if it is a model of both $\mathcal{T}$ and $\mathcal{A}$. We say that $\mathcal{K}$ is *consistent* (or *satisfiable*) if it has a model. We use the terminology introduced for TBoxes for KBs as well. For example, KBs $\mathcal{K}_1$ and $\mathcal{K}_2$ are *logically equivalent* if they have the same models (or, equivalently, entail each other).

We next introduce query answering for KBs, beginning with conjunctive queries [7–9]. An *atom* is of the form $A(x)$ or $r(x, y)$, where x, y are from a set of *variables* $\mathsf{N_V}$, A is a concept name, and r a role name. A *conjunctive query* (or CQ) is an expression of the form $q(\boldsymbol{x}) = \exists \boldsymbol{y} \, \varphi(\boldsymbol{x}, \boldsymbol{y})$, where $\boldsymbol{x}$ and $\boldsymbol{y}$ are disjoint sequences of variables and φ is a conjunction of atoms that only contain variables from $\boldsymbol{x} \cup \boldsymbol{y}$—we (ab)use set-theoretic notation for sequences where convenient. We often write $A(x) \in \boldsymbol{q}$ and $r(x, y) \in \boldsymbol{q}$ to indicate that $A(x)$ and $r(x, y)$ are conjuncts of φ. We call a CQ $\boldsymbol{q}$ *rooted* (rCQ) if every $y \in \boldsymbol{y}$ is connected to some $x \in \boldsymbol{x}$ by a path in the undirected graph whose nodes are the variables in $\boldsymbol{q}$ and edges are the pairs $\{u, v\}$ with $r(u, v) \in \boldsymbol{q}$, for some r. A *union of CQs* (UCQ) is a disjunction $q(\boldsymbol{x}) = \bigvee_i q_i(\boldsymbol{x})$ of CQs $q_i(\boldsymbol{x})$ with the same *answer variables* $\boldsymbol{x}$; it is *rooted* (rUCQ) if all the q_i are rooted. If the sequence $\boldsymbol{x}$ is empty, $q(\boldsymbol{x})$ is called a *Boolean* CQ or UCQ.

Given a UCQ $q(\boldsymbol{x}) = \bigvee_i q_i(\boldsymbol{x})$ and a KB $\mathcal{K}$, a sequence $\boldsymbol{a}$ of individual names from $\mathcal{K}$ of the same length as $\boldsymbol{x}$ is called a *certain answer to* $q(\boldsymbol{x})$ *over* $\mathcal{K}$ if, for every model $\mathcal{I}$ of $\mathcal{K}$, there exist a CQ q_i in $\boldsymbol{q}$ and a map (homomorphism) h of its variables to $\Delta^{\mathcal{I}}$ such that

- if x is the j-th element of $\boldsymbol{x}$ and a the j-th element of $\boldsymbol{a}$, then $h(x) = a^{\mathcal{I}}$;
- $A(z) \in \boldsymbol{q}$ implies $h(z) \in A^{\mathcal{I}}$, and $r(z, z') \in \boldsymbol{q}$ implies $(h(z), h(z')) \in r^{\mathcal{I}}$.

If this is the case, we write $\mathcal{K} \models q(\boldsymbol{a})$. For a Boolean UCQ $\boldsymbol{q}$, we also say that the certain answer over $\mathcal{K}$ is 'yes' if $\mathcal{K} \models \boldsymbol{q}$ and 'no' otherwise. *CQ or UCQ answering* means to decide, given a CQ or UCQ $q(\boldsymbol{x})$, a KB $\mathcal{K}$ and a tuple $\boldsymbol{a}$ from $\mathsf{ind}(\mathcal{K})$, whether $\mathcal{K} \models q(\boldsymbol{a})$.

3   Inseparability

Since there is no single inseparability relation between ontologies that is appropriate for all applications, we start by identifying basic properties that any semantic notion of inseparability between TBoxes or KBs should satisfy. We also introduce notation that will be used throughout the survey and discuss a few applications of inseparability.

For uniformity, we assume that the term 'ontology' refers to both TBoxes and KBs.

Definition 1 (inseparability). Let S be the set of ontologies (either TBoxes or KBs) formulated in a description logic $\mathcal{L}$. A map that assigns to each signature Σ an equivalence relation $\equiv_\Sigma$ on S is an *inseparability relation on* S if the following conditions hold:

(i) if $\mathcal{O}_1$ and $\mathcal{O}_2$ are logically equivalent, then $\mathcal{O}_1 \equiv_\Sigma \mathcal{O}_2$, for all signatures Σ and $\mathcal{O}_1, \mathcal{O}_2 \in S$;

(ii) $\Sigma_1 \subseteq \Sigma_2$ implies $\equiv_{\Sigma_1} \supseteq \equiv_{\Sigma_2}$, for all finite signatures Σ_1 and Σ_2.

By condition (i), an inseparability relations does not depend on the syntactic presentation of an ontology, but only on its semantics. Condition (ii) formalizes the requirement that if the set of relevant symbols increases ($\Sigma_2 \supseteq \Sigma_1$), then more ontologies become separable. Depending on the intended application, additional properties may also be required. For example, we refer the reader to [10] for a detailed discussion of robustness properties that are relevant for applications to modularity. We illustrate inseparability relations by three very basic examples.

Example 2

(1) Let S be the set of ontologies formulated in a description logic $\mathcal{L}$, and let $\mathcal{O}_1 \equiv_{\mathsf{equiv}} \mathcal{O}_2$ if and only if $\mathcal{O}_1$ and $\mathcal{O}_2$ are logically equivalent, for any $\mathcal{O}_1, \mathcal{O}_2 \in S$. Then $\equiv_{\mathsf{equiv}}$ is an inseparability relation that does not depend on the concrete signature. It is the finest inseparability relation possible. The inseparability relations considered in this survey are more coarse.

(2) Let S be the set of KBs in a description logic $\mathcal{L}$, and let $\mathcal{K}_1 \equiv_{\mathsf{sat}} \mathcal{K}_2$ if and only if $\mathcal{K}_1$ and $\mathcal{K}_2$ are equisatisfiable, for any $\mathcal{K}_1, \mathcal{K}_2 \in S$. Then $\equiv_{\mathsf{sat}}$ is another inseparability relation that does not depend on the concrete signature. It has two equivalence classes—the satisfiable KBs and the unsatisfiable KBs—and is not sufficiently fine-grained for most applications.

(3) Let S be the set of TBoxes in a description logic $\mathcal{L}$, and let $\mathcal{T}_1 \equiv_\Sigma^{\mathsf{hierarchy}} \mathcal{T}_2$ if and only if

$$\mathcal{T}_1 \models A \sqsubseteq B \quad \Longleftrightarrow \quad \mathcal{T}_2 \models A \sqsubseteq B, \quad \text{for all concept names } A, B \in \Sigma.$$

Then each relation $\equiv_\Sigma^{\mathsf{hierarchy}}$ is an inseparability relation. It distinguishes between two TBoxes if and only if they do not entail the same subsumption hierarchy over the concept names in Σ, and it is appropriate for applications that are only concerned with subsumption hierarchies such as producing a systematic catalog of vocabulary items, which is in fact the prime use of SNOMED CT[1].

As discussed in the introduction, the inseparability relations considered in this chapter are more sophisticated than those in Example 2. Details are given in

---

[1] http://www.ihtsdo.org/snomed-ct.

the subsequent sections. We remark that some versions of query inseparability that we are going to consider are, strictly speaking, not covered by Definition 1 since two signatures are involved (one for the query and one for the data). However, it is easy to extend Definition 1 accordingly.

We now present some important applications of inseparability.

Versioning. Maintaining and updating ontologies is very difficult without tools that support versioning. One can distinguish *three* approaches to versioning [11]: versioning based on syntactic difference (syntactic diff), versioning based on structural difference (structural diff), and versioning based on logical difference (logical diff). The *syntactic diff* underlies most existing version control systems used in software development [12] such as RCS, CVS, SCCS. It works with text files and represents the difference between versions as blocks of text present in one version but not in the other. As observed in [13], ontology versioning cannot rely on a purely syntactic diff operation since many syntactic differences (e.g., the order of ontology axioms) do not affect the semantics. The *structural diff* extends the syntactic diff by taking into account information about the structure of ontologies. Its main characteristic is that it regards ontologies as structured objects, such as an *is-a* taxonomy [13], a set of RDF triples [14] or a set of class defining axioms [15,16]. Though helpful, the structural diff still has no unambiguous semantic foundation and is syntax dependent. Moreover, it is tailored towards applications of ontologies that are based on the induced concept hierarchy (or some mild extension thereof), but does not capture other applications such as querying data under ontologies. In contrast, the *logical diff* [17,18] completely abstracts away from the presentation of the ontology and regards two versions of an ontology as identical if they are inseparable with respect to an appropriate inseparability relation such as concept inseparability or query inseparability. The result of the logical diff is then presented in terms of witnesses for separability.

Ontology Refinement. When extending an ontology with new concept inclusions or other statements, one usually wants to preserve the semantics of a large part Σ of its vocabulary. For example, when extending SNOMED CT with 50 additional concept names on top of the more than 300K existing ones, one wants to ensure that the meaning of unrelated parts of the vocabulary does not change. This preservation problem is formalized by demanding that the original ontology $\mathcal{O}_{original}$ and the extended ontology $\mathcal{O}_{original} \cup \mathcal{O}_{add}$ are Σ-inseparable (for an appropriate notion of inseparability) [19]. It should be noted that ontology refinement can be regarded as a versioning problem as discussed above, where $\mathcal{O}_{original}$ and $\mathcal{O}_{original} \cup \mathcal{O}_{add}$ are versions of an ontology that have to be compared.

Ontology Reuse. A frequent operation in ontology engineering is to import an existing ontology $\mathcal{O}_{im}$ into an ontology $\mathcal{O}_{host}$ that is currently being developed, with the aim of reusing the vocabulary of $\mathcal{O}_{im}$. Consider, for example, a host ontology $\mathcal{O}_{host}$ describing research projects that imports an ontology $\mathcal{O}_{im}$, which defines medical terms Σ to be used in the definition of research projects in $\mathcal{O}_{host}$. Then one typically wants to use the medical terms Σ exactly as defined in $\mathcal{O}_{im}$.

However, using those terms to define concepts in $\mathcal{O}_{host}$ might have unexpected consequences also for the terms in $\mathcal{O}_{im}$, that is, it might 'damage' the modeling of those terms. To avoid this, one wants to ensure that $\mathcal{O}_{host} \cup \mathcal{O}_{im}$ and $\mathcal{O}_{im}$ are Σ-inseparable [20]. Again, this can be regarded as a versioning problem for the ontology $\mathcal{O}_{im}$.

Modularity. Modular ontologies and the extraction of modules are an important ontology engineering challenge [21,22]. Understanding Σ-inseparability of ontologies is crucial for most approaches to this problem. For example, a very natural and popular definition of a module $\mathcal{M}$ of an ontology $\mathcal{O}$ demands that $\mathcal{M} \subseteq \mathcal{O}$ and that $\mathcal{M}$ is Σ-inseparable from $\mathcal{O}$ for the signature Σ of $\mathcal{M}$ (called self-contained module). Under this definition, the ontology $\mathcal{O}$ can be safely replaced by the module $\mathcal{M}$ in the sense specified by the inseparability relation and as far as the signature Σ of $\mathcal{M}$ is concerned. A stronger notion of module (called depleting module [23]) demands that $\mathcal{M} \subseteq \mathcal{O}$ and that $\mathcal{O} \setminus \mathcal{M}$ is Σ-inseparable from the empty ontology for the signature Σ of $\mathcal{M}$. The intuition is that the ontology statements outside of $\mathcal{M}$ should not say anything non-tautological about signature items in the module $\mathcal{M}$.

Ontology Mappings. The construction of mappings (or alignments) between ontologies is an important challenge in ontology engineering and integration [24]. Given two ontologies $\mathcal{O}_1$ and $\mathcal{O}_2$ in different signatures Σ_1 and Σ_2, the problem is to align the vocabulary items in Σ_1 with those in Σ_2 using a TBox $\mathcal{T}_{12}$ that states logical relationships between Σ_1 and Σ_2. For example, $\mathcal{T}_{12}$ could consist of statements of the form $A_1 \equiv A_2$ or $A_1 \sqsubseteq A_2$, where A_1 is a concept name in Σ_1 and A_2 is a concept name in Σ_2. When constructing such mappings, we typically do not want one ontology to interfere with the semantics of the other ontology via the mapping [25–27]. This condition can (and has been) formalized using inseparability. In fact, the non-interference requirement can be given by the condition that $\mathcal{O}_i$ and $\mathcal{O}_1 \cup \mathcal{O}_2 \cup \mathcal{T}_{12}$ are Σ_i inseparable, for $i = 1, 2$.

Knowledge Base Exchange. This application is a natural extension of *data exchange* [28], where the task is to transform a data instance $\mathcal{D}_1$ structured under a source schema Σ_1 into a data instance $\mathcal{D}_2$ structured under a target schema Σ_2 given a mapping $\mathcal{M}_{12}$ relating Σ_1 and Σ_2. In knowledge base exchange [29], we are interested in translating a KB $\mathcal{K}_1$ in a source signature Σ_1 to a KB $\mathcal{K}_2$ in a target signature Σ_2 according to a mapping given by a TBox $\mathcal{T}_{12}$ that consists of CIs and RIs in $\Sigma_1 \cup \Sigma_2$ defining concept and role names in Σ_2 in terms of concepts and roles in Σ_1. A good solution to this problem can be viewed as a KB $\mathcal{K}_2$ that it is inseparable from $\mathcal{K}_1 \cup \mathcal{T}_{12}$ with respect to a suitable Σ_2-inseparability relation.

Forgetting and Uniform Interpolation. When adapting an ontology to a new application, it is often useful to eliminate those symbols in its signature that are not relevant for the new application while retaining the semantics of the remaining ones. Another reason for eliminating symbols is predicate hiding, i.e., an ontology is to be published, but some part of it should be concealed from the public because it is confidential [30]. Moreover, one can view the elimination of symbols as an approach to ontology summary: the smaller and more focussed ontology summarizes what the original ontology says about the remaining signature items. The idea of eliminating symbols from a theory has been studied in AI under the name of forgetting a signature Σ [31]. In mathematical logic and modal logic, forgetting has been investigated under the dual notion of uniform interpolation [32–37]. Under both names, the problem has been studied extensively in DL research [38–44]. Using inseparability, we can formulate the condition that the result $\mathcal{O}_{\text{forget}}$ of eliminating Σ from $\mathcal{O}$ should not change the semantics of the remaining symbols by demanding that $\mathcal{O}$ and $\mathcal{O}_{\text{forget}}$ are $\text{sig}(\mathcal{O}) \setminus \Sigma$-inseparable for the signature $\text{sig}(\mathcal{O})$ of $\mathcal{O}$.

4 Concept Inseparability

We consider inseparability relations that distinguish TBoxes if and only if they do not entail the same concept inclusions in a selected signature.[2] The resulting *concept inseparability* relations are appropriate for applications that focus on TBox reasoning. We start by defining concept inseparability and the related notions of concept entailment and concept conservative extensions. We give illustrating examples and discuss the relationship between the three notions and their connection to logical equivalence. We then take a detailed look at concept inseparability in $\mathcal{ALC}$ and in $\mathcal{EL}$. In both cases, we first establish a model-theoretic characterization and then show how this characterization can be used to decide concept entailment with the help of automata-theoretic techniques. We also briefly discuss extensions of $\mathcal{ALC}$ and the special case of $\mathcal{EL}$ with acyclic TBoxes.

Definition 2 (concept inseparability, entailment and conservative extension). Let $\mathcal{T}_1$ and $\mathcal{T}_2$ be TBoxes formulated in some DL $\mathcal{L}$, and let Σ be a signature. Then

- the Σ-*concept difference between* $\mathcal{T}_1$ *and* $\mathcal{T}_2$ is the set $\text{cDiff}_\Sigma(\mathcal{T}_1, \mathcal{T}_2)$ of all Σ-concept inclusions (and role inclusions, if admitted by $\mathcal{L}$) α that are formulated in $\mathcal{L}$ and satisfy $\mathcal{T}_2 \models \alpha$ and $\mathcal{T}_1 \not\models \alpha$;
- $\mathcal{T}_1$ Σ-*concept entails* $\mathcal{T}_2$ if $\text{cDiff}_\Sigma(\mathcal{T}_1, \mathcal{T}_2) = \varnothing$;
- $\mathcal{T}_1$ and $\mathcal{T}_2$ are Σ-*concept inseparable* if $\mathcal{T}_1$ Σ-concept entails $\mathcal{T}_2$ and vice versa;
- $\mathcal{T}_2$ is a *concept conservative extension* of $\mathcal{T}_1$ if $\mathcal{T}_2 \supseteq \mathcal{T}_1$ and $\mathcal{T}_1$ and $\mathcal{T}_2$ are $\text{sig}(\mathcal{T}_1)$-inseparable.

We illustrate this definition by a number of examples.

Example 3 (concept entailment vs. logical entailment). If $\Sigma \supseteq \text{sig}(\mathcal{T}_1 \cup \mathcal{T}_2)$, then Σ-concept entailment is equivalent to logical entailment, that is, $\mathcal{T}_1$ Σ-concept entails $\mathcal{T}_2$ iff $\mathcal{T}_1 \models \mathcal{T}_2$. We recommend the reader to verify that this is a straightforward consequence of the definitions (it is crucial to observe that, because of our assumption on Σ, the concept inclusions in $\mathcal{T}_2$ qualify as potential members of $\text{cDiff}_\Sigma(\mathcal{T}_1, \mathcal{T}_2)$).

---

[2] For DLs that admit role inclusions, one additionally considers entailment of these.

Example 4 (definitorial extension). An important way to extend an ontology is to introduce definitions of new concept names. Let $\mathcal{T}_1$ be a TBox, say formulated in $\mathcal{ALC}$, and let $\mathcal{T}_2 = \{A \equiv C\} \cup \mathcal{T}_1$, where A is a fresh concept name. Then $\mathcal{T}_2$ is called a *definitorial extension* of $\mathcal{T}_1$. Clearly, unless $\mathcal{T}_1$ is inconsistent, we have $\mathcal{T}_1 \not\models \mathcal{T}_2$. However, $\mathcal{T}_2$ is a concept-conservative extension of $\mathcal{T}_1$. For the proof, assume that $\mathcal{T}_1 \not\models \alpha$ and $\mathrm{sig}(\alpha) \subseteq \mathrm{sig}(\mathcal{T}_1)$. We show that $\mathcal{T}_2 \not\models \alpha$. There is a model $\mathcal{I}_1$ of $\mathcal{T}_1$ such that $\mathcal{I} \not\models \alpha$. Modify $\mathcal{I}$ by setting $A^{\mathcal{I}} = C^{\mathcal{I}}$. Then, since $A \notin \mathrm{sig}(\mathcal{T}_1)$, the new $\mathcal{I}$ is still a model of $\mathcal{T}_1$ and we still have $\mathcal{I} \not\models \alpha$. Moreover, $\mathcal{I}$ satisfies $A \equiv C$, and thus is a model of $\mathcal{T}_2$. Consequently, $\mathcal{T}_2 \not\models \alpha$.

The notion of concept inseparability depends on the DL in which the separating concept inclusions can be formulated. Note that, in Definition 12, we assume that this DL is the one in which the original TBoxes are formulated. Throughout this chapter, we will thus make sure that the DL we work with is always clear from the context. We illustrate the difference that the choice of the 'separating DL' can make by two examples.

Example 5. Consider the $\mathcal{ALC}$ TBoxes

$$\mathcal{T}_1 = \{A \sqsubseteq \exists r.\top\} \quad \text{and} \quad \mathcal{T}_2 = \{A \sqsubseteq \exists r.B \sqcap \exists r.\neg B\}$$

and the signature $\Sigma = \{A, r\}$. If we view $\mathcal{T}_1$ and $\mathcal{T}_2$ as $\mathcal{ALCQ}$ TBoxes and consequently allow concept inclusions formulated in $\mathcal{ALCQ}$ to separate them, then $A \sqsubseteq (\geq 2r.\top) \in \mathrm{cDiff}_\Sigma(\mathcal{T}_1, \mathcal{T}_2)$, and so $\mathcal{T}_1$ and $\mathcal{T}_2$ are Σ-concept separable. However, $\mathcal{T}_1$ and $\mathcal{T}_2$ are Σ-concept inseparable when we only allow separation in terms of $\mathcal{ALC}$-concept inclusions. Intuitively, this is the case because, in $\mathcal{ALC}$, one cannot count the number of r-successors of an individual. We will later introduce the model-theoretic machinery required to prove such statements in a formal way.

Example 6. Consider the $\mathcal{EL}$ TBoxes

$\mathcal{T}_1 = \{\mathsf{Human} \sqsubseteq \exists\mathsf{eats}.\top, \ \mathsf{Plant} \sqsubseteq \exists\mathsf{grows\_in}.\mathsf{Area}, \ \mathsf{Vegetarian} \sqsubseteq \mathsf{Healthy}\}$,

$\mathcal{T}_2 = \mathcal{T}_1 \cup \{\mathsf{Human} \sqsubseteq \exists\mathsf{eats}.\mathsf{Food}, \ \mathsf{Food} \sqcap \mathsf{Plant} \sqsubseteq \mathsf{Vegetarian}\}$.

It can be verified that

$$\mathsf{Human} \sqcap \forall\mathsf{eats}.\mathsf{Plant} \sqsubseteq \exists\mathsf{eats}.\mathsf{Vegetarian}$$

is entailed by $\mathcal{T}_2$ but not by $\mathcal{T}_1$. If we view $\mathcal{T}_1$ and $\mathcal{T}_2$ as $\mathcal{ALC}$ TBoxes, then $\mathcal{T}_2$ is thus not a concept conservative extension of $\mathcal{T}_1$. However, we will show later that if we view $\mathcal{T}_1$ and $\mathcal{T}_2$ as $\mathcal{EL}$ TBoxes, then $\mathcal{T}_2$ is a concept conservative extension of $\mathcal{T}_1$ (i.e., $\mathcal{T}_1$ and $\mathcal{T}_2$ are $\mathrm{sig}(\mathcal{T}_1)$-inseparable in terms of $\mathcal{EL}$-concept inclusions).

As remarked in the introduction, conservative extensions are a special case of both inseparability and entailment. The former is by definition and the latter since $\mathcal{T}_2$ is a concept conservative extension of $\mathcal{T}_1 \subseteq \mathcal{T}_2$ iff $\mathcal{T}_1$ $\mathrm{sig}(\mathcal{T}_1)$-entails $\mathcal{T}_2$. Before turning our attention to specific DLs, we discuss a bit more the relationship between entailment and inseparability. On the one hand, inseparability is defined in terms of entailment and thus inseparability can be decided by two entailment checks. One might wonder about the converse direction, i.e., whether entailment can be reduced in some natural way to inseparability. This question is related to the following robustness condition.

Definition 3 (robustness under joins). A DL $\mathcal{L}$ is *robust under joins* for concept inseparability if, for all $\mathcal{L}$ TBoxes $\mathcal{T}_1$ and $\mathcal{T}_2$ and signatures Σ with $\mathsf{sig}(\mathcal{T}_1) \cap \mathsf{sig}(\mathcal{T}_2) \subseteq \Sigma$, the following are equivalent:

(i) $\mathcal{T}_1$ Σ-concept entails $\mathcal{T}_2$ in $\mathcal{L}$;
(ii) $\mathcal{T}_1$ and $\mathcal{T}_1 \cup \mathcal{T}_2$ are Σ-concept inseparable in $\mathcal{L}$.

Observe that the implication $(ii) \Rightarrow (i)$ is trivial. The converse holds for many DLs such as $\mathcal{ALC}$, $\mathcal{ALCI}$ and $\mathcal{EL}$; see [10] for details. However, there are also standard DLs such as $\mathcal{ALCH}$ for which robustness under joins fails.

Theorem 1. *If a DL $\mathcal{L}$ is robust under joins for concept inseparability, then concept entailment in $\mathcal{L}$ can be polynomially reduced to concept inseparability in $\mathcal{L}$.*

Proof. Assume that we want to decide whether $\mathcal{T}_1$ Σ-concept entails $\mathcal{T}_2$. By replacing every non-Σ-symbol X shared by $\mathcal{T}_1$ and $\mathcal{T}_2$ with a fresh symbol X_1 in $\mathcal{T}_1$ and a distinct fresh symbol X_2 in $\mathcal{T}_2$, we can achieve that $\Sigma \supseteq \mathsf{sig}(\mathcal{T}_1) \cap \mathsf{sig}(\mathcal{T}_2)$ without changing (non-)Σ-concept entailment of $\mathcal{T}_2$ by $\mathcal{T}_1$. We then have, by robustness under joins, that $\mathcal{T}_1$ Σ-concept entails $\mathcal{T}_2$ iff $\mathcal{T}_1$ and $\mathcal{T}_1 \cup \mathcal{T}_2$ are Σ-concept inseparable.   □

For DLs $\mathcal{L}$ that are not robust under joins for concept inseparability (such as $\mathcal{ALCH}$) it has not yet been investigated whether there exist natural polynomial reductions of concept entailment to concept inseparability.

4.1 Concept Inseparability for $\mathcal{ALC}$

We first give a model-theoretic characterization of concept entailment in $\mathcal{ALC}$ in terms of bisimulations and then show how this characterization can be used to obtain an algorithm for deciding concept entailment based on automata-theoretic techniques. We also discuss the complexity, which is 2ExpTime-complete, and the size of minimal counterexamples that witness inseparability.

Bisimulations are a central tool for studying the expressive power of $\mathcal{ALC}$ and of modal logics; see for example [45,46]. By a *pointed interpretation* we mean a pair $(\mathcal{I}, d)$, where $\mathcal{I}$ is an interpretation and $d \in \Delta^{\mathcal{I}}$.

Definition 4 (Σ-bisimulation). Let Σ be a finite signature and $(\mathcal{I}_1, d_1)$ and $(\mathcal{I}_2, d_2)$ pointed interpretations. A relation $S \subseteq \Delta^{\mathcal{I}_1} \times \Delta^{\mathcal{I}_2}$ is a *Σ-bisimulation* between $(\mathcal{I}_1, d_1)$ and $(\mathcal{I}_2, d_2)$ if $(d_1, d_2) \in S$ and, for all $(d, d') \in S$, the following conditions are satisfied:

(base) $d \in A^{\mathcal{I}_1}$ iff $d' \in A^{\mathcal{I}_2}$, for all $A \in \Sigma \cap \mathsf{N_C}$;

(zig) if $(d, e) \in r^{\mathcal{I}_1}$, then there exists $e' \in \Delta^{\mathcal{I}_2}$ such that $(d', e') \in r^{\mathcal{I}_2}$ and $(e, e') \in S$, for all $r \in \Sigma \cap \mathsf{N_R}$;

(zag) if $(d', e') \in r^{\mathcal{I}_2}$, then there exists $e \in \Delta^{\mathcal{I}_1}$ such that $(d, e) \in r^{\mathcal{I}_1}$ and $(e, e') \in S$, for all $r \in \Sigma \cap \mathsf{N_R}$.

We say that $(\mathcal{I}_1, d_1)$ and $(\mathcal{I}_2, d_2)$ are Σ-*bisimilar* and write $(\mathcal{I}_1, d_1) \sim_\Sigma^{\mathsf{bisim}} (\mathcal{I}_2, d_2)$ if there exists a Σ-bisimulation between them.

We now recall the main connection between bisimulations and $\mathcal{ALC}$. Say that $(\mathcal{I}_1, d_1)$ and $(\mathcal{I}_2, d_2)$ are $\mathcal{ALC}_\Sigma$-*equivalent*, in symbols $(\mathcal{I}_1, d_1) \equiv_\Sigma^{\mathcal{ALC}} (\mathcal{I}_2, d_2)$, in case $d_1 \in C^{\mathcal{I}_1}$ iff $d_2 \in C^{\mathcal{I}_2}$ for all Σ-concepts C in $\mathcal{ALC}$. An interpretation $\mathcal{I}$ is of *finite outdegree* if the set $\{d' \mid (d, d') \in \bigcup_{r \in \mathsf{N_R}} r^{\mathcal{I}}\}$ is finite, for any $d \in \Delta^{\mathcal{I}}$.

Theorem 2. *Let $(\mathcal{I}_1, d_1)$ and $(\mathcal{I}_2, d_2)$ be pointed interpretations and Σ a signature. Then $(\mathcal{I}_1, d_1) \sim_\Sigma^{\mathsf{bisim}} (\mathcal{I}_2, d_2)$ implies $(\mathcal{I}_1, d_1) \equiv_\Sigma^{\mathcal{ALC}} (\mathcal{I}_2, d_2)$. The converse holds if $\mathcal{I}_1$ and $\mathcal{I}_2$ are of finite outdegree.*

Example 7. The following classical example shows that without the condition of finite outdegree, the converse direction does not hold.

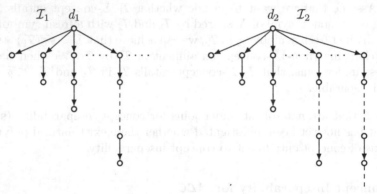

Here, $(\mathcal{I}_1, d_1)$ is a pointed interpretation with an r-chain of length n starting from d_1, for each $n \geq 1$. $(\mathcal{I}_2, d_2)$ coincides with $(\mathcal{I}_1, d_1)$ except that it also contains an infinite r-chain starting from d_2. Let $\Sigma = \{r\}$. It can be proved that $(\mathcal{I}_1, d_1) \equiv_\Sigma^{\mathcal{ALC}} (\mathcal{I}_2, d_2)$. However, $(\mathcal{I}_1, d_1) \not\sim_\Sigma^{\mathsf{bisim}} (\mathcal{I}_2, d_2)$ due to the infinite chain in $(\mathcal{I}_2, d_2)$.

As a first application of Theorem 2, we note that $\mathcal{ALC}$ cannot distinguish between an interpretation and its *unraveling* into a tree. An interpretation $\mathcal{I}$ is called a *tree interpretation* if $r^{\mathcal{I}} \cap s^{\mathcal{I}} = \varnothing$ for any $r \neq s$ and the directed graph $(\Delta^{\mathcal{I}}, \bigcup_{r \in \mathsf{N_R}} r^{\mathcal{I}})$ is a (possibly infinite) tree. The root of $\mathcal{I}$ is denoted by $\rho^{\mathcal{I}}$. By the unraveling technique [45], one can show that every pointed interpretation $(\mathcal{I}, d)$ is Σ-bisimilar to a pointed tree interpretation $(\mathcal{I}^*, \rho^{\mathcal{I}^*})$, for any signature Σ. Indeed, suppose $(\mathcal{I}, d)$ is given. The domain $\Delta^{\mathcal{I}^*}$ of $\mathcal{I}^*$ is the set of words $w = d_0 r_0 d_1 \cdots r_n d_n$ such that $d_0 = d$ and $(d_i, d_{i+1}) \in r_i^{\mathcal{I}}$ for all $i < n$ and roles names r_i. We set $\mathsf{tail}(d_0 r_0 d_1 \cdots r_n d_n) = d_n$ and define the interpretation $A^{\mathcal{I}^*}$ and $r^{\mathcal{I}^*}$ of concept names A and role names r by setting:

- $w \in A^{\mathcal{I}^*}$ if $\mathsf{tail}(w) \in A^{\mathcal{I}^*}$;
- $(w, w') \in r^{\mathcal{I}^*}$ if $w' = wrd'$.

The following lemma can be proved by a straightforward induction.

Lemma 1. *The relation* $S = \{(w, \mathsf{tail}(w)) \mid w \in \Delta^{\mathcal{I}^*}\}$ *is a* Σ*-bisimulation between* $(\mathcal{I}^*, \rho^{\mathcal{I}^*})$ *and* $(\mathcal{I}, d)$, *for any signature* Σ.

We now characterize concept entailment (and thus also concept inseparability and concept conservative extensions) in $\mathcal{ALC}$ using bisimulations, following [40].

Theorem 3. *Let* $\mathcal{T}_1$ *and* $\mathcal{T}_2$ *be* $\mathcal{ALC}$ *TBoxes and* Σ *a signature. Then* $\mathcal{T}_1$ Σ-*concept entails* $\mathcal{T}_2$ *iff, for any model* $\mathcal{I}_1$ *of* $\mathcal{T}_1$ *and any* $d_1 \in \Delta^{\mathcal{I}_1}$, *there exist a model* $\mathcal{I}_2$ *of* $\mathcal{T}_2$ *and* $d_2 \in \Delta^{\mathcal{I}_2}$ *such that* $(\mathcal{I}_1, d_1) \sim_{\Sigma}^{\mathsf{bisim}} (\mathcal{I}_2, d_2)$.

For $\mathcal{I}_1$ of finite outdegree, one can prove this result directly by employing compactness arguments and Theorem 2. For the general case, we refer to [40]. We illustrate Theorem 3 by sketching a proof of the statement from Example 5 (in a slightly more general form).

Example 8. Consider the $\mathcal{ALC}$ TBoxes

$$\mathcal{T}_1 = \{A \sqsubseteq \exists r.\top\} \cup \mathcal{T} \quad \text{and} \quad \mathcal{T}_2 = \{A \sqsubseteq \exists r.B \sqcap \exists r.\neg B\} \cup \mathcal{T},$$

where $\mathcal{T}$ is an $\mathcal{ALC}$ TBox and $B \notin \Sigma = \{A, r\} \cup \mathsf{sig}(\mathcal{T})$. We use Theorem 3 and Lemma 1 to show that $\mathcal{T}_1$ Σ-concept entails $\mathcal{T}_2$. Suppose $\mathcal{I}$ is a model of $\mathcal{T}_1$ and $d \in \Delta^{\mathcal{I}}$. Using tree unraveling, we construct a tree model $\mathcal{I}^*$ of $\mathcal{T}_1$ with $(\mathcal{I}, d) \sim_{\Sigma}^{\mathsf{bisim}} (\mathcal{I}^*, \rho^{\mathcal{I}^*})$. As bisimulations and $\mathcal{ALC}$ TBoxes are oblivious to duplication of successors, we find a tree model $\mathcal{J}$ of $\mathcal{T}_1$ such that $e \in A^{\mathcal{J}}$ implies $\#\{d \mid (e, d) \in r^{\mathcal{J}}\} \geq 2$ for all $e \in \Delta^{\mathcal{J}}$ and $(\mathcal{I}^*, \rho^{\mathcal{I}^*}) \sim_{\Sigma}^{\mathsf{bisim}} (\mathcal{J}, \rho^{\mathcal{J}})$. By reinterpreting $B \notin \Sigma$, we can find $\mathcal{J}'$ that coincides with $\mathcal{J}$ except that now we ensure that $e \in A^{\mathcal{J}}$ implies $e \in (\exists r.B \sqcap \exists r.\neg B)^{\mathcal{J}'}$ for all $e \in \Delta^{\mathcal{J}}$. But then $\mathcal{J}'$ is a model of $\mathcal{T}_2$ and $(\mathcal{I}, d) \sim_{\Sigma}^{\mathsf{bisim}} (\mathcal{J}', \rho^{\mathcal{J}})$, as required.
    Below, we illustrate possible interpretations $\mathcal{I}^*$, $\mathcal{J}$ and $\mathcal{J}'$ satisfying the above conditions, for a given interpretation $\mathcal{I}$.

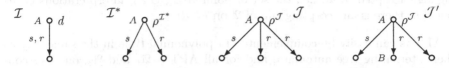

Theorem 3 is a useful starting point for constructing decision procedures for concept entailment in $\mathcal{ALC}$ and related problems. This can be done from first principles as in [19, 47]. Here we present an approach that uses tree automata. We use amorphous alternating parity tree automata [48], which actually run on unrestricted interpretations rather than on trees. They still belong to the family of *tree* automata as they are in the tradition of more classical forms of such automata and cannot distinguish between an interpretation and its unraveling into a tree (which indicates a connection to bisimulations).

Definition 5 (APTA). An (*amorphous*) *alternating parity tree automaton* (or APTA for short) is a tuple $\mathcal{A} = (Q, \Sigma_N, \Sigma_E, q_0, \delta, \Omega)$, where Q is a finite set of *states*, $\Sigma_N \subseteq N_C$ is the finite *node alphabet*, $\Sigma_E \subseteq N_R$ is the finite *edge alphabet*, $q_0 \in Q$ is the *initial state*, $\delta : Q \to \text{mov}(\mathcal{A})$ is the transition function with $\text{mov}(\mathcal{A}) = \{\text{true}, \text{false}, A, \neg A, q, q \wedge q', q \vee q', \langle r \rangle q, [r]q \mid A \in \Sigma_N, q, q' \in Q, r \in \Sigma_E\}$ the set of *moves* of the automaton, and $\Omega : Q \to \mathbb{N}$ is the *priority function*.

Intuitively, the move q means that the automaton sends a copy of itself in state q to the element of the interpretation that it is currently processing, $\langle r \rangle q$ means that a copy in state q is sent to an r-successor of the current element, and $[r]q$ means that a copy in state q is sent to every r-successor.

It will be convenient to use unrestricted modal logic formulas in negation normal form when specifying the transition function of APTAs. The more restricted form required by Definition 5 can then be attained by introducing intermediate states. We next introduce the semantics of APTAs.

In what follows, a Σ-*labelled tree* is a pair (T, ℓ) with T a tree and $\ell : T \to \Sigma$ a node labelling function. A *path* π in a tree T is a subset of T such that $\varepsilon \in \pi$ and for each $x \in \pi$ that is not a leaf in T, π contains one child of x.

Definition 6 (run). Let $(\mathcal{I}, d_0)$ be a pointed $\Sigma_N \cup \Sigma_E$-interpretation and let $\mathfrak{A} = (Q, \Sigma_N, \Sigma_E, q_0, \delta, \Omega)$ be an APTA. A *run* of $\mathfrak{A}$ on $(\mathcal{I}, d_0)$ is a $Q \times \Delta^{\mathcal{I}}$-labelled tree (T, ℓ) such that $\ell(\varepsilon) = (q_0, d_0)$ and for every $x \in T$ with $\ell(x) = (q, d)$:

- $\delta(q) \neq \text{false}$;
- if $\delta(q) = A$ ($\delta(q) = \neg A$), then $d \in A^{\mathcal{I}}$ ($d \notin A^{\mathcal{I}}$);
- if $\delta(q) = q' \wedge q''$, then there are children y, y' of x with $\ell(y) = (q', d)$ and $\ell(y') = (q'', d)$; item if $\delta(q) = q' \vee q''$, then there is a child y of x such that $\ell(y) = (q', d)$ or $\ell(y') = (q'', d)$;
- if $\delta(q) = \langle r \rangle q'$, then there is a $(d, d') \in r^{\mathcal{I}}$ and a child y of x such that $\ell(y) = (q', d')$;
- if $\delta(q) = [r]q'$ and $(d, d') \in r^{\mathcal{I}}$, then there is a child y of x with $\ell(y) = (q', d')$.

A run (T, ℓ) is *accepting* if, for every path π of T, the maximal $i \in \mathbb{N}$ with $\{x \in \pi \mid \ell(x) = (q, d) \text{ with } \Omega(q) = i\}$ infinite is even. We use $L(\mathfrak{A})$ to denote the language accepted by $\mathfrak{A}$, i.e., the set of pointed $\Sigma_N \cup \Sigma_E$-interpretations $(\mathcal{I}, d)$ such that there is an accepting run of $\mathfrak{A}$ on $(\mathcal{I}, d)$.

APTAs can easily be complemented in polynomial time in the same way as other alternating tree automata, and for all APTAs $\mathfrak{A}_1$ and $\mathfrak{A}_2$, one can construct in polynomial time an APTA that accepts $L(\mathfrak{A}_1) \cap L(\mathfrak{A}_2)$. The emptiness problem for APTAs is ExpTime-complete [48].

We now describe how APTAs can be used to decide concept entailment in $\mathcal{ALC}$. Let $\mathcal{T}_1$ and $\mathcal{T}_2$ be $\mathcal{ALC}$ TBoxes and Σ a signature. By Theorem 3, $\mathcal{T}_1$ does not Σ-concept entail $\mathcal{T}_2$ iff there is a model $\mathcal{I}_1$ of $\mathcal{T}_1$ and a $d_1 \in \Delta^{\mathcal{I}_1}$ such that $(\mathcal{I}_1, d_1) \not\sim_\Sigma^{\text{bisim}} (\mathcal{I}_2, d_2)$ for all models $\mathcal{I}_2$ of $\mathcal{T}_2$ and $d_2 \in \Delta^{\mathcal{I}_2}$. We first observe that this still holds when we restrict ourselves to *rooted interpretations*, that is, to pointed interpretations $(\mathcal{I}_i, d_i)$ such that every $e \in \Delta^{\mathcal{I}_i}$ is reachable from d_i by some sequence of role names. In fact, whenever $(\mathcal{I}_1, d_1) \not\sim_\Sigma^{\mathsf{bisim}} (\mathcal{I}_2, d_2)$, then also $(\mathcal{I}_1^r, d_1) \not\sim_\Sigma^{\mathsf{bisim}} (\mathcal{I}_2^r, d_2)$ where $\mathcal{I}_i^r$ is the restriction of $\mathcal{I}_i$ to the elements reachable from d_i. Moreover, if $\mathcal{I}_1$ and $\mathcal{I}_2$ are models of $\mathcal{T}_1$ and $\mathcal{T}_2$, respectively, then the same is true for $\mathcal{I}_1^r$ and $\mathcal{I}_2^r$. Rootedness is important because APTAs can obviously not speak about unreachable parts of a pointed interpretation. We now construct two APTAs $\mathfrak{A}_1$ and $\mathfrak{A}_2$ such that for all rooted $\mathsf{sig}(\mathcal{T}_1)$-interpretations $(\mathcal{I}, d)$,

1. $(\mathcal{I}, d) \in L(\mathfrak{A}_1)$ iff $\mathcal{I} \models \mathcal{T}_1$;
2. $(\mathcal{I}, d) \in L(\mathfrak{A}_2)$ iff there exist a model $\mathcal{J}$ of $\mathcal{T}_2$ and an $e \in \Delta^{\mathcal{J}}$ such that $(\mathcal{I}, d) \sim_\Sigma^{\mathsf{bisim}} (\mathcal{J}, e)$.

Defining $\mathfrak{A}$ as $\mathfrak{A}_1 \cap \overline{\mathfrak{A}_2}$, we then have $L(\mathfrak{A}) = \varnothing$ iff $\mathcal{T}_1$ Σ-concept entails $\mathcal{T}_2$. It is easy to construct the automaton $\mathfrak{A}_1$. We only illustrate the idea by an example. Assume that $\mathcal{T}_1 = \{A \sqsubseteq \neg \forall r.B\}$. We first rewrite $\mathcal{T}_1$ into the equivalent TBox $\{\top \sqsubseteq \neg A \sqcup \exists r. \neg B\}$ and then use an APTA with only state q_0 and

$$\delta(q_0) = \bigwedge_{s \in N_R} [s] q_0 \wedge (\neg A \vee \langle r \rangle \neg B).$$

The acceptance condition is trivial, that is, $\Omega(q_0) = 0$. The construction of $\mathfrak{A}_2$ is more interesting. We require the notion of a type, which occurs in many constructions for $\mathcal{ALC}$. Let $\mathsf{cl}(\mathcal{T}_2)$ denote the set of concepts used in $\mathcal{T}_2$, closed under subconcepts and single negation. A type t is a set $t \subseteq \mathsf{cl}(\mathcal{T}_2)$ such that, for some model $\mathcal{I}$ of $\mathcal{T}_2$ and some $d \in \Delta^{\mathcal{I}}$, we have $t = \{C \in \mathsf{cl}(\mathcal{T}_2) \mid d \in C^{\mathcal{I}}\}$. Let $\mathsf{TP}(\mathcal{T}_2)$ denote the set of all types for $\mathcal{T}_2$. For $t, t' \in \mathsf{TP}(\mathcal{T}_2)$ and a role name r, we write $t \leadsto_r t'$ if (i) $\forall r.C \in t$ implies $C \in t'$ and (ii) $C \in t'$ implies $\exists r.C \in t$ whenever $\exists r.C \in \mathsf{cl}(\mathcal{T})$. Now we define $\mathfrak{A}_2$ to have state set $Q = \mathsf{TP}(\mathcal{T}_2) \uplus \{q_0\}$ and the following transitions:

$$\delta(q_0) = \bigvee \mathsf{TP}(\mathcal{T}_2),$$
$$\delta(t) = \bigwedge_{A \in t \cap N_C \cap \Sigma} A \wedge \bigwedge_{A \in (N_C \cap \Sigma) \setminus t} \neg A$$
$$\wedge \bigwedge_{r \in \Sigma \cap N_R} [r] \bigvee \{t' \in \mathsf{TP}(\mathcal{T}_2) \mid t \leadsto_r t'\}$$
$$\wedge \bigwedge_{\exists r.C \in t, r \in \Sigma} \langle r \rangle \bigvee \{t' \in \mathsf{TP}(\mathcal{T}_2) \mid t \leadsto_r t', C \in t'\}.$$

Here, the empty conjunction represents true and the empty disjunction represents false. The acceptance condition is again trivial, but note that this might change with complementation. The idea is that $\mathfrak{A}_2$ (partially) guesses a model $\mathcal{J}$ that is Σ-bisimilar to the input interpretation $\mathcal{I}$, represented as types. Note that $\mathfrak{A}_2$ verifies only the Σ-part of $\mathcal{J}$ on $\mathcal{I}$, and that it might label the same element with different types (which can then only differ in their non-Σ-parts). A detailed proof that the above automaton works as expected is provided in [40]. In summary, we obtain the upper bound in the following theorem.

Theorem 4. *In $\mathcal{ALC}$, concept entailment, concept inseparability, and concept conservative extensions are $2\mathrm{ExpTime}$-complete.*

The sketched APTA-based decision procedure actually yields an upper bound that is slightly stronger than what is stated in Theorem 4: the algorithm for concept entailment (and concept conservative extensions) actually runs in time $2^{p(|\mathcal{T}_1|\cdot 2^{|\mathcal{T}_2|})}$ for some polynomial $p()$ and is thus only single exponential in $|\mathcal{T}_1|$. For simplicity, in the remainder of the chapter we will typically not explicitly report on such fine-grained upper bounds that distinguish between different inputs.

The lower bound stated in Theorem 4 is proved (for concept conservative extensions) in [19] using a rather intricate reduction of the word problem of exponentially space bounded alternating Turing machines (ATMs). An interesting issue that is closely related to computational hardness is to analyze the size of the smallest concept inclusions that witness non-Σ-concept entailment of a TBox $\mathcal{T}_2$ by a TBox $\mathcal{T}_1$, that is, of the members of $\mathsf{cDiff}_\Sigma(\mathcal{T}_1, \mathcal{T}_2)$. It is shown in [19] for the case of concept conservative extensions in $\mathcal{ALC}$ (and thus also for concept entailment) that smallest witness inclusions can be triple exponential in size, but not larger. An example that shows why witness inclusions can get large is given in Sect. 4.3.

4.2 Concept Inseparability for Extensions of $\mathcal{ALC}$

We briefly discuss results on concept inseparability for extensions of $\mathcal{ALC}$ and give pointers to the literature.

In principle, the machinery and results that we have presented for $\mathcal{ALC}$ can be adapted to many extensions of $\mathcal{ALC}$, for example, with number restrictions, inverse roles, and role inclusions. To achieve this, the notion of bisimulation has to be adapted to match the expressive power of the considered DL and the automata construction has to be modified. In particular, amorphous automata as used above are tightly linked to the expressive power of $\mathcal{ALC}$ and have to be replaced by traditional alternating tree automata (running on trees with fixed outdegree) which requires a slightly more technical automaton construction.

As an illustration, we only give some brief examples. To obtain an analogue of Theorem 2 for $\mathcal{ALCI}$, one needs to extend bisimulations that additionally respect successors reachable by an inverse role; to obtain such a result for $\mathcal{ALCQ}$, we need bisimulations that respect the number of successors [45,46,49]. Corresponding versions of Theorem 3 can then be proved using techniques from [46,49].

Example 9. Consider the $\mathcal{ALCQ}$ TBoxes

$$\mathcal{T}_1 = \{A \sqsubseteq \; \geq 2\, r.\top\} \cup \mathcal{T} \quad \text{and} \quad \mathcal{T}_2 = \{A \sqsubseteq \exists r.B \sqcap \exists r.\neg B\} \cup \mathcal{T},$$

where $\mathcal{T}$ is an $\mathcal{ALCQ}$ TBox that does not use the concept name B. Suppose $\Sigma = \{A, r\} \cup \mathsf{sig}(\mathcal{T})$. Then $\mathcal{T}_1$ and $\mathcal{T}_2$ are Σ-concept inseparable in $\mathcal{ALCQ}$. Formally, this can be shown using the characterizations from [49].

The above approach has not been fully developed in the literature. However, using more elementary methods, the following complexity result has been established in [47].

Theorem 5. *In $\mathcal{ALCQI}$, concept entailment, concept inseparability, and concept conservative extensions are 2ExpTime-complete.*

It is also shown in [47] that, in $\mathcal{ALCQI}$, smallest counterexamples are still triple exponential, and that further adding nominals to $\mathcal{ALCQI}$ results in undecidability.

Theorem 6. *In $\mathcal{ALCQIO}$, concept entailment, concept inseparability, and concept conservative extensions are undecidable.*

For a number of prominent extensions of $\mathcal{ALC}$, concept inseparability has not yet been investigated in much detail. This particularly concerns extensions with transitive roles [50]. We note that it is not straightforward to lift the above techniques to DLs with transitive roles; see [51] where conservative extensions in modal logics with transitive frames are studied and [36] in which modal logics with bisimulation quantifiers (which are implicit in Theorem 3) are studied, including cases with transitive frame classes. As illustrated in Sect. 8, extensions of $\mathcal{ALC}$ with the universal role are also an interesting subject to study.

4.3   Concept Inseparability for $\mathcal{EL}$

We again start with model-theoretic characterizations and then proceed to decision procedures, complexity, and the length of counterexamples. In contrast to $\mathcal{ALC}$, we use simulations, which intuitively are 'half a bisimulation', much like $\mathcal{EL}$ is 'half of $\mathcal{ALC}$'. The precise definition is as follows.

Definition 7 (Σ-simulation). Let Σ be a finite signature and $(\mathcal{I}_1, d_1)$, $(\mathcal{I}_2, d_2)$ pointed interpretations. A relation $S \subseteq \Delta^{\mathcal{I}_1} \times \Delta^{\mathcal{I}_2}$ is a Σ-*simulation* from $(\mathcal{I}_1, d_1)$ to $(\mathcal{I}_2, d_2)$ if $(d_1, d_2) \in S$ and, for all $(d, d') \in S$, the following conditions are satisfied:

(base$^\ell$) if $d \in A^{\mathcal{I}_1}$, then $d' \in A^{\mathcal{I}_2}$, for all $A \in \Sigma \cap \mathsf{N_C}$;
(zig) if $(d, e) \in r^{\mathcal{I}_1}$, then there exists $e' \in \Delta^{\mathcal{I}_2}$ such that $(d', e') \in r^{\mathcal{I}_2}$ and $(e, e') \in S$, for all $r \in \Sigma \cap \mathsf{N_R}$.

We say that $(\mathcal{I}_2, d_2)$ Σ-*simulates* $(\mathcal{I}_1, d_1)$ and write $(\mathcal{I}_1, d_1) \leq_\Sigma^{\mathsf{sim}} (\mathcal{I}_2, d_2)$ if there exist a Σ-simulation from $(\mathcal{I}_1, d_1)$ to $(\mathcal{I}_2, d_2)$. We say that $(\mathcal{I}_1, d_1)$ and $(\mathcal{I}_2, d_2)$ are Σ-*equisimilar*, in symbols $(\mathcal{I}_1, d_1) \sim_\Sigma^{\mathsf{esim}} (\mathcal{I}_2, d_2)$, if both $(\mathcal{I}_1, d_1) \leq_\Sigma^{\mathsf{sim}} (\mathcal{I}_2, d_2)$ and $(\mathcal{I}_2, d_2) \leq_\Sigma^{\mathsf{sim}} (\mathcal{I}_1, d_1)$.

A pointed interpretation $(\mathcal{I}_1, d_1)$ is $\mathcal{EL}_\Sigma$-*contained in* $(\mathcal{I}_2, d_2)$, in symbols $(\mathcal{I}_1, d_1) \leq_\Sigma^{\mathcal{EL}} (\mathcal{I}_2, d_2)$, if $d_1 \in C^{\mathcal{I}_1}$ implies $d_2 \in C^{\mathcal{I}_2}$, for all $\mathcal{EL}_\Sigma$-concepts C. We call pointed interpretations $(\mathcal{I}_1, d_1)$ and $(\mathcal{I}_2, d_2)$ $\mathcal{EL}_\Sigma$-*equivalent*, in symbols $(\mathcal{I}_1, d_1) \equiv_\Sigma^{\mathcal{EL}} (\mathcal{I}_2, d_2)$, in case $(\mathcal{I}_1, d_1) \leq_\Sigma^{\mathcal{EL}} (\mathcal{I}_2, d_2)$ and $(\mathcal{I}_2, d_2) \leq_\Sigma^{\mathcal{EL}} (\mathcal{I}_1, d_1)$. The following was shown in [52,53].

Theorem 7. *Let $(\mathcal{I}_1, d_1)$ and $(\mathcal{I}_2, d_2)$ be pointed interpretations and Σ a signature. Then $(\mathcal{I}_1, d_1) \leq_\Sigma^{\mathsf{sim}} (\mathcal{I}_2, d_2)$ implies $(\mathcal{I}_1, d_1) \leq_\Sigma^{\mathcal{EL}} (\mathcal{I}_2, d_2)$. The converse holds if $\mathcal{I}_1$ and $\mathcal{I}_2$ are of finite outdegree.*

The interpretations given in Example 7 can be used to show that the converse direction in Theorem 7 does not hold in general (since $(\mathcal{I}_2, d_2) \not\preceq^{\mathsf{sim}}_\Sigma (\mathcal{I}_1, d_1)$). It is instructive to see pointed interpretations that are equisimilar but not bisimilar.

Example 10. Consider the interpretations $\mathcal{I}_1 = (\{d_1, e_1\}, A^{\mathcal{I}_1} = \{e_1\}, r^{\mathcal{I}_1} = \{(d_1, e_1)\})$ and $\mathcal{I}_2 = (\{d_2, e_2, e_3\}, A^{\mathcal{I}_2} = \{e_2\}, r^{\mathcal{I}_2} = \{(d_2, e_2), (d_2, e_3)\})$ and let $\Sigma = \{r, A\}$. Then $(\mathcal{I}_1, d_1)$ and $(\mathcal{I}_2, d_2)$ are Σ-equisimilar but not Σ-bisimilar.

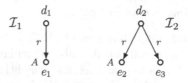

Similar to Theorem 3, Σ-equisimilarity can be used to give a model-theoretic characterization of concept entailment (and thus also concept inseparability and concept conservative extensions) in $\mathcal{EL}$ [53].

Theorem 8. *Let $\mathcal{T}_1$ and $\mathcal{T}_2$ be $\mathcal{EL}$ TBoxes and Σ a signature. Then $\mathcal{T}_1$ Σ-concept entails $\mathcal{T}_2$ iff, for any model $\mathcal{I}_1$ of $\mathcal{T}_1$ and any $d_1 \in \Delta^{\mathcal{I}_1}$, there exist a model $\mathcal{I}_2$ of $\mathcal{T}_2$ and $d_2 \in \Delta^{\mathcal{I}_2}$ such that $(\mathcal{I}_1, d_1) \sim^{\mathsf{esim}}_\Sigma (\mathcal{I}_2, d_2)$.*

We illustrate Theorem 8 by proving that the TBoxes $\mathcal{T}_1$ and $\mathcal{T}_2$ from Example 6 are Σ-concept inseparable in $\mathcal{EL}$.

Example 11. Recall that $\Sigma = \mathsf{sig}(\mathcal{T}_1)$ and

$\mathcal{T}_1 = \{\mathsf{Human} \sqsubseteq \exists\,\mathsf{eats}.\top,\ \mathsf{Plant} \sqsubseteq \exists\,\mathsf{grows\_in}.\mathsf{Area},\ \mathsf{Vegetarian} \sqsubseteq \mathsf{Healthy}\}$,

$\mathcal{T}_2 = \mathcal{T}_1 \cup \{\mathsf{Human} \sqsubseteq \exists\,\mathsf{eats}.\mathsf{Food},\ \mathsf{Food} \sqcap \mathsf{Plant} \sqsubseteq \mathsf{Vegetarian}\}$.

Let $\mathcal{I}$ be a model of $\mathcal{T}_1$ and $d \in \Delta^{\mathcal{I}}$. We may assume that $\mathsf{Food}^{\mathcal{I}} = \varnothing$. Define $\mathcal{I}'$ by adding, for every $e \in \mathsf{Human}^{\mathcal{I}}$, a fresh individual $\mathsf{new}(e)$ to $\Delta^{\mathcal{I}}$ with $(e, \mathsf{new}(e)) \in \mathsf{eats}^{\mathcal{I}'}$ and $\mathsf{new}(e) \in \mathsf{Food}^{\mathcal{I}'}$. Clearly, $\mathcal{I}'$ is a model of $\mathcal{T}_2$. We show that $(\mathcal{I}, d)$ and $(\mathcal{I}', d)$ are Σ-equisimilar. The identity $\{(e, e) \mid e \in \Delta^{\mathcal{I}}\}$ is obviously a Σ-simulation from $(\mathcal{I}, d)$ to $(\mathcal{I}', d)$. Conversely, pick for each $e \in \mathsf{Human}^{\mathcal{I}'}$ an $\mathsf{old}(e) \in \Delta^{\mathcal{I}}$ with $(e, \mathsf{old}(e)) \in \mathsf{eats}^{\mathcal{I}}$, which must exist by the first CI of $\mathcal{T}_1$. It can be verified that

$$S = \{(e, e) \mid e \in \Delta^{\mathcal{I}}\} \cup \{(\mathsf{new}(e), \mathsf{old}(e)) \mid e \in \Delta^{\mathcal{I}}\}$$

is a Σ-simulation from $(\mathcal{I}', d)$ to $(\mathcal{I}, d)$. Note that $(\mathcal{I}, d)$ and $(\mathcal{I}', d)$ are not guaranteed to be Σ-bisimilar.

As in the $\mathcal{ALC}$ case, Theorem 8 gives rise to a decision procedure for concept entailment based on tree automata. However, we can now get the complexity down to EXPTIME.[3] To achieve this, we define the automaton $\mathfrak{A}_2$ in a more careful way than for $\mathcal{ALC}$, while we do not touch the construction of $\mathfrak{A}_1$.

---

[3] An alternative elementary proof is given in [52].

Let $\mathsf{sub}(\mathcal{T}_2)$ denote the set of concepts that occur in $\mathcal{T}_2$, closed under subconcepts. For any $C \in \mathsf{sub}(\mathcal{T}_2)$, we use $\mathsf{con}_{\mathcal{T}}(C)$ to denote the set of concepts $D \in \mathsf{sub}(\mathcal{T}_2)$ such that $\mathcal{T} \models C \sqsubseteq D$. We define the APTA based on the set of states

$$Q = \{q_0\} \uplus \{q_C, \overline{q}_C \mid C \in \mathsf{sub}(\mathcal{T}_2)\},$$

where q_0 is the starting state. The transitions are as follows:

$$\delta(q_0) = \bigwedge_{C \in \mathsf{sub}(\mathcal{T}_2)} (q_C \vee \overline{q}_C) \wedge \bigwedge_{r \in \Sigma} [r]q_0,$$

$$\delta(q_A) = A \wedge \bigwedge_{C \in \mathsf{con}_{\mathcal{T}}(A)} q_C \text{ and } \delta(\overline{q}_A) = \neg A \text{ for all } A \in \mathsf{sub}(\mathcal{T}_2) \cap \mathsf{N}_\mathsf{C} \cap \Sigma,$$

$$\delta(q_{C \sqcap D}) = q_C \wedge q_D \wedge \bigwedge_{E \in \mathsf{con}_{\mathcal{T}}(C \sqcap D)} q_E \text{ and}$$

$$\delta(\overline{q}_{C \sqcap D}) = \overline{q}_C \vee \overline{q}_D \text{ for all } C \sqcap D \in \mathsf{sub}(\mathcal{T}_2),$$

$$\delta(q_{\exists r.C}) = \langle r \rangle q_C \wedge \bigwedge_{D \in \mathsf{con}_{\mathcal{T}}(\exists r.C)} q_D \text{ and}$$

$$\delta(\overline{q}_{\exists r.C}) = [r]\overline{q}_C \text{ for all } \exists r.C \in \mathsf{sub}(\mathcal{T}_2) \text{ with } r \in \Sigma,$$

$$\delta(q_\top) = \bigwedge_{C \in \mathsf{con}_{\mathcal{T}}(\top)} q_C \text{ and } \delta(\overline{q}_\top) = \mathsf{false}.$$

Observe that, in each case, the transition for $\overline{q}_C$ is the dual of the transition for q_C, except that the latter has an additional conjunction pertaining to $\mathsf{con}_{\mathcal{T}}$. As before, we set $\Omega(q) = 0$ for all $q \in Q$. An essential difference between the above APTA $\mathfrak{A}_2$ and the one that we had constructed for $\mathcal{ALC}$ is that the latter had to look at sets of subconcepts (in the form of a type) while the automaton above always considers only a single subconcept at the time. A proof that the above automaton works as expected can be extracted from [53].

Theorem 9. *In $\mathcal{EL}$, concept entailment, concept inseparability, and concept conservative extensions are* ExpTime-*complete.*

The lower bound in Theorem 9 is proved (for concept conservative extension) in [52] using a reduction of the word problem of polynomially space bounded ATMs. It can be extracted from the proofs in [52] that smallest concept inclusions that witness failure of concept entailment (or concept conservative extensions) are at most double exponentially large, measured in the size of the input TBoxes.[4] The following example shows a case where they are also at least double exponentially large.

Example 12. For each $n \geq 1$, we give TBoxes $\mathcal{T}_1$ and $\mathcal{T}_2$ whose size is polynomial in n and such that $\mathcal{T}_2$ is not a concept conservative extension of $\mathcal{T}_1$, but the elements of $\mathsf{cDiff}_\Sigma(\mathcal{T}_1, \mathcal{T}_2)$ are of size at least 2^{2^n} for $\Sigma = \mathsf{sig}(\mathcal{T}_1)$. It is instructive to start with the definition of $\mathcal{T}_2$, which is as follows:

---

[4] This should not be confused with the size of uniform interpolants, which can even be triple exponential in $\mathcal{EL}$ [54].

$$A \sqsubseteq \overline{X}_0 \sqcap \cdots \sqcap \overline{X}_{n-1},$$
$$\textstyle\prod_{\sigma \in \{r,s\}} \exists \sigma.(\overline{X}_i \sqcap X_0 \sqcap \cdots \sqcap X_{i-1}) \sqsubseteq X_i, \qquad \text{for } i < n,$$
$$\textstyle\prod_{\sigma \in \{r,s\}} \exists \sigma.(X_i \sqcap X_0 \sqcap \cdots \sqcap X_{i-1}) \sqsubseteq \overline{X}_i, \qquad \text{for } i < n,$$
$$\textstyle\prod_{\sigma \in \{r,s\}} \exists \sigma.(\overline{X}_i \sqcap \overline{X}_j) \sqsubseteq \overline{X}_i, \qquad \text{for } j < i < n,$$
$$\textstyle\prod_{\sigma \in \{r,s\}} \exists \sigma.(X_i \sqcap \overline{X}_j) \sqsubseteq X_i, \qquad \text{for } j < i < n,$$
$$X_0 \sqcap \cdots \sqcap X_{n-1} \sqsubseteq B.$$

The concept names $X_0, \ldots, X_{n-1}$ and $\overline{X}_0, \ldots, \overline{X}_{n-1}$ are used to represent a binary counter: if X_i is true, then the i-th bit is positive and if $\overline{X}_i$ is true, then it is negative. These concept names will not be used in $\mathcal{T}_1$ and thus cannot occur in $\mathsf{cDiff}_\Sigma(\mathcal{T}_1, \mathcal{T}_2)$ for the signature Σ of $\mathcal{T}_1$. Observe that Lines 2–5 implement incrementation of the counter. We are interested in consequences of $\mathcal{T}_2$ that are of the form $C_{2^n} \sqsubseteq B$, where

$$C_0 = A, \qquad C_i = \exists r.C_{i-1} \sqcap \exists s.C_{i-1},$$

which we would like to be the smallest elements of $\mathsf{cDiff}_\Sigma(\mathcal{T}_1, \mathcal{T}_2)$. Clearly, C_{2^n} is of size at least 2^{2^n}. Ideally, we would like to employ a trivial TBox $\mathcal{T}_1$ that uses only signature $\Sigma = \{A, B, r, s\}$ and has no interesting consequences (only tautologies). If we do exactly this, though, there are some undesired (single exponentially) 'small' CIs in $\mathsf{cDiff}_\Sigma(\mathcal{T}_1, \mathcal{T}_2)$, in particular $C'_n \sqsubseteq B$, where

$$C'_0 = A, \qquad C'_i = A \sqcap \exists r.C_{i-1} \sqcap \exists s.C_{i-1}.$$

Intuitively, the multiple use of A messes up our counter, making bits both true and false at the same time and resulting in all concept names X_i to become true already after travelling n steps along r. We thus have to achieve that these CIs are already consequences of $\mathcal{T}_1$. To this end, we define $\mathcal{T}_1$ as

$$\exists \sigma.A \sqsubseteq A', \qquad A' \sqcap A \sqsubseteq B', \qquad \exists \sigma.B' \sqsubseteq B', \qquad B' \sqsubseteq B$$

where σ ranges over $\{r, s\}$, and include these concept assertions also in $\mathcal{T}_2$ to achieve $\mathcal{T}_1 \subseteq \mathcal{T}_2$ as required for conservative extensions.

4.4 Concept Inseparability for Acyclic $\mathcal{EL}$ TBoxes

We show that concept inseparability for *acyclic* $\mathcal{EL}$ TBoxes can be decided in polynomial time and discuss interesting applications to versioning and the logical diff of TBoxes. We remark that TBoxes used in practice are often acyclic, and that, in fact, many biomedical ontologies such as SNOMED CT are acyclic $\mathcal{EL}$ TBoxes or mild extensions thereof.

Concept inseparability of acyclic $\mathcal{EL}$ TBoxes is still far from being a trivial problem. For example, it can be shown that smallest counterexamples from $\mathsf{cDiff}_\Sigma(\mathcal{T}_1, \mathcal{T}_2)$ can be exponential in size [11]. However, acyclic $\mathcal{EL}$ TBoxes enjoy the pleasant property that if $\mathsf{cDiff}_\Sigma(\mathcal{T}_1, \mathcal{T}_2)$ is non-empty, then it must contain a concept inclusion of the form $C \sqsubseteq A$ or $A \sqsubseteq C$, with A a concept name. This is a consequence of the following result, established in [11].

Theorem 10. *Suppose T_1 and T_2 are acyclic $\mathcal{EL}$ TBoxes and Σ a signature. If $C \sqsubseteq D \in \mathsf{cDiff}_\Sigma(T_1, T_2)$, then there exist subconcepts C' of C and D' of D such that $C' \sqsubseteq D' \in \mathsf{cDiff}_\Sigma(T_1, T_2)$, and C' or D' is a concept name.*

Theorem 10 implies that every logical difference between T_1 and T_2 is associated with a concept name from Σ (that must occur in T_2). This opens up an interesting perspective for representing the logical difference between TBoxes since, in contrast to $\mathsf{cDiff}_\Sigma(T_1, T_2)$, the set of all concept names A that are associated with a logical difference $C \sqsubseteq A$ or $A \sqsubseteq C$ is finite. One can thus summarize for the user the logical difference between two TBoxes T_1 and T_2 by presenting her with the list of all such concept names A.

Let T_1 and T_2 be acyclic $\mathcal{EL}$ TBoxes and Σ a signature. We define the set of *left-hand* Σ-concept difference witnesses $\mathsf{cWtn}^{\mathsf{lhs}}_\Sigma(T_1, T_2)$ (or *right-hand* Σ-concept difference witnesses $\mathsf{cWtn}^{\mathsf{rhs}}_\Sigma(T_1, T_2)$) as the set of all $A \in \Sigma \cap \mathsf{N_C}$ such that there exists a concept C with $A \sqsubseteq C \in \mathsf{cDiff}_\Sigma(T_1, T_2)$ (or $C \sqsubseteq A \in \mathsf{cDiff}_\Sigma(T_1, T_2)$, respectively). Note that, by Theorem 10, T_1 Σ-concept entails T_2 iff $\mathsf{cWtn}^{\mathsf{lhs}}_\Sigma(T_1, T_2) = \mathsf{cWtn}^{\mathsf{rhs}}_\Sigma(T_1, T_2) = \varnothing$. In the following, we explain how both sets can be computed in polynomial time. The constructions are from [11].

The tractability of computing $\mathsf{cWtn}^{\mathsf{lhs}}_\Sigma(T_1, T_2)$ follows from Theorem 7 and the fact that $\mathcal{EL}$ has canonical models. More specifically, for every $\mathcal{EL}$ TBox T and $\mathcal{EL}$ concept C one can construct in polynomial time a canonical pointed interpretation $(\mathcal{I}_{T,C}, d)$ such that, for any $\mathcal{EL}$ concept D, we have $d \in D^{\mathcal{I}_{T,C}}$ iff $T \models C \sqsubseteq D$. Then Theorem 7 yields for any $A \in \Sigma$ that

$$A \in \mathsf{cWtn}^{\mathsf{lhs}}_\Sigma(T_1, T_2) \quad \Longleftrightarrow \quad (\mathcal{I}_2, d_2) \not\leq^{\mathsf{sim}}_\Sigma (\mathcal{I}_1, d_1)$$

where $(\mathcal{I}_i, d_i)$ are canonical pointed interpretations for T_i and A, $i = 1, 2$. Since the existence of a simulation between polynomial size pointed interpretations can be decided in polynomial time [55], we have proved the following result.

Theorem 11. *For $\mathcal{EL}$ TBoxes T_1 and T_2 and a signature Σ, $\mathsf{cWtn}^{\mathsf{lhs}}_\Sigma(T_1, T_2)$ can be computed in polynomial time.*

We now consider $\mathsf{cWtn}^{\mathsf{rhs}}_\Sigma(T_1, T_2)$, that is, Σ-CIs of the form $C \sqsubseteq A$. To check, for a concept name $A \in \Sigma$, whether $A \in \mathsf{cWtn}^{\mathsf{rhs}}_\Sigma(T_1, T_2)$, ideally we would like to compute all concepts C such that $T_1 \not\models C \sqsubseteq A$ and then check whether $T_2 \models C \sqsubseteq A$. Unfortunately, there are infinitely many such concepts C. Note that if $T_2 \models C \sqsubseteq A$ and C' is more specific than C in the sense that $\models C' \sqsubseteq C$, then $T_2 \models C' \sqsubseteq A$. If there is a *most specific* concept C_A among all C with $T_1 \not\models C \sqsubseteq A$, it thus suffices to compute this C_A and check whether $T_2 \models C_A \sqsubseteq A$. Intuitively, though, such a C_A is only guaranteed to exist when we admit infinitary concepts. The solution is to represent C_A not as a concept, but as a TBox. We only demonstrate this approach by an example and refer the interested reader to [11] for further details.

Example 13. (a) Suppose that $T_1 = \{A \equiv \exists r.A_1\}$, $T_2 = \{A \equiv \exists r.A_2\}$ and $\Sigma = \{A, A_1, A_2, r\}$. A concept C_A such that $T_1 \not\models C_A \sqsubseteq A$ should have neither A nor $\exists r.A_1$ as top level conjuncts. This can be captured by the CIs

$$X_A \sqsubseteq A_1 \sqcap A_2 \sqcap \exists r.(A \sqcap A_2 \sqcap \exists r.X_\Sigma), \tag{1}$$
$$X_\Sigma \sqsubseteq A \sqcap A_1 \sqcap A_2 \sqcap \exists r.X_\Sigma, \tag{2}$$

where X_A and X_Σ are fresh concept names and X_A represents the most specific concept C_A with $\mathcal{T}_1 \not\models C_A \sqsubseteq A$. We have $\mathcal{T}_2 \cup \{(1),(2)\} \models X_A \sqsubseteq A$ and thus $A \in \mathsf{cWtn}_\Sigma(\mathcal{T}_1, \mathcal{T}_2)$.

(b) Consider next $\mathcal{T}_1 = \{A \equiv \exists r.A_1 \sqcap \exists r.A_2\}$, $\mathcal{T}_2 = \{A \equiv \exists r.A_2\}$ and $\Sigma = \{A, A_1, A_2, r\}$. A concept C_A with $\mathcal{T}_1 \not\models C_A \sqsubseteq A$ should not have both $\exists r.A_1$ and $\exists r.A_2$ as top level conjuncts. Thus the most specific C_A should contain exactly one of these top level conjuncts, which gives rise to a choice. We use the CIs

$$X_A^1 \sqsubseteq A_1 \sqcap A_2 \sqcap \exists r.(A \sqcap A_2 \sqcap \exists r.X_\Sigma), \tag{3}$$
$$X_A^2 \sqsubseteq A_1 \sqcap A_2 \sqcap \exists r.(A \sqcap A_1 \sqcap \exists r.X_\Sigma), \tag{4}$$

where, intuitively, the disjunction of X_A^1 and X_A^2 represents the most specific C_A. We have $\mathcal{T}_2 \cup \{(3),(4)\} \models X_A^1 \sqsubseteq A$ and thus $A \in \mathsf{cWtn}_\Sigma(\mathcal{T}_1, \mathcal{T}_2)$.

The following result is proved by generalizing the examples given above.

Theorem 12. *For $\mathcal{EL}$ TBoxes $\mathcal{T}_1$ and $\mathcal{T}_2$ and signatures Σ, $\mathsf{cWtn}_\Sigma^{\mathsf{rhs}}(\mathcal{T}_1, \mathcal{T}_2)$ can be computed in polynomial time.*

The results stated above can be generalized to extensions of acyclic $\mathcal{EL}$ with role inclusions and domain and range restrictions and have been implemented in the CEX tool for computing logical difference [11].

An alternative approach to computing right-hand Σ-concept difference witnesses based on checking for the existence of a simulations between polynomial size hypergraphs has been introduced in [56]. It has recently been extended [57] to the case of unrestricted $\mathcal{EL}$ TBoxes; the hypergraphs then become exponential in the size of the input.

5   Model Inseparability

We consider inseparability relations according to which two TBoxes are indistinguishable w.r.t. a signature Σ in case their models coincide when restricted to Σ. A central observation is that two TBoxes are Σ-model inseparable iff they cannot be distinguished by entailment of a second-order (SO) sentence in Σ. As a consequence, model inseparability implies concept inseparability for any DL $\mathcal{L}$ and is thus language independent and very robust. It is particularly useful when a user is not committed to a certain DL or is interested in more than just terminological reasoning.

We start this section with introducing model inseparability and the related notions of model entailment and model conservative extensions. We then look at the relationship between these notions and also compare model inseparability to concept inseparability. We next discuss complexity. It turns out that model inseparability is undecidable for almost all DLs, including $\mathcal{EL}$, with the exception of some *DL-Lite* dialects. Interestingly, by restricting the signature Σ to be a set concept names, one can often restore decidability. We then move to model inseparability in the case in which one TBox is empty, which is of particular interest for applications in ontology reuse and module extraction. While this restricted case is still undecidable in $\mathcal{EL}$, it is decidable for acyclic $\mathcal{EL}$ TBoxes. We close the section by discussing approximations of model inseparability that play an important role in module extraction.

Two interpretation $\mathcal{I}$ and $\mathcal{J}$ *coincide for* a signature Σ, written $\mathcal{I} =_\Sigma \mathcal{J}$, if $\Delta^{\mathcal{I}} = \Delta^{\mathcal{J}}$ and $X^{\mathcal{I}} = X^{\mathcal{J}}$ for all $X \in \Sigma$. Our central definitions are now as follows.

Definition 8 (model inseparability, entailment and conservative extensions). Let $\mathcal{T}_1$ and $\mathcal{T}_2$ be TBoxes and let Σ be a signature. Then

- the Σ-*model difference* between $\mathcal{T}_1$ and $\mathcal{T}_2$ is the set $\mathsf{mDiff}_\Sigma(\mathcal{T}_1, \mathcal{T}_2)$ of all models $\mathcal{I}$ of $\mathcal{T}_1$ such that there does not exist a model $\mathcal{J}$ of $\mathcal{T}_2$ with $\mathcal{J} =_\Sigma \mathcal{I}$;
- $\mathcal{T}_1$ Σ-*model entails* $\mathcal{T}_2$ if $\mathsf{mDiff}_\Sigma(\mathcal{T}_1, \mathcal{T}_2) = \varnothing$;
- $\mathcal{T}_1$ and $\mathcal{T}_2$ are Σ-*model inseparable* if $\mathcal{T}_1$ Σ-model entails $\mathcal{T}_2$ and vice versa;
- $\mathcal{T}_2$ is a *model conservative extension* of $\mathcal{T}_1$ if $\mathcal{T}_2 \supseteq \mathcal{T}_1$ and $\mathcal{T}_1$ and $\mathcal{T}_2$ are $\mathsf{sig}(\mathcal{T}_1)$-model inseparable.

Similarly to concept entailment (Example 3), model entailment coincides with logical entailment when $\Sigma \supseteq \mathsf{sig}(\mathcal{T}_1 \cup \mathcal{T}_2)$. We again recommend to the reader to verify this to become acquainted with the definitions. Also, one can show as in the proof from Example 4 that definitorial extensions are always model conservative extensions.

Regarding the relationship between concept inseparability and model inseparability, we note that the latter implies the former. The proof of the following result goes through for any DL $\mathcal{L}$ that enjoys a coincidence lemma (that is, for any DL, and even when $\mathcal{L}$ is the set of all second-order sentences).

Theorem 13. *Let $\mathcal{T}_1$ and $\mathcal{T}_2$ be TBoxes formulated in some DL $\mathcal{L}$ and Σ a signature such that $\mathcal{T}_1$ Σ-model entails $\mathcal{T}_2$. Then $\mathcal{T}_1$ Σ-concept entails $\mathcal{T}_2$.*

Proof. Suppose $\mathcal{T}_1$ Σ-model entails $\mathcal{T}_2$, and let α be a Σ-inclusion in $\mathcal{L}$ such that $\mathcal{T}_2 \models \alpha$. We have to show that $\mathcal{T}_1 \models \alpha$. Let $\mathcal{I}$ be a model of $\mathcal{T}_1$. There is a model $\mathcal{J}$ of $\mathcal{T}_2$ such that $\mathcal{J} =_\Sigma \mathcal{I}$. Then $\mathcal{J} \models \alpha$, and so $\mathcal{I} \models \alpha$ since $\mathsf{sig}(\alpha) \subseteq \Sigma$. $\qquad\qquad\square$

As noted, Theorem 13 even holds when $\mathcal{L}$ is the set of all SO-sentences. Thus, if $\mathcal{T}_1$ Σ-model entails $\mathcal{T}_2$ then, for every SO-sentence φ in the signature Σ, $\mathcal{T}_2 \models \varphi$ implies $\mathcal{T}_1 \models \varphi$. It is proved in [10] that, in fact, the latter exactly characterizes Σ-model entailment.

The following example shows that concept inseparability in $\mathcal{ALCQ}$ does not imply model inseparability (similar examples can be given for any DL and even for full first-order logic [10]).

Example 14. Consider the $\mathcal{ALCQ}$ TBoxes and signature from Example 9:

$$\mathcal{T}_1 = \{A \sqsubseteq \,\geq 2r.\top\} \qquad \mathcal{T}_2 = \{A \sqsubseteq \exists r.B \sqcap \exists r.\neg B\} \qquad \Sigma = \{A, r\}.$$

We have noted in Example 5 that $\mathcal{T}_1$ and $\mathcal{T}_2$ are Σ-concept inseparable. However, it is easy to see that the following interpretation is in $\mathsf{mDiff}_\Sigma(\mathcal{T}_1, \mathcal{T}_2)$.

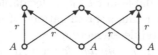

We note that the relationship between model-based notions of conservative extension and language-dependent notions of conservative extensions was also extensively discussed in the literature on software specification [58–62].

We now consider the relationship between model entailment and model inseparability. As in the concept case, model inseparability is defined in terms of model entailment and can be decided by two model entailment checks. Conversely, model entailment can be polynomially reduced to model inseparability (in constast to concept inseparability, where this depends on the DL under consideration).

Lemma 2. *In any DL $\mathcal{L}$, model entailment can be polynomially reduced to model inseparability.*

Proof. Assume that we want to decide whether $\mathcal{T}_1$ Σ-model entails $\mathcal{T}_2$ holds. By replacing every non-Σ-symbol X shared by $\mathcal{T}_1$ and $\mathcal{T}_2$ with a fresh symbol X_1 in $\mathcal{T}_1$ and a distinct fresh symbol X_2 in $\mathcal{T}_2$, we can achieve that $\Sigma \supseteq \mathrm{sig}(\mathcal{T}_1) \cap \mathrm{sig}(\mathcal{T}_2)$ without changing the original (non-)Σ-model entailment of $\mathcal{T}_2$ by $\mathcal{T}_1$. We then have that $\mathcal{T}_1$ Σ-model entails $\mathcal{T}_2$ iff $\mathcal{T}_1$ and $\mathcal{T}_1 \cup \mathcal{T}_2$ are Σ-model inseparable. □

The proof of Lemma 2 shows that any DL $\mathcal{L}$ is *robust under joins for model inseparability*, defined analogously to robustness under joins for concept inseparability; see Definition 3.

5.1 Undecidability of Model Inseparability

Model-inseparability is computationally much harder than concept inseparability. In fact, it is undecidable already for $\mathcal{EL}$ TBoxes [63]. Here, we give a short and transparent proof showing that model conservative extensions are undecidable in $\mathcal{ALC}$. The proof is by reduction of the following undecidable $\mathbb{N} \times \mathbb{N}$ *tiling problem* [64–66]: given a finite set $\mathfrak{T}$ of *tile types* T, each with four colors *left(T)*, *right(T)*, *up(T)* and *down(T)*, decide whether $\mathfrak{T}$ *tiles* the grid $\mathbb{N} \times \mathbb{N}$ in the sense that there exists a function (called a *tiling*) τ from $\mathbb{N} \times \mathbb{N}$ to $\mathfrak{T}$ such that

- $up(\tau(i,j)) = down(\tau(i, j+1))$ and
- $right(\tau(i,j)) = left(\tau(i+1, j))$.

If we think of a tile as a physical 1×1-square with a color on each of its four edges, then a tiling τ of $\mathbb{N} \times \mathbb{N}$ is just a way of placing tiles, each of a type from $\mathfrak{T}$, to cover the $\mathbb{N} \times \mathbb{N}$ grid, with no rotation of the tiles allowed and such that the colors on adjacent edges are identical.

Theorem 14. *In $\mathcal{ALC}$, model conservative extensions are undecidable.*

Proof. Given a set $\mathfrak{T}$ of tile types, we regard each $T \in \mathfrak{T}$ as a concept name and let x and y be role names. Let $\mathcal{T}_1$ be the TBox with the following CIs:

$$\top \sqsubseteq \bigsqcup_{T \in \mathfrak{T}} T,$$
$$T \sqcap T' \sqsubseteq \bot, \quad \text{for } T \neq T',$$
$$T \sqcap \exists x.T' \sqsubseteq \bot, \quad \text{for } \mathsf{right}(T) \neq \mathsf{left}(T'),$$
$$T \sqcap \exists y.T' \sqsubseteq \bot, \quad \text{for } \mathsf{up}(T) \neq \mathsf{down}(T'),$$
$$\top \sqsubseteq \exists x.\top \sqcap \exists y.\top.$$

Let $\mathcal{T}_2 = \mathcal{T}_1 \cup \mathcal{T}$, where $\mathcal{T}$ consists of a single CI:

$$\top \sqsubseteq \exists u.(\exists x.B \sqcap \exists x.\neg B) \sqcup \exists u.(\exists y.B \sqcap \exists y.\neg B) \sqcup \exists u.(\exists x.\exists y.B \sqcap \exists y.\exists x.\neg B),$$

where u is a fresh role name and B is a fresh concept name. Let $\Sigma = \mathsf{sig}(\mathcal{T}_1)$. One can show that $\mathcal{T}$ can be satisfied in a model $\mathcal{J} =_\Sigma \mathcal{I}$ iff in $\mathcal{I}$ either x is not functional or y is not functional or $x \circ y \neq y \circ x$. It is not hard to see then that $\mathfrak{T}$ tiles $\mathbb{N} \times \mathbb{N}$ iff $\mathcal{T}_1$ and $\mathcal{T}_2$ are not Σ-model inseparable. $\qquad\square$

The only standard DLs for which model inseparability is known to be decidable are certain *DL-Lite* dialects. In fact, it is shown in [23] that Σ-model entailment between TBoxes in the extensions of *DL-Lite*$_{core}$ with Boolean operators and unqualified number restrictions is decidable. The computational complexity remains open and for the extension *DL-Lite*$_{core}^{\mathcal{H}}$ of *DL-Lite*$_{core}$ with role hierarchies, even decidability is open. The decidability proof given in [23] is by reduction to the two-sorted first-order theory of Boolean algebras (BA) combined with Presburger arithmetic (PA) for representing cardinalities of sets. The decidability of this theory, called BAPA, has been first proved in [67]. Here we do not go into the decidability proof, but confine ourselves to giving an instructive example which shows that *uncountable* models have to be considered when deciding model entailment in *DL-Lite*$_{core}$ extended with unqualified number restrictions [23].

Example 15. The TBox $\mathcal{T}_1$ states, using auxiliary role names r and s, that the extension of the concept name B is infinite:

$$\mathcal{T}_1 = \{\top \sqsubseteq \exists r.\top, \ \exists r^-.\top \sqsubseteq \exists s.\top, \ \exists s^-.\top \sqsubseteq B,$$
$$B \sqsubseteq \exists s.\top, \ (\geq 2\,s^-.\top) \sqsubseteq \bot, \ \exists r^-.\top \sqcap \exists s^-.\top \sqsubseteq \bot\}.$$

The TBox $\mathcal{T}_2$ states that p is an injective function from A to B:

$$\mathcal{T}_2 = \{A \equiv \exists p.\top, \ \exists p^-.\top \sqsubseteq B, \ (\geq 2\,p.\top) \sqsubseteq \bot, \ (\geq 2\,p^-.\top) \sqsubseteq \bot\}.$$

Let $\Sigma = \{A, B\}$. There exists an uncountable model $\mathcal{I}$ of $\mathcal{T}_1$ with uncountable $A^{\mathcal{I}}$ and at most countable $B^{\mathcal{I}}$. Thus, there is no injection from $A^{\mathcal{I}}$ to $B^{\mathcal{I}}$, and so $\mathcal{I} \in \mathsf{mDiff}_\Sigma(\mathcal{T}_1, \mathcal{T}_2)$ and $\mathcal{T}_1$ does not Σ-model entail $\mathcal{T}_2$. Observe, however, that if

$\mathcal{I}$ is a countably infinite model of $\mathcal{T}_1$, then there is always an injection from $A^{\mathcal{I}}$ to $B^{\mathcal{I}}$. Thus, in this case there exists a model $\mathcal{I}'$ of $\mathcal{T}_2$ with $\mathcal{I}' =_{\Sigma} \mathcal{I}$. It follows that uncountable models of $\mathcal{T}_1$ are needed to prove that $\mathcal{T}_1$ does not Σ-model entail $\mathcal{T}_2$.

An interesting way to make Σ-model inseparability decidable is to require that Σ contains only concept names. We show that, in this case, one can use the standard filtration technique from modal logic to show that there always exists a counterexample to Σ-model inseparability of at most exponential size (in sharp contrast to Example 15).

Lemma 3. *Suppose $\mathcal{T}_1$ and $\mathcal{T}_2$ are $\mathcal{ALC}$ TBoxes and Σ contains concept names only. If $\mathsf{mDiff}_{\Sigma}(\mathcal{T}_1, \mathcal{T}_2) \neq \varnothing$, then there is an interpretation $\mathcal{I}$ in $\mathsf{mDiff}_{\Sigma}(\mathcal{T}_1, \mathcal{T}_2)$ such that $|\Delta^{\mathcal{I}}| \leq 2^{|\mathcal{T}_1| + |\mathcal{T}_2|}$.*

Proof. Assume $\mathcal{I} \in \mathsf{mDiff}_{\Sigma}(\mathcal{T}_1, \mathcal{T}_2)$. Define an equivalence relation $\sim \subseteq \Delta^{\mathcal{I}} \times \Delta^{\mathcal{I}}$ by setting $d \sim d'$ iff, for all $C \in \mathsf{sub}(\mathcal{T}_1 \cup \mathcal{T}_2)$, we have $d \in C^{\mathcal{I}_1}$ iff $d' \in C^{\mathcal{I}_2}$. Let $[d] = \{d' \in \Delta^{\mathcal{I}} \mid d' \sim d\}$. Define an interpretation $\mathcal{I}'$ by taking

$$\Delta^{\mathcal{I}'} = \{[d] \mid d \in \Delta^{\mathcal{I}}\},$$
$$A^{\mathcal{I}'} = \{[d] \mid d \in A^{\mathcal{I}}\} \text{ for all } A \in \mathsf{sub}(\mathcal{T}_1),$$
$$r^{\mathcal{I}'} = \{([d], [d']) \mid \exists e \in [d] \exists e' \in [d'] (e, e') \in r^{\mathcal{I}}\} \text{ for all role names } r.$$

It is not difficult to show that $d \in C^{\mathcal{I}}$ iff $[d] \in C^{\mathcal{I}'}$ for all $d \in \Delta^{\mathcal{I}}$ and $C \in \mathsf{sub}(\mathcal{T}_1)$. Thus $\mathcal{I}'$ is a model of $\mathcal{T}_1$. We now show that there does not exist a model $\mathcal{J}'$ of $\mathcal{T}_2$ with $\mathcal{I}' =_{\Sigma} \mathcal{J}'$. For a proof by contradiction, assume that such a $\mathcal{J}'$ exists. We define a model $\mathcal{J}$ of $\mathcal{T}_2$ with $\mathcal{J} =_{\Sigma} \mathcal{I}$, and thus derive a contradiction to the assumption that $\mathcal{I} \in \mathsf{mDiff}_{\Sigma}(\mathcal{T}_1, \mathcal{T}_2)$. To this end, let $A^{\mathcal{J}} = A^{\mathcal{I}}$ for all $A \in \Sigma$ and set

$$A^{\mathcal{J}} = \{d \mid [d] \in A^{\mathcal{J}}\} \text{ for all } A \notin \Sigma,$$
$$r^{\mathcal{J}} = \{(d, d') \mid ([d], [d']) \in r^{\mathcal{J}}\} \text{ for all role names } r.$$

Note that the role names (which are all not in Σ), are interpreted in a 'maximal' way. It can be proved that $d \in C^{\mathcal{J}}$ iff $[d] \in C^{\mathcal{J}'}$ for all $d \in \Delta^{\mathcal{I}}$ and $C \in \mathsf{sub}(\mathcal{T}_2)$. Thus $\mathcal{J}$ is a model of $\mathcal{T}_2$ and we have derived a contradiction. □

Using the bounded model property established in Lemma 3, one can prove a $\mathrm{coNExp}^{\mathrm{NP}}$ upper bound for model inseparability. A matching lower bound and several extensions of this result are proved in [63].

Theorem 15. *In $\mathcal{ALC}$, Σ-model inseparability is $\mathrm{coNExp}^{\mathrm{NP}}$-complete when Σ is restricted to sets of concept names.*

Proof. We sketch the proof of the upper bound. It is sufficient to show that one can check in $\mathrm{NExp}^{\mathrm{NP}}$ whether $\mathsf{mDiff}_{\Sigma}(\mathcal{T}_1, \mathcal{T}_2) \neq \varnothing$. By Lemma 3, one can do this by guessing a model $\mathcal{I}$ of $\mathcal{T}_1$ of size at most $2^{|\mathcal{T}_1| + |\mathcal{T}_2|}$ and then calling an oracle to verify that there is no model $\mathcal{J}$ of $\mathcal{T}_2$ with $\mathcal{J} =_{\Sigma} \mathcal{I}$. The oracle runs in NPsince we can give it the guessed $\mathcal{I}$ as an input, thus we are asking for a model of $\mathcal{T}_2$ of size polynomial in the size of the oracle input.

The lower bound is proved in [63] by a reduction of satisfiability in circumscribed $\mathcal{ALC}$ KBs, which is known to be $\text{coNExp}^{\text{NP}}$-hard. □

In [63], Theorem 15 is generalized to $\mathcal{ALCI}$. We conjecture that it can be further extended to most standard DLs that admit the finite model property. For DLs without the finite model property such as $\mathcal{ALCQI}$, we expect that BAPA-based techniques, as used for circumscription in [68], can be employed to obtain an analog of Theorem 15.

5.2 Model Inseparability from the Empty TBox

We now consider model inseparability in the case where one TBox is empty. To motivate this important case, consider the application of ontology reuse, where one wants to import a TBox $\mathcal{T}_{\text{im}}$ into a TBox $\mathcal{T}$ that is currently being developed. Recall that the result of importing $\mathcal{T}_{\text{im}}$ in $\mathcal{T}$ is the union $\mathcal{T} \cup \mathcal{T}_{\text{im}}$ and that, when importing $\mathcal{T}_{\text{im}}$ into $\mathcal{T}$, the TBox $\mathcal{T}$ is not supposed to interfere with the modeling of the symbols from $\mathcal{T}_{\text{im}}$. We can formalize this requirement by demanding that

- $\mathcal{T} \cup \mathcal{T}_{\text{im}}$ and $\mathcal{T}_{\text{im}}$ are Σ-model inseparable for $\Sigma = \text{sig}(\mathcal{T}_{\text{im}})$.

In this scenario, one has to be prepared for the imported TBox $\mathcal{T}_{\text{im}}$ to be revised. Thus, one would like to design the importing TBox $\mathcal{T}$ such that *any* TBox $\mathcal{T}_{\text{im}}$ can be imported into $\mathcal{T}$ without undesired interaction as long as the signature of $\mathcal{T}_{\text{im}}$ is not changed. Intuitively, $\mathcal{T}$ provides a *safe interface for importing ontologies* that only share symbols from some fixed signature Σ with $\mathcal{T}$. This idea led to the definition of safety for a signature in [20]:

Definition 9. Let $\mathcal{T}$ be an $\mathcal{L}$ TBox. We say that $\mathcal{T}$ is *safe for a signature Σ under model inseparability* if $\mathcal{T} \cup \mathcal{T}_{\text{im}}$ is $\text{sig}(\mathcal{T}_{\text{im}})$-model inseparable from $\mathcal{T}_{\text{im}}$ for all $\mathcal{L}$ TBoxes $\mathcal{T}_{\text{im}}$ with $\text{sig}(\mathcal{T}) \cap \text{sig}(\mathcal{T}_{\text{im}}) \subseteq \Sigma$.

As one quantifies over all TBoxes $\mathcal{T}_{\text{im}}$ in Definition 9, safety for a signature seems hard to deal with algorithmically. Fortunately, it turns out that the quantification can be avoided. This is related to the following robustness property.[5]

Definition 10. A DL $\mathcal{L}$ is said to be *robust under replacement for model inseparability* if, for all $\mathcal{L}$ TBoxes $\mathcal{T}_1$ and $\mathcal{T}_2$ and signatures Σ, the following condition is satisfied: if $\mathcal{T}_1$ and $\mathcal{T}_2$ are Σ model inseparable, then $\mathcal{T}_1 \cup \mathcal{T}$ and $\mathcal{T}_2 \cup \mathcal{T}$ are Σ-model inseparable for all $\mathcal{L}$ TBoxes $\mathcal{T}$ with $\text{sig}(\mathcal{T}) \cap \text{sig}(\mathcal{T}_1 \cup \mathcal{T}_2) \subseteq \Sigma$.

The following has been observed in [20]. It again applies to any standard DL, and in fact even to second-order logic.

---

[5] Similar robustness properties and notions of equivalence have been discussed in logic programming, we refer the reader to [69–71] and references therein. We will discuss this robustness property further in Sect. 8.

Theorem 16. *In any DL $\mathcal{L}$, model inseparability is robust under replacement.*

Using robustness under replacement, it can be proved that safety for a signature is nothing but inseparability from the empty TBox, in this way eliminating the quantification over TBoxes used in the original definition. This has first been observed in [20]. The connection to robustness under replacement is from [63].

Theorem 17. *A TBox $\mathcal{T}$ is safe for a signature Σ under model-inseparability iff $\mathcal{T}$ is Σ-model inseparable from the empty TBox.*

Proof. Assume first that $\mathcal{T}$ is not Σ-model inseparable from $\varnothing$. Then $\mathcal{T} \cup \mathcal{T}_{im}$ is not Σ-model inseparable from $\mathcal{T}_{im}$, where $\mathcal{T}_{im}$ is the trivial Σ-TBox $\mathcal{T}_{im} = \{A \sqsubseteq A \mid A \in \Sigma \cap \mathsf{N_C}\} \cup \{\exists r.\top \sqsubseteq \top \mid r \in \Sigma \cap \mathsf{N_R}\}$. Hence $\mathcal{T}$ is not safe for Σ. Now assume $\mathcal{T}$ is Σ-model inseparable from $\varnothing$ and let $\mathcal{T}_{im}$ be a TBox such that $\mathsf{sig}(\mathcal{T}) \cap \mathsf{sig}(\mathcal{T}_{im}) \subseteq \Sigma$. Then it follows from robustness under replacement that $\mathcal{T} \cup \mathcal{T}_{im}$ is $\mathsf{sig}(\mathcal{T}_{im})$-model inseparable from $\mathcal{T}_{im}$. $\qquad\square$

By Theorem 17, deciding safety of a TBox $\mathcal{T}$ for a signature Σ under model inseparability amounts to checking Σ-model inseparability from the empty TBox. We thus consider the latter problem as an important special case of model inseparability. Unfortunately, even in $\mathcal{EL}$, model inseparability from the empty TBox is undecidable [63].

Theorem 18. *In $\mathcal{EL}$, model inseparability from the empty TBox is undecidable.*

We now consider acyclic $\mathcal{EL}$ TBoxes as an important special case. As we have mentioned before, many large-scale TBoxes are in fact acyclic $\mathcal{EL}$ TBoxes or mild extensions thereof. Interestingly, model inseparability of acyclic TBoxes from the empty TBox can be decided in polynomial time [63]. The approach is based on a characterization of model inseparability from the empty TBox in terms of certain syntactic and semantic *dependencies*. The following example shows two cases of how an acyclic $\mathcal{EL}$ TBox can fail to be model inseparable from the empty TBox. These two cases will then give rise to two types of syntactic dependencies.

Example 16

(a) Let $\mathcal{T} = \{A \sqsubseteq \exists r.B, B \sqsubseteq \exists s.E\}$ and $\Sigma = \{A, s\}$. Then $\mathcal{T}$ is not Σ-model inseparable from the empty TBox: for the interpretation $\mathcal{I}$ with $\Delta^{\mathcal{I}} = \{d\}$, $A^{\mathcal{I}} = \{d\}$, and $s^{\mathcal{I}} = \varnothing$, there is no model $\mathcal{J}$ of $\mathcal{T}$ with $\mathcal{J} =_{\Sigma} \mathcal{I}$.
(b) Let $\mathcal{T} = \{A_1 \sqsubseteq \exists r.B_1, A_2 \sqsubseteq \exists r.B_2, A \equiv B_1 \sqcap B_2\}$ and $\Sigma = \{A_1, A_2, A\}$. Then $\mathcal{T}$ is not Σ-model inseparable from the empty TBox: for the interpretation $\mathcal{I}$ with $\Delta^{\mathcal{I}} = \{d\}$, $A_1^{\mathcal{I}} = A_2^{\mathcal{I}} = \{d\}$, and $A^{\mathcal{I}} = \varnothing$, there is no model $\mathcal{J}$ of $\mathcal{T}$ with $\mathcal{J} =_{\Sigma} \mathcal{I}$.

Intuitively, in part (a) of Example 16, the reason for separability from the empty TBox is that we can start with a Σ-concept name that occurs on some left-hand side (which is A) and then deduce from it that another Σ-symbol (which is s) must be non-empty. Part (b) is of a slightly different nature.

We start with a set of Σ-concept names (which is $\{A_1, A_2\}$) and from that deduce a set of concepts that implies another Σ-concept (which is A) via a concept definition, right-to-left. It turns out that it is convenient to distinguish between these two cases also in general. We first introduce some notation. For an acyclic TBox $\mathcal{T}$, let

- $\mathsf{lhs}(\mathcal{T})$ denote the set of concept names A such that there is some CI $A \equiv C$ or $A \sqsubseteq C$ in $\mathcal{T}$;
- $\mathsf{def}(\mathcal{T})$ denote the set of concept names A such that there is a definition $A \equiv C$ in $\mathcal{T}$;
- $\mathsf{depend}_{\mathcal{T}}^{\equiv}(A)$ be defined exactly as $\mathsf{depend}_{\mathcal{T}}(A)$ in Sect. 2, except that only concept definitions $A \equiv C$ are considered while concept inclusions $A \sqsubseteq C$ are disregarded.

Definition 11. Let $\mathcal{T}$ be an acyclic $\mathcal{EL}$ TBox, Σ a signature, and $A \in \Sigma$. We say that

- A has a direct Σ-dependency in $\mathcal{T}$ if $\mathsf{depend}_{\mathcal{T}}(A) \cap \Sigma \neq \varnothing$;
- A has an indirect Σ-dependency in $\mathcal{T}$ if $A \in \mathsf{def}(\mathcal{T}) \cap \Sigma$ and there are $A_1, \ldots, A_n \in \mathsf{lhs}(\mathcal{T}) \cap \Sigma$ such that $A \notin \{A_1, \ldots, A_n\}$ and

$$\mathsf{depend}_{\mathcal{T}}^{\equiv}(A) \setminus \mathsf{def}(\mathcal{T}) \subseteq \bigcup_{1 \leq i \leq n} \mathsf{depend}_{\mathcal{T}}(A_i).$$

We say that $\mathcal{T}$ contains an (in)direct Σ-dependency if there is an $A \in \Sigma$ that has an (in)direct Σ-dependency in $\mathcal{T}$.

It is proved in [63] that, for every acyclic $\mathcal{EL}$ TBox $\mathcal{T}$ and signature Σ, $\mathcal{T}$ is Σ-model inseparable from the empty TBox iff $\mathcal{T}$ has neither direct nor indirect Σ-dependencies. It can be decided in PTIME in a straightforward way whether a given $\mathcal{EL}$ TBox contains a direct Σ-dependency. For indirect Σ-dependencies, this is less obvious since we start with a set of concept names from $\mathsf{lhs}(\mathcal{T}) \cap \Sigma$. Fortunately, it can be shown that if a concept name $A \in \Sigma$ has an indirect Σ-dependency in $\mathcal{T}$ induced by concept names $A_1, \ldots, A_n \in \mathsf{lhs}(\mathcal{T}) \cap \Sigma$, then A has an indirect Σ-dependency in $\mathcal{T}$ induced by the set of concept names $(\mathsf{lhs}(\mathcal{T}) \cap \Sigma) \setminus \{A\}$. We thus only need to consider the latter set.

Theorem 19. In $\mathcal{EL}$, model inseparability of acyclic TBoxes from the empty TBox is in PTIME.

Also in [63], Theorem 19 is extended from $\mathcal{EL}$ to $\mathcal{ELI}$, and it is shown that, in $\mathcal{ALC}$ and $\mathcal{ALCI}$, model inseparability from the empty TBox is Π_2^p-complete for acyclic TBoxes.

5.3 Locality-Based Approximations

We have seen in the previous section that model inseparability from the empty TBox is of great practical value in the context of ontology reuse, that it is undecidable even in $\mathcal{EL}$, and that decidability can (sometimes) be regained by restricting TBoxes to be acyclic. In the non-acyclic case, one option is to resort to approximations from above. This leads to the (semantic) notion of $\varnothing$-locality and its syntactic companion $\bot$-locality. We discuss the former in this section and the latter in Sect. 8.

A TBox $\mathcal{T}$ is called $\varnothing$-*local w.r.t. a signature* Σ if, for every interpretation $\mathcal{I}$, there exists a model $\mathcal{J}$ of $\mathcal{T}$ such that $\mathcal{I} =_\Sigma \mathcal{J}$ and $A^{\mathcal{J}} = r^{\mathcal{J}} = \varnothing$, for all $A \in \mathsf{N_C} \setminus \Sigma$ and $r \in \mathsf{N_R} \setminus \Sigma$; in other words, every interpretation of Σ-symbols can be trivially extended to a model of $\mathcal{T}$ by interpreting non-Σ symbols as the empty set. Note that, if $\mathcal{T}$ is $\varnothing$-local w.r.t. Σ, then it is Σ-model inseparable from the empty TBox and thus, by Theorem 17, safe for Σ under model inseparability. The following example shows that the converse does not hold.

Example 17. Let $\mathcal{T} = \{A \sqsubseteq B\}$ and $\Sigma = \{A\}$. Then $\mathcal{T}$ is Σ-model inseparable from $\varnothing$, but $\mathcal{T}$ is not $\varnothing$-local w.r.t. Σ.

In contrast to model inseparability, $\varnothing$-locality is decidable also in $\mathcal{ALC}$ and beyond, and is computationally not harder than standard reasoning tasks such as satisfiability. The next procedure for checking $\varnothing$-locality was given in [72].

Theorem 20. *Let* $\mathcal{T}$ *be an* $\mathcal{ALCQI}$ *TBox and* Σ *a signature. Suppose* $\mathcal{T}|_{\Sigma=\varnothing}$ *is obtained from* $\mathcal{T}$ *by replacing all concepts of the form* A, $\exists r.C$, $\exists r^-.C$, $(\geq n\, r.C)$ *and* $(\geq n\, r^-.C)$ *with* $\bot$ *whenever* $A \notin \Sigma$ *and* $r \notin \Sigma$. *Then* $\mathcal{T}$ *is* $\varnothing$-*local w.r.t.* Σ *iff* $\mathcal{T}|_{\Sigma=\varnothing}$ *is logically equivalent to the empty TBox.*

While Theorem 20 is stated here for $\mathcal{ALCQI}$—the most expressive DL considered in this chapter—the original result in [73] is more general and applies to $\mathcal{SHOIQ}$ knowledge bases. There is also a dual notion of Δ-locality [20], in which non-Σ symbols are interpreted as the entire domain and which can also be reduced to logical equivalence.

We also remark that, unlike model inseparability from the empty TBox, model inseparability cannot easily be reduced to logical equivalence in the style of Theorem 20.

Example 18. Let $\mathcal{T} = \{A \sqsubseteq B \sqcup C\}$, $\mathcal{T}' = \{A \sqsubseteq B\}$ and $\Sigma = \{A, B\}$. Then the TBoxes $\mathcal{T}|_{\Sigma=\varnothing}$ and $\mathcal{T}'|_{\Sigma=\varnothing}$ are logically equivalent, yet $\mathcal{T}$ is not Σ-model inseparable from $\mathcal{T}'$.

$\varnothing$-locality and its syntactic companion $\bot$-locality are prominently used in ontology modularization [20,73–75]. A subset $\mathcal{M}$ of $\mathcal{T}$ is called a $\varnothing$-*local* Σ-*module* of $\mathcal{T}$ if $\mathcal{T} \setminus \mathcal{M}$ is $\varnothing$-local w.r.t. Σ. It can be shown that every $\varnothing$-local Σ-module $\mathcal{M}$ of $\mathcal{T}$ is self-contained (that is, $\mathcal{M}$ is Σ-model inseparable from $\mathcal{T}$) and depleting (that is, $\mathcal{T} \setminus \mathcal{M}$ is $\Sigma \cup \mathsf{sig}(\mathcal{M})$-model inseparable from the empty TBox). In addition, $\varnothing$-local modules are also *subsumer-preserving*, that is, for every $A \in \Sigma \cap \mathsf{N_C}$ and $B \in \mathsf{N_C}$, if $\mathcal{T} \models A \sqsubseteq B$ then $\mathcal{M} \models A \sqsubseteq B$. This property is particular useful in modular reasoning [76–78].

A $\varnothing$-local module of a given ontology $\mathcal{T}$ for a given signature Σ can be computed in a straightforward way as follows. Starting with $\mathcal{M} = \varnothing$, iteratively add to $\mathcal{M}$ every $\alpha \in \mathcal{T}$ such that $\alpha|_{\Sigma \cup \mathsf{sig}(\mathcal{M})=\varnothing}$ is not a tautology until $\mathcal{T} \setminus \mathcal{M}$ is $\varnothing$-local w.r.t. $\Sigma \cup \mathsf{sig}(\mathcal{M})$. The resulting module might be larger than necessary because this procedure actually generates a $\varnothing$-local $\Sigma \cup \mathsf{sig}(\mathcal{M})$-module rather than only a Σ-module and because $\varnothing$-locality overapproximates model inseparability, but in most practical cases results in reasonably small modules [74].

6   Query Inseparability for KBs

In this section, we consider inseparability of KBs rather than of TBoxes. One main application of KBs is to provide access to the data stored in their ABox by means of database-style queries, also taking into account the knowledge from the TBox to compute more complete answers. This approach to querying data is known as *ontology-mediated querying* [79], and it is a core part of the *ontology-based data access* (OBDA) paradigm [80]. In many applications of KBs, a reasonable notion of inseparability is thus the one where both KBs are required to give the same answers to all relevant queries that a user might pose. Of course, such an inseparability relation depends on the class of relevant queries and on the signature that we are allowed to use in the query. We will consider the two most important query languages, which are conjunctive queries (CQs) and unions thereof (UCQs), and their rooted fragments, rCQs and rUCQs.

We start the section by introducing query inseparability of KBs and related notions of query entailment and query conservative extensions. We then discuss the connection to the logical equivalence of KBs, how the choice of a query language impacts query inseparability, and the relation between query entailment and query inseparability. Next, we give model-theoretic characterizations of query inseparability which are based on model classes that are complete for query answering and on (partial or full) homomorphisms. We then move to decidability and complexity, starting with $\mathcal{ALC}$ and then proceeding to *DL-Lite*, $\mathcal{EL}$, and Horn-$\mathcal{ALC}$. In the case of $\mathcal{ALC}$, inseparability in terms of CQs turns out to be undecidable while inseparability in terms of UCQs is decidable in 2ExpTime (and the same is true for the rooted versions of these query languages). In the mentioned Horn DLs, CQ inseparability coincides with UCQ inseparability and is decidable, with the complexity ranging from PTime for $\mathcal{EL}$ via ExpTime for *DL-Lite*$_{core}^{\mathcal{H}}$, *DL-Lite*$_{horn}^{\mathcal{H}}$, and Horn-$\mathcal{ALC}$ to 2ExpTime for Horn-$\mathcal{ALCI}$.

Definition 12 (query inseparability, entailment and conservative extensions). Let $\mathcal{K}_1$ and $\mathcal{K}_2$ be KBs, Σ a signature, and $\mathcal{Q}$ a class of queries. Then

- the Σ-$\mathcal{Q}$ *difference* between $\mathcal{K}_1$ and $\mathcal{K}_2$ is the set $\mathsf{qDiff}_{\Sigma}^{\mathcal{Q}}(\mathcal{K}_1, \mathcal{K}_2)$ of all $\boldsymbol{q}(\boldsymbol{a})$ such that $\boldsymbol{q}(\boldsymbol{x}) \in \mathcal{Q}_{\Sigma}$, $\boldsymbol{a} \subseteq \mathsf{ind}(\mathcal{A}_2)$, $\mathcal{K}_2 \models \boldsymbol{q}(\boldsymbol{a})$ and $\mathcal{K}_1 \not\models \boldsymbol{q}(\boldsymbol{a})$;
- $\mathcal{K}_1$ Σ-$\mathcal{Q}$ *entails* $\mathcal{K}_2$ if $\mathsf{qDiff}_{\Sigma}^{\mathcal{Q}}(\mathcal{K}_1, \mathcal{K}_2) = \varnothing$;
- $\mathcal{K}_1$ and $\mathcal{K}_2$ are Σ-$\mathcal{Q}$ *inseparable* if $\mathcal{K}_1$ Σ-$\mathcal{Q}$ entails $\mathcal{K}_2$ and vice versa;
- $\mathcal{K}_2$ is a $\mathcal{Q}$-*conservative extension* of $\mathcal{K}_1$ if $\mathcal{K}_2 \supseteq \mathcal{K}_1$, and $\mathcal{K}_1$ and $\mathcal{K}_2$ are $\mathsf{sig}(\mathcal{K}_1)$-$\mathcal{Q}$ inseparable.

If $\boldsymbol{q}(\boldsymbol{a}) \in \mathsf{qDiff}_{\Sigma}^{\mathcal{Q}}(\mathcal{K}_1, \mathcal{K}_2)$, then we say that $\boldsymbol{q}(\boldsymbol{a})$ Σ-$\mathcal{Q}$ *separates* $\mathcal{K}_1$ and $\mathcal{K}_2$.

Note that slight variations of the definition of query inseparability are possible; for example, one can allow signatures to also contain individual names and then consider only query answers that consist of these names [81].

Query inseparability is a coarser relationship between KBs than logical equivalence even when $\Sigma \supseteq \text{sig}(\mathcal{K}_1) \cup \text{sig}(\mathcal{K}_2)$. Recall that this is in sharp contrast to concept and model inseparability, for which we observed that they coincide with logic equivalence under analogous assumptions on Σ.

Example 19 (query inseparability and logical equivalence). Let $\mathcal{K}_i = (\mathcal{T}_i, \mathcal{A}_i)$, $i = 1, 2$, where $\mathcal{A}_1 = \{A(c)\}$, $\mathcal{T}_1 = \{A \sqsubseteq B\}$, $\mathcal{A}_2 = \{A(c), B(c)\}$, and $\mathcal{T}_2 = \varnothing$. Then $\mathcal{K}_1$ and $\mathcal{K}_2$ are Σ-UCQ inseparable for any signature Σ but clearly $\mathcal{K}_1$ and $\mathcal{K}_2$ are not logically equivalent.

This example shows that there are drastic logical differences between KBs that cannot be detected by UCQs. This means that, when we aim to replace a KB with a query inseparable one, we have significant freedom to modify the KB. In the example above, we went from a KB with a non-empty TBox to a KB with an empty TBox, which should be easier to deal with when queries have to be answered efficiently.

We now compare the notions of $\mathcal{Q}$ inseparability induced by different choices of the query language $\mathcal{Q}$. A first observation is that, for Horn DLs such as $\mathcal{EL}$, there is no difference between UCQ inseparability and CQ inseparability. The same applies to rCQs and rUCQs. This follows from the fact that KBs formulated in a Horn DL have a universal model, that is, a single model that gives the same answers to queries as the KB itself—see Sect. 6.1 for more details.[6]

Theorem 21. *Let $\mathcal{K}_1$ and $\mathcal{K}_2$ be KBs formulated in a Horn DL, and let Σ be a signature. Then*

(i) $\mathcal{K}_1$ Σ-UCQ entails $\mathcal{K}_2$ iff $\mathcal{K}_1$ Σ-CQ entails $\mathcal{K}_2$;
(ii) $\mathcal{K}_1$ Σ-rUCQ entails $\mathcal{K}_2$ iff $\mathcal{K}_1$ Σ-rCQ entails $\mathcal{K}_2$.

The equivalences above do not hold for DLs that are not Horn, as shown by the following example:

Example 20. Let $\mathcal{K}_i = (\mathcal{T}_i, \mathcal{A})$, for $i = 1, 2$, be the $\mathcal{ALC}$ KBs where $\mathcal{T}_1 = \varnothing$, $\mathcal{T}_2 = \{A \sqsubseteq B_1 \sqcup B_2\}$, and $\mathcal{A} = \{A(c)\}$. Let $\Sigma = \{A, B_1, B_2\}$. Then $\mathcal{K}_1$ Σ-CQ entails $\mathcal{K}_2$, but the UCQ (actually rUCQ) $q(x) = B_1(x) \vee B_2(x)$ shows that $\mathcal{K}_1$ does not Σ-UCQ entail $\mathcal{K}_2$.

As in the case of concept and model inseparability (of TBoxes), it is instructive to consider the connection between query entailment and query inseparability. As before, query inseparability is defined in terms of query entailment. The converse direction is harder to analyze. Recall that, for concept and model

---

[6] In fact, when we say 'Horn DL', we mean a DL in which every KB has a universal model.

inseparability, we employed robustness under joins to reduce entailment to inseparability. Robustness under joins is defined as follows for Σ-Q-inseparability: if $\Sigma \supseteq sig(\mathcal{K}_1) \cap sig(\mathcal{K}_2)$, then $\mathcal{K}_1$ Σ-Q entails $\mathcal{K}_2$ iff $\mathcal{K}_1$ and $\mathcal{K}_1 \cup \mathcal{K}_2$ are Σ-Q inseparable. Unfortunately, this property does not hold.

Example 21. Let $\mathcal{K}_i = (\mathcal{T}_i, \mathcal{A})$, $i = 1, 2$, be Horn-$\mathcal{ALC}$ KBs with

$$\mathcal{T}_1 = \{A \sqsubseteq \exists r.B \sqcap \exists r.\neg B\}, \quad \mathcal{T}_2 = \{A \sqsubseteq \exists r.B \sqcap \forall r.B\}, \quad \mathcal{A} = \{A(c)\}.$$

Let $\Sigma = \{A, B, r\}$. Then, for any class of queries Q introduced above, $\mathcal{K}_1$ Σ-Q entails $\mathcal{K}_2$ but $\mathcal{K}_1$ and $\mathcal{K}_1 \cup \mathcal{K}_2$ are not Σ-Q inseparable since $\mathcal{K}_1 \cup \mathcal{K}_2$ is not satisfiable.

Robustness under joins has not yet been studied systematically for query inseparability. While Example 21 shows that query inseparability does not enjoy robustness under joins in Horn-$\mathcal{ALC}$, it is open whether the same is true in $\mathcal{EL}$ and the *DL-Lite* family. Interestingly, there is a (non-trivial) polynomial time reduction of query entailment to query inseparability that works for many Horn DLs and does not rely on robustness under joins [81].

Theorem 22. *Σ-CQ entailment of KBs is polynomially reducible to Σ-CQ inseparability of KBs for any Horn DL containing $\mathcal{EL}$ or DL-Lite$_{core}^{\mathcal{H}}$, and contained in Horn-$\mathcal{ALCHI}$.*

6.1 Model-Theoretic Criteria for Query Inseparability

We now provide model-theoretic characterizations of query inseparability. Recall that query inseparability is defined in terms of certain answers and that, given a KB $\mathcal{K}$ and a query $q(x)$, a tuple $a \subseteq ind(\mathcal{K})$ is a certain answer to $q(x)$ over $\mathcal{K}$ iff, for every model $\mathcal{I}$ of $\mathcal{K}$, we have $\mathcal{I} \models q(a)$. It is well-known that, in many cases, it is actually not necessary to consider *all* models $\mathcal{I}$ of $\mathcal{K}$ to compute certain answers. We say that a class M of models of $\mathcal{K}$ is *complete for $\mathcal{K}$ and a class Q of queries* if, for every $q(x) \in Q$, we have $\mathcal{K} \models q(a)$ iff $\mathcal{I} \models q(a)$ for all $\mathcal{I} \in M$.

In the following, we give some important examples of model classes for which KBs are complete.

Example 22. Given an $\mathcal{ALC}$ KB $\mathcal{K} = (\mathcal{T}, \mathcal{A})$, we denote by $M_{tree}^b(\mathcal{K})$ the class of all models $\mathcal{I}$ of $\mathcal{K}$ that can be constructed by choosing, for each $a \in ind(\mathcal{A})$, a tree interpretation $\mathcal{I}_a$ (see Sect. 4.1) of outdegree bounded by $|\mathcal{T}|$ and with root u, taking their disjoint union, and then adding the pair (a, b) to $r^{\mathcal{I}}$ whenever $r(a, b) \in \mathcal{A}$. It is known that $M_{tree}^b(\mathcal{K})$ is complete for $\mathcal{K}$ and UCQs, and thus for any class of queries considered in this chapter [7]. If $\mathcal{K}$ is formulated in Horn-$\mathcal{ALC}$ or in $\mathcal{EL}$, then there is even a single model $\mathcal{C}_{\mathcal{K}}$ in $M_{tree}^b(\mathcal{K})$ such that $\{\mathcal{C}_{\mathcal{K}}\}$ is complete for $\mathcal{K}$ and UCQs, the *universal* (or *canonical*) *model* of $\mathcal{K}$ [82].

If the KB is formulated in an extension of $\mathcal{ALC}$, the class of models needs to be adapted appropriately. The only such extension we are going to consider is $\mathcal{ALCHI}$ and its fragment DL-Lite$_{core}^{\mathcal{H}}$. In this case, one needs a more liberal definition of tree interpretation where role edges can point both downwards and upwards and multi edges are allowed. We refer to the resulting class of models as M^b_{utree}.

It is well-known from model theory [83] that, for any CQ $q(x)$ and any tuple $a \subseteq \text{ind}(\mathcal{K})$, we have $\mathcal{I} \models q(a)$ for all $\mathcal{I} \in M$ iff $\prod M \models q(a)$, where $\prod M$ is the *direct product* of interpretations in M. More precisely, if $M = \{\mathcal{I}_i \mid i \in I\}$, for some set I, then $\prod M = (\Delta^{\prod M}, \cdot^{\prod M})$, where

- $\Delta^{\prod M} = \prod_{i \in I} \Delta^{\mathcal{I}_i}$ is the Cartesian product of the $\Delta^{\mathcal{I}_i}$;
- $a^{\prod M} = (a^{\mathcal{I}_i})_{i \in I}$, for any individual name a;
- $A^{\prod M} = \{(d_i)_{i \in I} \mid d_i \in A^{\mathcal{I}_i} \text{ for all } i \in I\}$, for any concept name A;
- $r^{\prod M} = \{(d_i, e_i)_{i \in I} \mid (d_i, e_i) \in r^{\mathcal{I}_i} \text{ for all } i \in I\}$, for any role name r.

It is to be noted that in general $\prod M$ is *not* a model of $\mathcal{K}$, even if every interpretation in M is.

Example 23. Two interpretations $\mathcal{I}_1$ and $\mathcal{I}_2$ are shown below together with their direct product $\mathcal{I}_1 \times \mathcal{I}_2$ (all the arrows are assumed to be labelled with r):

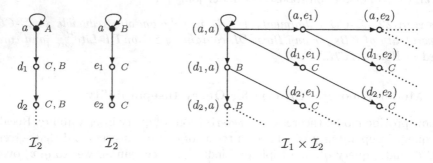

Now, consider the CQ $q_1(x) = \exists y, z\, (r(x,y) \wedge r(y,z) \wedge B(y) \wedge C(z))$. We clearly have $\mathcal{I}_1 \models q_1(a)$, $\mathcal{I}_2 \models q_1(a)$, and $\mathcal{I}_1 \times \mathcal{I}_2 \models q_1(a)$. On the other hand, for the Boolean CQ $q_2 = \exists x, y, z\, (r(x,y) \wedge r(y,z) \wedge C(y) \wedge B(z))$, we have $\mathcal{I}_1 \models q_2$ but $\mathcal{I}_2 \not\models q_2$, and so $\mathcal{I}_1 \times \mathcal{I}_2 \not\models q_2$.

Another well-known model-theoretic notion that we need for our characterizations is that of homomorphism. Let $\mathcal{I}_1$ and $\mathcal{I}_2$ be interpretations, and Σ a signature. A function $h \colon \Delta^{\mathcal{I}_2} \to \Delta^{\mathcal{I}_1}$ is a *Σ-homomorphism from $\mathcal{I}_2$ to $\mathcal{I}_1$* if

- $h(a^{\mathcal{I}_2}) = a^{\mathcal{I}_1}$ for all $a \in \mathsf{N}_\mathsf{I}$ interpreted by $\mathcal{I}_2$,
- $d \in A^{\mathcal{I}_2}$ implies $h(d) \in A^{\mathcal{I}_1}$ for all $d \in \Delta^{\mathcal{I}_2}$ and Σ-concept names A,
- $(d, e) \in r^{\mathcal{I}_2}$ implies $(h(d), h(e)) \in r^{\mathcal{I}_1}$ for all $d, e \in \Delta^{\mathcal{I}_2}$ and Σ-role names r.

It is readily seen that if $\mathcal{I}_2 \models q(a)$, for a Σ-CQ $q(x)$, and there is a Σ-homomorphism from $\mathcal{I}_2$ to $\mathcal{I}_1$, then $\mathcal{I}_1 \models q(a)$. Furthermore, if we regard $q(a)$ as an interpretation whose domain consists of the elements in a (substituted for the answer variables) and of the quantified variables in $q(x)$, and whose interpretation function is given by its atoms, then $\mathcal{I}_2 \models q(a)$ iff there exists a Σ-homomorphism from $q(a)$ to $\mathcal{I}_2$.

To give model-theoretic criteria for CQ entailment and UCQ entailment, we actually start with partial Σ-homomorphisms, which we replace by full homomorphisms in a second step. Let n be a natural number. We say that $\mathcal{I}_2$ is $n\Sigma$-homomorphically embeddable into $\mathcal{I}_1$ if, for any subinterpretation $\mathcal{I}_2'$ of $\mathcal{I}_2$ with $|\Delta^{\mathcal{I}_2'}| \leq n$, there is a Σ-homomorphism from $\mathcal{I}_2'$ to $\mathcal{I}_1$.[7] If $\mathcal{I}_2$ is $n\Sigma$-homomorphically embeddable into $\mathcal{I}_1$ for any $n > 0$, then we say that $\mathcal{I}_2$ is finitely Σ-homomorphically embeddable into $\mathcal{I}_1$.

Theorem 23. *Let $\mathcal{K}_1$ and $\mathcal{K}_2$ be KBs, Σ a signature, and $M_i^{\mathcal{Q}}$ a class of interpretations that is complete for $\mathcal{K}_i$ and the class of queries $\mathcal{Q}$, for $i = 1, 2$ and $\mathcal{Q} \in \{CQ, UCQs\}$. Then*

(i) $\mathcal{K}_1$ Σ-UCQ entails $\mathcal{K}_2$ iff, for any $n > 0$ and $\mathcal{I}_1 \in M_1^{UCQ}$, there exists $\mathcal{I}_2 \in M_2^{UCQ}$ that is $n\Sigma$-homomorphically embeddable into $\mathcal{I}_1$.
(ii) $\mathcal{K}_1$ Σ-CQ entails $\mathcal{K}_2$ iff $\prod M_2^{CQ}$ is finitely Σ-homomorphically embeddable into $\prod M_1^{CQ}$.

As finite Σ-homomorphic embeddability is harder to deal with algorithmically than full Σ-homomorphic embeddability, it would be convenient to replace finite Σ-homomorphic embeddability with Σ-homomorphic embeddability in Theorem 23. We first observe that this is not possible in general:

Example 24. Let $\mathcal{K}_i = (\mathcal{T}_i, \mathcal{A})$, $i = 1, 2$, be *DL-Lite$_{core}$* KBs where $\mathcal{A} = \{A(c)\}$, and

$$\mathcal{T}_1 = \{A \sqsubseteq \exists s.\top, \ \exists s^-.\top \sqsubseteq \exists r.\top, \ \exists r^-.\top \sqsubseteq \exists r.\top\},$$

$$\mathcal{T}_2 = \{A \sqsubseteq \exists s.\top, \ \exists s^-.\top \sqsubseteq \exists r^-.\top, \ \exists r.\top \sqsubseteq \exists r^-.\top\}.$$

Let $\Sigma = \{A, r\}$. Recall that the class of models $\{\mathcal{C}_{\mathcal{K}_i}\}$ is complete for $\mathcal{K}_i$ and UCQs, where $\mathcal{C}_{\mathcal{K}_i}$ is the canonical model of $\mathcal{K}_i$:

The KBs $\mathcal{K}_1$ and $\mathcal{K}_2$ are Σ-UCQ inseparable, but $\mathcal{C}_{\mathcal{K}_2}$ is not Σ-homomorphically embeddable into $\mathcal{C}_{\mathcal{K}_1}$.

The example above uses inverse roles and it turns out that these are indeed needed to construct counterexamples against the version of Theorem 23 where finite homomorphic embeddability is replaced with full embeddability. The following result showcases this. It concentrates on *Horn-ALC* and on *ALC*, which do not admit inverse roles, and establishes characterizations of query entailment based on full homomorphic embeddings.

---

[7] $\mathcal{I}_2'$ is a *subinterpretation* of $\mathcal{I}_2$ if $\Delta^{\mathcal{I}_2'} \subseteq \Delta^{\mathcal{I}_2}$, $A^{\mathcal{I}_2'} = A^{\mathcal{I}_2} \cap \Delta^{\mathcal{I}_2'}$ and $r^{\mathcal{I}_2'} = r^{\mathcal{I}_2} \cap (\Delta^{\mathcal{I}_2'} \times \Delta^{\mathcal{I}_2'})$, for all concept names A and role names r.

Theorem 24

(i) Let $\mathcal{K}_1$ and $\mathcal{K}_2$ be Horn-$\mathcal{ALC}$ KBs. Then $\mathcal{K}_1$ Σ-CQ entails $\mathcal{K}_2$ iff $\mathcal{C}_{\mathcal{K}_2}$ is Σ-homomorphically embeddable into $\mathcal{C}_{\mathcal{K}_1}$.

(ii) Let $\mathcal{K}_1$ and $\mathcal{K}_2$ be $\mathcal{ALC}$ KBs. Then $\mathcal{K}_1$ Σ-UCQ entails $\mathcal{K}_2$ iff, for every $\mathcal{I}_1 \in M^b_{tree}(\mathcal{K}_1)$, there exists $\mathcal{I}_2 \in M^b_{tree}(\mathcal{K}_2)$ such that $\mathcal{I}_2$ is Σ-homomorphically embeddable into $\mathcal{I}_1$.

Claim (i) of Theorem 24 is proved in [81] using a game-theoretic characterization (which we discuss below). The proof of (ii) is given in [84]. One first proves using an automata-theoretic argument that one can work without loss of generality with models in $M^b_{tree}(\mathcal{K}_1)$ in which the tree interpretations $\mathcal{I}_a$ attached to the ABox individuals a are regular. Second, since nodes in $\mathcal{I}_a$ are related to their children using role names only (as opposed to inverse roles), Σ-homomorphisms on tree interpretations correspond to Σ-simulations (see Sects. 4.3 and 6.3). Finally, using this observation one can construct the required Σ-homomorphism as the union of finite Σ-homomorphisms on finite initial parts of the tree interpretations $\mathcal{I}_a$.

Note that Theorem 24 omits the case of $\mathcal{ALC}$ KBs and CQ entailment, for which we are not aware of a characterization in terms of full homomorphic embeddability.

Another interesting aspect of Example 24 is that the canonical model of $\mathcal{K}_2$ contains elements that are not reachable along a path of Σ-roles. In fact, just like inverse roles, this is a crucial feature for the example to work. We illustrate this by considering *rooted* UCQs (rUCQs). Recall that in an rUCQ, every variables has to be connected to an answer variable. For answering a Σ-rUCQ, Σ-disconnected parts of models such as in Example 24 can essentially be ignored since the query cannot 'see' them. As a consequence, we can sometimes replace finite homomorphic embeddability with full homomorphic embeddability. We give an example characterization to illustrate this. Call an interpretation $\mathcal{I}$ Σ-*connected* if, for every $u \in \Delta^{\mathcal{I}}$, there is a path $r_1^{\mathcal{I}}(a, u_1), \ldots, r_n^{\mathcal{I}}(u_n, u)$ with an individual a and $r_i \in \Sigma$. An interpretation $\mathcal{I}_2$ is con-Σ-*homomorphically embeddable into* $\mathcal{I}_1$ if the maximal Σ-connected subinterpretation $\mathcal{I}_2'$ of $\mathcal{I}_2$ is Σ-homomorphically embeddable into $\mathcal{I}_1$.

Theorem 25. *Let $\mathcal{K}_1$ and $\mathcal{K}_2$ be $\mathcal{ALCHI}$ KBs and Σ a signature. Then $\mathcal{K}_1$ Σ-rUCQ entails $\mathcal{K}_2$ iff for any $\mathcal{I}_1 \in M^b_{utree}(\mathcal{K}_1)$, there exists $\mathcal{I}_2 \in M^b_{utree}(\mathcal{K}_2)$ that is con-Σ-homomorphically embeddable into $\mathcal{I}_1$.*

Theorem 25 is proved for $\mathcal{ALC}$ in [85]. The extension to $\mathcal{ALCHI}$ is straightforward. The model-theoretic criteria given above are a good starting point for designing decision procedures for query inseparability. But can they be checked effectively? We first consider this question for $\mathcal{ALC}$ and then move to Horn DLs.

6.2   Query Inseparability of $\mathcal{ALC}$ KBs

We begin with CQ entailment and inseparability in $\mathcal{ALC}$ and show that both problems are undecidable even for very restricted classes of KBs. The same is true for rCQs. We then show that, in contrast to the CQ case, UCQ inseparability in $\mathcal{ALC}$ is decidable in 2ExpTime.

The following example illustrates the notion of CQ-inseparability of $\mathcal{ALC}$ KBs.

Example 25. Suppose $\mathcal{T}_1 = \varnothing$, $\mathcal{T}_2 = \{E \sqsubseteq A \sqcup B\}$, $\mathcal{A}$ looks like on the left-hand side of the picture below, and $\Sigma = \{r, A, B\}$. Then we can separate $\mathcal{K}_2 = (\mathcal{T}_2, \mathcal{A})$ from $\mathcal{K}_1 = (\mathcal{T}_1, \mathcal{A})$ by the Σ-CQ $q(x)$ shown on the right-hand side of the picture since clearly $(\mathcal{T}_1, \mathcal{A}) \not\models q(a)$, whereas $(\mathcal{T}_2, \mathcal{A}) \models q(a)$. To see the latter, we first observe that, in any model $\mathcal{I}$ of $\mathcal{K}_2$, we have $(i)\ c \in A^{\mathcal{I}}$ or $(ii)\ c \in B^{\mathcal{I}}$. In case (i), $\mathcal{I} \models q(a)$ because of the path $r(a, c), r(c, d)$; and if (ii) holds, then $\mathcal{I} \models q(a)$ because of the path $r(a, b), r(b, c)$ (cf. [86, Example 4.2.5]).

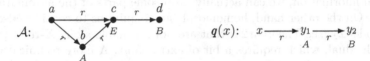

Theorem 26. *Let $\mathcal{Q} \in \{CQ, rCQ\}$.*

(i) Σ-$\mathcal{Q}$ entailment of an $\mathcal{ALC}$ KB by an $\mathcal{EL}$ KB is undecidable.
(ii) Σ-$\mathcal{Q}$ inseparability of an $\mathcal{ALC}$ and an $\mathcal{EL}$ KBs is undecidable.

The proof of this theorem given in [85] uses a reduction of an undecidable tiling problem. As usual in encodings of tilings, it is not hard to synchronize tile colours along one dimension. The following example gives a hint of how this can be achieved in the second dimension.

Example 26. Suppose a KB $\mathcal{K}$ has the two models $\mathcal{I}_i$, $i = 1, 2$, that are formed by the points on the path between a and e_i on the right-hand side of the picture below (this can be easily achieved using an inclusion of the form $D \sqsubseteq D_1 \sqcup D_2$), with a being an ABox point with a loop and the e_i being the only instances of a concept C. Let q be the CQ on the left-hand side of the picture. Then we can have $\mathcal{K} \models q(a)$ only if $d_1, d_3 \in A^{\mathcal{I}_2}$ and $d_2, d_4 \in B^{\mathcal{I}_2}$, with the fat black and grey arrows indicating homomorphisms from q to the $\mathcal{I}_i$ (the grey one sends x_0–x_2 to a using the ABox loop). This trick can be used to pass the tile colours from one row to another.

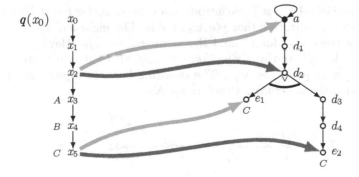

As we saw in Example 20, UCQs distinguish between more KBs than CQs, that is, UCQ inseparability is a different and in fact more fine-grained notion than CQ inseparability. This has the remarkable effect that decidability is regained [84].

Theorem 27. *In $\mathcal{ALC}$, Σ-$\mathcal{Q}$ entailment and Σ-$\mathcal{Q}$ inseparability of KBs are* 2ExpTime*-complete, for $\mathcal{Q} \in \{UCQ, rUCQ\}$.*

The proof of the upper bound in Theorem 27 uses tree automata and relies on the characterization of Theorem 24 (ii) [84]. In principle, the automata construction is similar to the one given in the proof sketch of Theorem 4. The main differences between the two constructions is that we have to replace bisimulations with homomorphisms. Since homomorphisms preserve only positive and existential information, we can actually drop some parts of the automaton construction. On the other hand, homomorphisms require us to consider also parts of the model $\mathcal{I}_2$ (see Theorem 24) that are not reachable along Σ-roles from an ABox individual, which requires a bit of extra effort. A more technical issue is that the presence of an ABox seems to not go together so well with amorphous automata and thus one resorts to more traditional tree automata along with a suitable encoding of the ABox and of the model $\mathcal{I}_1$ as a labeled tree. The lower bound is proved by an ATM reduction.

6.3   Query Inseparability of KBs in Horn DLs

We first consider DLs without inverse roles and then DLs that admit inverse roles. In both cases, we sketch decision procedures that are based on games played on canonical models $\mathcal{C}_\mathcal{K}$ as mentioned in Example 22. It is well known from logic programming and databases [87] that such models can be constructed by the *chase procedure*. We illustrate the chase (in a somewhat different but equivalent form) by the following example.

Example 27. Consider the *DL-Lite*$_{core}^{\mathcal{H}}$ KB $\mathcal{K}_2 = (\mathcal{T}_2, \mathcal{A}_2)$ with $\mathcal{A}_2 = \{A(a)\}$ and

$$\mathcal{T}_2 = \{A \sqsubseteq B, \ A \sqsubseteq \exists p.\top, \ \exists p^-.\top \sqsubseteq \exists r^-.\top, \ \exists r.\top \sqsubseteq \exists q^-.\top, \ \exists q.\top \sqsubseteq \exists q^-.\top,$$
$$\exists r.\top \sqsubseteq \exists s^-.\top, \ \exists s.\top \sqsubseteq \exists t^-.\top, \ \exists t.\top \sqsubseteq \exists s^-.\top, \ t^- \sqcap s \sqsubseteq \bot\}.$$

We first construct a 'closure' of the ABox $\mathcal{A}_2$ under the inclusions in $\mathcal{T}_2$. For instance, to satisfy $A \sqsubseteq \exists p.\top$, we introduce a *witness* w_p *for* p and draw an arrow $\rightsquigarrow$ from a to w_p indicating that $p(a, w_p)$ holds. The inclusion $\exists p^-.\top \sqsubseteq \exists r^-.\top$ requires a witness w_{r^-} for r^- and the arrow $w_p \rightsquigarrow w_{r^-}$. Having reached the witness w_{q^-} for q^- and applying $\exists q.\top \sqsubseteq \exists q^-.\top$ to it, we 'reuse' w_{q^-} and simply draw a loop $w_{q^-} \rightsquigarrow w_{q^-}$. The resulting finite interpretation $\mathcal{G}_2$ shown below is called the *generating structure for* $\mathcal{K}_2$:

Note that $\mathcal{G}_2$ is *not* a model of $\mathcal{K}_2$ because $(w_{s^-}, w_{t^-}) \in (t^-)^{\mathcal{G}_2} \cap s^{\mathcal{G}_2}$. We can obtain a model of $\mathcal{K}_2$ by *unravelling* the witness part of the generating structure $\mathcal{G}_2$ into an infinite tree (in general, forest). The resulting interpretation $\mathcal{I}_2$ shown below is a canonical model of $\mathcal{K}_2$.

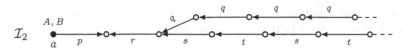

The generating structure underlying the canonical model $\mathcal{C}_\mathcal{K}$ of a Horn KB $\mathcal{K}$ defined above will be denoted by $\mathcal{G}_\mathcal{K}$. By Theorem 24, if $\mathcal{K}_1$ and $\mathcal{K}_2$ are KBs formulated in a Horn DL, then $\mathcal{K}_1$ Σ-CQ entails $\mathcal{K}_2$ iff $\mathcal{C}_{\mathcal{K}_2}$ is $n\Sigma$-homomorphically embeddable into $\mathcal{C}_{\mathcal{K}_2}$ for any $n > 0$.

In what follows, we require the following upper bounds on the size of generating structures for Horn KBs [81]:

Theorem 28

(i) *The generating structure for any consistent Horn-$\mathcal{ALCHI}$ KB $(\mathcal{T}, \mathcal{A})$ can be constructed in time $|\mathcal{A}| \cdot 2^{p(|\mathcal{T}|)}$, where p is some fixed polynomial;*

(ii) *The generating structure for any consistent KB $(\mathcal{T}, \mathcal{A})$ formulated in a DL from the $\mathcal{EL}$ or DL-Lite family can be constructed in time $|\mathcal{A}| \cdot p(|\mathcal{T}|)$, where p is some fixed polynomial.*

We now show that checking whether a canonical model is $n\Sigma$-homomorphically embeddable into another canonical model can be established by playing games on their underlying generating structures. For more details, the reader is referred to [81].

Suppose $\mathcal{K}_1$ and $\mathcal{K}_2$ are (consistent) Horn KBs, $\mathcal{C}_1$ and $\mathcal{C}_2$ are their canonical models, and Σ a signature. First, we reformulate the definition of $n\Sigma$-homomorphic embedding in game-theoretic terms. The states of our game are of the form $(\pi \mapsto \sigma)$, where $\pi \in \Delta^{\mathcal{C}_2}$ and $\sigma \in \Delta^{\mathcal{C}_1}$. Intuitively, $(\pi \mapsto \sigma)$ means that 'π is to be Σ-homomorphically mapped to σ'. The game is played by player 1 and player 2 starting from some initial state $(\pi_0 \mapsto \sigma_0)$. The aim of player 1 is to demonstrate that there exists a Σ-homomorphism from (a finite subinterpretation of) $\mathcal{C}_2$ into $\mathcal{C}_1$ with π_0 mapped to σ_0, while player 2 wants to show that there is no such homomorphism. In each round $i > 0$ of the game, player 2 challenges player 1 with some $\pi_i \in \Delta^{\mathcal{C}_2}$ that is related to π_{i-1} by some Σ-role. Player 1, in turn, has to respond with some $\sigma_i \in \Delta^{\mathcal{C}_1}$ such that the already constructed partial Σ-homomorphism can be extended with $(\pi_i \mapsto \sigma_i)$, in particular:

- $\pi_i \in A^{\mathcal{C}_2}$ implies $\sigma_i \in A^{\mathcal{C}_1}$, for any Σ-concept name A, and
- $(\pi_{i-1}, \pi_i) \in r^{\mathcal{C}_2}$ implies $(\sigma_{i-1}, \sigma_i) \in r^{\mathcal{C}_1}$, for any Σ-role r.

A play of length n starting from a state $\mathfrak{s}_0$ is any sequence $\mathfrak{s}_0, \ldots, \mathfrak{s}_n$ of states obtained as described above. For any ordinal $\lambda \le \omega$, we say that player 1 has a λ-*winning strategy* in the game starting from $\mathfrak{s}_0$ if, for any play $\mathfrak{s}_0, \ldots, \mathfrak{s}_n$ with $n < \lambda$ that is played according to this strategy, player 1 has a response to any challenge of player 2 in the final state $\mathfrak{s}_n$.

It is easy to see that if, for any $\pi_0 \in \Delta^{\mathcal{C}_2}$, there is $\sigma_0 \in \Delta^{\mathcal{C}_1}$ such that player 1 has an ω-winning strategy in this game starting from $(\pi_0 \to \sigma_0)$, then there is a Σ-homomorphism from $\mathcal{C}_2$ into $\mathcal{C}_1$, and the other way round. That $\mathcal{C}_2$ is *finitely* Σ-homomorphically embeddable into $\mathcal{C}_1$ is equivalent to the following condition:

– for any $\pi_0 \in \Delta^{\mathcal{C}_2}$ and any $n < \omega$, there exists $\sigma_0 \in \Delta^{\mathcal{C}_1}$ such that player 1 has an n-winning strategy in this game starting from $(\pi_0 \to \sigma_0)$.

Example 28. Suppose $\mathcal{C}_1$ and $\mathcal{C}_2$ look like in the picture below. An ω-winning strategy for player 1 starting from $(a \mapsto a)$ is shown by the dotted lines with the rounds of the game indicated by the numbers on the lines.

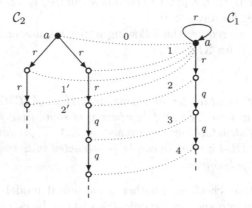

Note, however, that the game-theoretic criterion formulated above does not immediately yield any algorithm to decide finite homomorphic embeddability because both $\mathcal{C}_2$ and $\mathcal{C}_1$ can be infinite. It is readily seen that the canonical model $\mathcal{C}_2$ in the game can be replaced by the underlying generating structure $\mathcal{G}_2$, in which player 2 can only make challenges indicated by the generating relation $\rightsquigarrow$. The picture below illustrates the game played on $\mathcal{G}_2$ and $\mathcal{C}_1$ from Example 28.

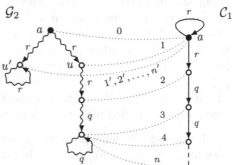

If the KBs are formulated in a Horn DL that does not allow *inverse roles*, then $\mathcal{C}_1$ can also be replaced with its generating structure $\mathcal{G}_1$ as illustrated by the picture below:

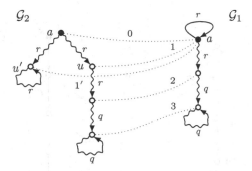

Reachability or simulation games on finite graphs such as the one discussed above have been extensively investigated in game theory [88,89]. In particular, it follows that checking the existence of n-winning strategies, for all $n < \omega$, can be done in polynomial time in the number of states and the number of the available challenges. Together with Theorem 28, this gives the upper bounds in the following theorem. Claim (i) was first observed in [52], while (ii) and the results on data complexity are from [81].

Theorem 29. Σ-CQ entailment and Σ-CQ inseparability of KBs are

(i) in PTIME for $\mathcal{EL}$;
(ii) EXPTIME-complete for Horn-$\mathcal{ALC}$.

Both problems are in PTIME for data complexity for both $\mathcal{EL}$ and Horn-$\mathcal{ALC}$.

Here, by 'data complexity' we mean that only the ABoxes of the two involved KBs are regarded as input, while the TBoxes are fixed. Analogously to data complexity in query answering, the rationale behind this setup is that, in data-centric applications, ABoxes tend to be huge compared to the TBoxes and thus it can result in more realistic complexities to assume that the latter are actually of constant size. The lower bound in Theorem 29 is proved by reduction of the word problem of polynomially space-bounded ATMs. We remind the reader at this point that, in all DLs studied in this section, CQ entailment coincides with UCQ entailment, and likewise for inseparability.

If inverse roles are available, then replacing canonical models with their generating structures in games often becomes less straightforward. We explain the issues using an example in $DL\text{-}Lite^{\mathcal{H}}_{core}$, where inverse roles interact in a problematic way with role inclusions. Similar effects can be observed in Horn-$\mathcal{ALCI}$, though, despite the fact that no role inclusions are available there.

Example 29. Consider the $DL\text{-}Lite^{\mathcal{H}}_{core}$ KBs $\mathcal{K}_1 = (\mathcal{T}_1, \{Q(u, u)\})$ with

$$\mathcal{T}_1 = \{\, A \sqsubseteq \exists s.\top,\ \exists s^-.\top \sqsubseteq \exists t.\top,\ \exists t^-.\top \sqsubseteq \exists s.\top,\ s \sqsubseteq q,\ t \sqsubseteq q,\ \exists q^-.\top \sqsubseteq \exists r.\top \,\}$$

and $\mathcal{K}_2$ from Example 27. Let $\Sigma = \{q, r, s, t\}$. The generating structure $\mathcal{G}_2$ for $\mathcal{K}_2$ and the canonical model $\mathcal{C}_1$ for $\mathcal{K}_1$, as well as a 4-winning strategy for player 1 in the game over $\mathcal{G}_2$ and $\mathcal{C}_1$ starting from the state (u_0, σ_4) are shown in the picture below:

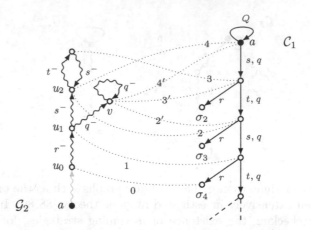

(In fact, for any $n > 0$, player 1 has an n-winning strategy starting from any $(u_0 \mapsto \sigma_m)$ provided that m is even and $m \geq n$.)

This game over $\mathcal{G}_2$ and $\mathcal{C}_1$ has its obvious counterparts over $\mathcal{G}_2$ and $\mathcal{G}_1$; one of them is shown on the left-hand side of the picture below. It is to be noted, however, that—unlike Example 28—the responses of player 1 are in the reverse direction of the $\rightsquigarrow$-arrows (which is possible because of the inverse roles).

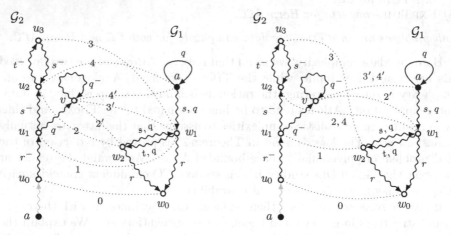

On the other hand, such reverse responses may create paths in $\mathcal{G}_1$ that do not have any real counterparts in $\mathcal{C}_1$, and so do not give rise to Σ-homomorphisms we need. An example is shown on the right-hand side of the picture above, where u_3 is mapped to w_2 and v to a in round 3, which is impossible to reproduce in the tree-shaped $\mathcal{C}_1$.

One way to ensure that, in the 'backwards game' over $\mathcal{G}_2$ and $\mathcal{G}_1$, all the challenges made by player 1 in any given state are responded by the same element of $\mathcal{G}_1$, is to use states of the form $(\varXi \mapsto w)$, where $\varXi$ is the set of elements of $\mathcal{G}_2$ to be mapped to an element w of $\mathcal{G}_1$. In our example above, we can use the state $(\{u_2, v\} \mapsto w_2)$, where the only challenge of player 2 is the set of $\rightsquigarrow$- successors of u_2 and v marked by Σ-roles, that is, $\Xi' = \{u_3, v\}$, to which player 1 responds with $(\Xi' \mapsto w_1)$.

By allowing more complex states, we increase their number and, as a consequence, the complexity of deciding finite Σ-homomorphic embeddability. The proof of the following theorem can be found in [81]:

Theorem 30. *Σ-CQ entailment and inseparability of KBs are*

(i) EXPTIME-*complete for DL-Lite$_{horn}^{\mathcal{H}}$ and DL-Lite$_{core}^{\mathcal{H}}$;*
(ii) 2EXPTIME-*complete for Horn-$\mathcal{ALCI}$ and Horn-$\mathcal{ALCHI}$.*

For all of these DLs, both problems are in PTIME *for data complexity.*

The lower bounds are once again proved using alternating Turing machines. We remark that CQ entailment and inseparability are in PTIME in *DL-Lite$_{core}$* and *DL-Lite$_{horn}$*. In *DL-Lite*, it is thus the combination of inverse roles and role hierarchies that causes hardness.

7   Query Inseparability of TBoxes

Query inseparability of KBs, as studied in the previous section, presupposes that the data (in the form of an ABox) is known to the user, as is the case for example in KB exchange [29]. In many query answering applications, though, the data is either not known during the TBox design or it changes so frequently that query inseparability w.r.t. one fixed data set is not a sufficiently robust notion. In such cases, one wants to decide query inseparability of *TBoxes* $\mathcal{T}_1$ and $\mathcal{T}_2$, defined by requiring that, for *all* ABoxes $\mathcal{A}$, the KBs $(\mathcal{T}_1, \mathcal{A})$ and $(\mathcal{T}_2, \mathcal{A})$ are query inseparable. To increase flexibility, we can also specify a signature of the ABoxes that we are considering, and we do not require that it coincides with the signature of the queries. In a sense, this change in the setup brings us closer to the material from Sects. 4 and 5, where also inseparability of TBoxes is studied. As in the KB case, the main classes of queries that we consider are CQs and UCQs as well as rCQs and rUCQs. To relate concept inseparability and query inseparability of TBoxes, we additionally consider a class of tree-shaped CQs.

The structure of this section is as follows. We start by discussing the impact that the choice of query language has on query inseparability of TBoxes. We then relate query inseparability of TBoxes to logical equivalence, concept inseparability, and model inseparability. For Horn DLs, query inseparability and concept inseparability are very closely related, while this is not the case for DLs with disjunction. Finally, we consider the decidability and complexity of query inseparability of TBoxes. Undecidability of CQ inseparability of $\mathcal{ALC}$ KBs transfers to the TBox case, and the same is true of upper complexity bounds in *DL-Lite*. New techniques are needed to establish decidability and complexity results for other Horn DLs such as $\mathcal{EL}$ and Horn-$\mathcal{ALC}$. A main observation underlying our algorithms is that it is sufficient to consider tree-shaped ABoxes when searching for witnesses of query separability of TBoxes.

Definition 13 (query inseparability, entailment and conservative extensions). Let $\mathcal{T}_1$ and $\mathcal{T}_2$ be TBoxes, $\Theta = (\Sigma_1, \Sigma_2)$ a pair of signatures, and $\mathcal{Q}$ a class of queries. Then

- the Θ-$\mathcal{Q}$ *difference* between $\mathcal{T}_1$ and $\mathcal{T}_2$ is the set $\mathsf{qDiff}_\Theta^{\mathcal{Q}}(\mathcal{T}_1, \mathcal{T}_2)$ of all pairs $(\mathcal{A}, \boldsymbol{q}(\boldsymbol{a}))$ such that $\mathcal{A}$ is a Σ_1-ABox and $\boldsymbol{q}(\boldsymbol{a}) \in \mathsf{qDiff}_{\Sigma_2}^{\mathcal{Q}}(\mathcal{K}_1, \mathcal{K}_2)$, where $\mathcal{K}_i = (\mathcal{T}_i, \mathcal{A})$ for $i = 1, 2$;
- $\mathcal{T}_1$ Θ-$\mathcal{Q}$ *entails* $\mathcal{T}_2$ if $\mathsf{qDiff}_\Theta^{\mathcal{Q}}(\mathcal{T}_1, \mathcal{T}_2) = \varnothing$;
- $\mathcal{T}_1$ and $\mathcal{T}_2$ are Θ-$\mathcal{Q}$ *inseparable* if $\mathcal{T}_1$ Θ-$\mathcal{Q}$ entails $\mathcal{T}_2$ and vice versa;
- $\mathcal{T}_2$ is a $\mathcal{Q}$ *conservative extension* of $\mathcal{T}_1$ iff $\mathcal{T}_2 \supseteq \mathcal{T}_1$ and $\mathcal{T}_1$ and $\mathcal{T}_2$ are Θ-$\mathcal{Q}$ inseparable for $\Sigma_1 = \Sigma_2 = \mathsf{sig}(\mathcal{T}_1)$.

If $(\mathcal{A}, \boldsymbol{q}(\boldsymbol{a})) \in \mathsf{qDiff}_\Theta^{\mathcal{Q}}(\mathcal{T}_1, \mathcal{T}_2)$, we say that $(\mathcal{A}, \boldsymbol{q}(\boldsymbol{a}))$ Θ-$\mathcal{Q}$ *separates* $\mathcal{T}_1$ and $\mathcal{T}_2$.

Note that Definition 13 does not require the separating ABoxes to be satisfiable with $\mathcal{T}_1$ and $\mathcal{T}_2$. Thus, a potential source of separability is that there is an ABox with which one of the TBoxes is satisfiable while the other is not. One could also define a (more brave) version of query inseparability where only ABoxes are considered that are satisfiable with $\mathcal{T}_1$ and $\mathcal{T}_2$. We will discuss this further in Sect. 7.2.

We now analyze the impact that the choice of query language has on query inseparability. To this end, we introduce a class of tree-shaped CQs that is closely related to $\mathcal{EL}$-concepts. Every $\mathcal{EL}$-concept C corresponds to a tree-shaped rCQ $\boldsymbol{q}_C(x)$ such that, for any interpretation $\mathcal{I}$ and $d \in \Delta^{\mathcal{I}}$, we have $d \in C^{\mathcal{I}}$ iff $\mathcal{I} \models \boldsymbol{q}_C(d)$. We denote $\boldsymbol{q}_C(x)$ by $C(x)$ and the Boolean CQ $\exists x\, \boldsymbol{q}_C(x)$ by $\exists x\, C(x)$. We use $\mathcal{Q}_{\mathcal{EL}}$ to denote the class of all queries of the former kind and $\mathcal{Q}_{\mathcal{EL}^u}$ for the class of queries of any of these two kinds. In the following theorem, the equivalence of (iii) with the other two conditions is of particular interest. Informally it says that, in $\mathcal{EL}$ and Horn-$\mathcal{ALC}$, tree-shaped queries are always sufficient to separate TBoxes.

Theorem 31. *Let $\mathcal{L}$ be a Horn DL, $\mathcal{T}_1$ and $\mathcal{T}_2$ TBoxes formulated in $\mathcal{L}$, and $\Theta = (\Sigma_1, \Sigma_2)$ a pair of signatures. Then the following conditions are equivalent:*

(i) $\mathcal{T}_1$ Θ-UCQ entails $\mathcal{T}_2$;
(ii) $\mathcal{T}_1$ Θ-CQ entails $\mathcal{T}_2$.

If $\mathcal{L}$ is $\mathcal{EL}$ or Horn-$\mathcal{ALC}$, then these conditions are also equivalent to

(iii) $\mathcal{T}_1$ Θ-$\mathcal{Q}_{\mathcal{EL}^u}$ entails $\mathcal{T}_2$.

The same is true when UCQs are replaced with rUCQs, CQs with rCQs, and $\mathcal{Q}_{\mathcal{EL}^u}$ with $\mathcal{Q}_{\mathcal{EL}}$ (simultaneously).

Proof. The first equivalence follows directly from the fact that KBs in Horn DLs are complete w.r.t. a single model (Example 22). We sketch the proof that Θ-$\mathcal{Q}_{\mathcal{EL}^u}$ entailment implies Θ-CQ entailment in $\mathcal{EL}$ and Horn-$\mathcal{ALC}$. Assume that there is a Σ_1-ABox $\mathcal{A}$ such that $\mathcal{K}_1$ does not Σ_2-CQ entail $\mathcal{K}_2$ for $\mathcal{K}_1 = (\mathcal{T}_1, \mathcal{A})$ and $\mathcal{K}_2 = (\mathcal{T}_2, \mathcal{A})$. Then $\mathcal{C}_{\mathcal{K}_2}$ is not finitely Σ-homomorphically embeddable into

$\mathcal{C}_{\mathcal{K}_1}$ (see Theorem 23). We thus find a finite subinterpretation $\mathcal{I}$ of an interpretation $\mathcal{I}_a$ in $\mathcal{C}_{\mathcal{K}_2}$ (see Example 22) that is not Σ-homomorphically embeddable into $\mathcal{C}_{\mathcal{K}_1}$. We can regard the Σ-reduct of $\mathcal{I}$ as a Σ-query in $\mathcal{Q}_{\mathcal{E}\mathcal{L}^u}$ which takes the form $C(x)$ if $\mathcal{I}$ contains a and $\exists x C(x)$ otherwise. This query witnesses that $\mathcal{K}_1$ does not Θ-$\mathcal{Q}_{\mathcal{E}\mathcal{L}^u}$ entail $\mathcal{K}_2$, as required.   □

For Horn DLs other than $\mathcal{E}\mathcal{L}$ and Horn-$\mathcal{A}\mathcal{L}\mathcal{C}$, the equivalence between (ii) and (iii) in Theorem 31 does often not hold in exactly the stated form. The reason is that additional constructors such as inverse roles and role inclusions require us to work with slightly different types of canonical models; see Example 22. However, the equivalence then holds for appropriate extensions of $\mathcal{Q}_{\mathcal{E}\mathcal{L}^u}$, for example, by replacing $\mathcal{E}\mathcal{L}$ concepts with $\mathcal{E}\mathcal{L}\mathcal{I}$ concepts when moving from Horn-$\mathcal{A}\mathcal{L}\mathcal{C}$ to Horn-$\mathcal{A}\mathcal{L}\mathcal{C}\mathcal{I}$.

Theorem 31 does not hold for non-Horn DLs. We have already observed in Example 20 that UCQ entailment and CQ entailment of KBs do not coincide in $\mathcal{A}\mathcal{L}\mathcal{C}$. Since the example uses the same ABox in both KBs, it also applies to the inseparability of TBoxes. In the following example, we prove that the equivalence between CQ inseparability and $\mathcal{Q}_{\mathcal{E}\mathcal{L}^u}$ inseparability fails in $\mathcal{A}\mathcal{L}\mathcal{C}$, too. The proof can actually be strengthened to show that, in $\mathcal{A}\mathcal{L}\mathcal{C}$, CQ inseparability is a stronger notion than inseparability by *acyclic* CQs (which generalize $\mathcal{Q}_{\mathcal{E}\mathcal{L}^u}$ by allowing multiple answer variables and edges in trees that are directed upwards).

Example 30. Let $\Sigma_1 = \{r\}$, $\Sigma_2 = \{r, A\}$, and $\Theta = (\Sigma_1, \Sigma_2)$. We construct an $\mathcal{A}\mathcal{L}\mathcal{C}$ TBox $\mathcal{T}_1$ as follows:

- to ensure that, for any Σ_1-ABox $\mathcal{A}$, the KB $(\mathcal{T}_1, \mathcal{A})$ is satisfiable iff $\mathcal{A}$ (viewed as an undirected graph with edges $\{\{a, b\} \mid r(a, b) \in \mathcal{A}\}$) is two-colorable, we take the CIs

$$B \sqsubseteq \forall r.\neg B, \quad \neg B \sqsubseteq \forall r.B;$$

- to ensure that, for any Σ_1-ABox $\mathcal{A}$, any model in $\boldsymbol{M}^b_{tree}(\mathcal{T}_1, \mathcal{A})$ has an infinite r-chain of nodes labeled with the concept name A whose root is not reachable from an ABox individual along Σ_2-roles we add the CIs

$$\exists r.\top \sqsubseteq \exists s.B', \quad B' \sqsubseteq A \sqcap \exists r.B'.$$

$\mathcal{T}_2$ is the extension of $\mathcal{T}_1$ with the CI

$$\exists r.\top \sqsubseteq A \sqcup \forall r.A.$$

Thus, models of $(\mathcal{T}_2, \mathcal{A})$ extend models of $(\mathcal{T}_1, \mathcal{A})$ by labelling certain individuals in $\mathcal{A}$ with A. The non-$\mathcal{A}$ part is not modified as we can assume that its elements are already labeled with A. Observe that $\mathcal{T}_1$ and $\mathcal{T}_2$ can be distinguished by the ABox $\mathcal{A} = \{r(a, b), r(b, a)\}$ and the CQ $\boldsymbol{q} = \exists x, y \, (A(x) \wedge r(x, y) \wedge r(y, x))$. Indeed, $a \in A^{\mathcal{I}}$ or $b \in A^{\mathcal{I}}$ holds in every model $\mathcal{I}$ of $(\mathcal{T}_2, \mathcal{A})$ but this is not the case for $(\mathcal{T}_1, \mathcal{A})$. We now argue that, for every Σ_1 ABox $\mathcal{A}$ and every Σ_2 concept C in $\mathcal{E}\mathcal{L}$, $(\mathcal{T}_2, \mathcal{A}) \models \exists x \, C(x)$ implies $(\mathcal{T}_1, \mathcal{A}) \models \exists x \, C(x)$ and $(\mathcal{T}_2, \mathcal{A}) \models C(a)$ implies $(\mathcal{T}_1, \mathcal{A}) \models C(a)$. Assume that $\mathcal{A}$ and C are given. As any model in $\boldsymbol{M}^b_{tree}(\mathcal{T}_1, \mathcal{A})$

has infinite r-chains labeled with A, we have $(\mathcal{T}_1, \mathcal{A}) \models \exists x\, C(x)$ for any Σ_2-concept C in $\mathcal{EL}$. Thus, we only have to consider the case $(\mathcal{T}_2, \mathcal{A}) \models C(a)$. If C does not contain A, then clearly $\mathcal{A} \models C(a)$, and so $(\mathcal{T}_1, \mathcal{A}) \models C(a)$, as required. If $\mathcal{A}$ is not 2-colorable, we also have $(\mathcal{T}_1, \mathcal{A}) \models C(a)$, as required. Otherwise C contains A and $\mathcal{A}$ is 2-colorable. But then it is easy to see that $(\mathcal{T}_2, \mathcal{A}) \not\models C(a)$ and we have derived a contradiction.

7.1   Relation to Other Notions of Inseparability

We now consider the relationship between query inseparability, model inseparability, and logical equivalence. Clearly, Σ-model inseparability entails Θ-$\mathcal{Q}$ inseparability for $\Theta = (\Sigma, \Sigma)$ and any class $\mathcal{Q}$ of queries. The same is true for logical equivalence, where we can even choose Θ freely. The converse direction is more interesting.

An ABox $\mathcal{A}$ is said to be *tree-shaped* if the directed graph $(\mathsf{ind}(\mathcal{A}), \{(a, b) \mid r(a, b) \in \mathcal{A}\})$ is a tree and $r(a, b) \in \mathcal{A}$ implies $s(a, b) \notin \mathcal{A}$ for any $a, b \in \mathsf{ind}(\mathcal{A})$ and $s \neq r$. We call $\mathcal{A}$ *undirected tree-shaped* (or *utree-shaped*) if the undirected graph $(\mathsf{ind}(\mathcal{A}), \{\{a, b\} \mid r(a, b) \in \mathcal{A}\})$ is a tree and $r(a, b) \in \mathcal{A}$ implies $s(a, b) \notin \mathcal{A}$ for any $a, b \in \mathsf{ind}(\mathcal{A})$ and $s \neq r$. Observe that every $\mathcal{EL}$ concept C corresponds to a tree-shaped ABox $\mathcal{A}_C$ and, conversely, every tree-shaped ABox $\mathcal{A}$ corresponds to an $\mathcal{EL}$-concept $C_{\mathcal{A}}$. In particular, for any TBox $\mathcal{T}$ and $\mathcal{EL}$ concept D, we have $\mathcal{T} \models C \sqsubseteq D$ iff $(\mathcal{T}, \mathcal{A}_C) \models D(\rho_C)$, ρ_C the root of $\mathcal{A}_C$.

Theorem 32. *Let $\mathcal{L} \in \{DL\text{-}Lite_{core}^{\mathcal{H}}, \mathcal{EL}\}$ and let $\Theta = (\Sigma_1, \Sigma_2)$ be a pair of signatures such that $\Sigma_i \supseteq \mathsf{sig}(\mathcal{T}_1) \cup \mathsf{sig}(\mathcal{T}_2)$ for $i \in \{1, 2\}$. Then the following conditions are equivalent:*

(i) $\mathcal{T}_1$ and $\mathcal{T}_2$ are logically equivalent;
(ii) $\mathcal{T}_1$ and $\mathcal{T}_2$ are Θ-rCQ inseparable.

Proof. We show $(ii) \Rightarrow (i)$ for $\mathcal{EL}$, the proof for $DL\text{-}Lite_{core}^{\mathcal{H}}$ is similar and omitted. Assume $\mathcal{T}_1$ and $\mathcal{T}_2$ are $\mathcal{EL}$ TBoxes that are not logically equivalent. Then there is $C \sqsubseteq D \in \mathcal{T}_2$ such that $\mathcal{T}_1 \not\models C \sqsubseteq D$ (or vice versa). We regard C as the tree-shaped Σ_1-ABox $\mathcal{A}_C$ with root ρ_C and D as the Σ_2-rCQ $D(x)$. Then $(\mathcal{T}_2, \mathcal{A}_C) \models D(\rho_C)$ but $(\mathcal{T}_1, \mathcal{A}_C) \not\models D(\rho_C)$. Thus $\mathcal{T}_1$ and $\mathcal{T}_2$ are Θ-rCQ separable. $\square$

Of course, Theorem 32 fails when the restriction of Θ is dropped. The following example shows that, even with this restriction, Theorem 32 does not hold for Horn-$\mathcal{ALC}$.

Example 31. Consider the Horn-$\mathcal{ALC}$ TBoxes

$$\mathcal{T}_1 = \{A \sqsubseteq \exists r.\neg A\} \quad \text{and} \quad \mathcal{T}_2 = \{A \sqsubseteq \exists r.\top\}.$$

Clearly, $\mathcal{T}_1$ and $\mathcal{T}_2$ are not logically equivalent. However, it is easy to see that they are Θ-UCQ inseparable for any Θ.

We now relate query inseparability and concept inseparability. In $\mathcal{ALC}$, these notions are incomparable. It already follows from Example 31 that UCQ inseparability does not imply concept inseparability. The following example shows that the converse implication does not hold either.

Example 32. Consider the $\mathcal{ALC}$ TBoxes $\mathcal{T}_1 = \varnothing$ and

$$\mathcal{T}_2 = \{B \sqcap \exists r.B \sqsubseteq A, \neg B \sqcap \exists r.\neg B \sqsubseteq A\}.$$

Using Theorem 3, one can show that $\mathcal{T}_1$ and $\mathcal{T}_2$ are Σ-concept inseparable, for $\Sigma = \{A, r\}$. However, $\mathcal{T}_1$ and $\mathcal{T}_2$ are not Θ-CQ inseparable for any $\Theta = (\Sigma_1, \Sigma_2)$ with $r \in \Sigma_1$ and $A \in \Sigma_2$ since for the ABox $\mathcal{A} = \{r(a, a)\}$ we have $(\mathcal{T}_1, \mathcal{A}) \not\models A(a)$ and $(\mathcal{T}_2, \mathcal{A}) \models A(a)$.

In Horn DLs, in contrast, concept inseparability and query inseparability are closely related. To explain why this is the case, consider $\mathcal{EL}$ as a paradigmatic example. Since $\mathcal{EL}$ concepts are positive and existential, an $\mathcal{EL}$ concept inclusion $C \sqsubseteq D$ which shows that two TBoxes $\mathcal{T}_1$ and $\mathcal{T}_2$ are not concept inseparable is almost the same as a witness $(\mathcal{A}, \boldsymbol{q}(\boldsymbol{a}))$ that query separates $\mathcal{T}_1$ and $\mathcal{T}_2$. In fact, both ABoxes and queries are positive and existential as well, but they need not be tree-shaped. Thus, a first puzzle piece is provided by Theorem 31 which implies that we need to consider only tree-shaped queries $\boldsymbol{q}$. This is complemented by the observation that it also suffices to consider only tree-shaped ABoxes $\mathcal{A}$. The latter is also an important foundation for designing decision procedures for query inseparability in Horn DLs. The following result was first proved in [52] for $\mathcal{EL}$. We state it here also for Horn-$\mathcal{ALC}$ [85] as this will be needed later on.

Theorem 33. *Let $\mathcal{T}_1$ and $\mathcal{T}_2$ be Horn-$\mathcal{ALC}$ TBoxes and $\Theta = (\Sigma_1, \Sigma_2)$. Then the following are equivalent:*

(i) $\mathcal{T}_1$ Θ-CQ entails $\mathcal{T}_2$;
(ii) for all utree-shaped Σ_1-ABoxes $\mathcal{A}$ and all $\mathcal{EL}$-concepts C in signature Σ_2:
    (a) if $(\mathcal{T}_2, \mathcal{A}) \models C(a)$, then $(\mathcal{T}_1, \mathcal{A}) \models C(a)$ where a is the root of $\mathcal{A}$;
    (b) if $(\mathcal{T}_2, \mathcal{A}) \models \exists x\, C(x)$, then $(\mathcal{T}_1, \mathcal{A}) \models \exists x\, C(x)$.

If $\mathcal{T}_1$ and $\mathcal{T}_2$ are $\mathcal{EL}$ TBoxes, then it is sufficient to consider tree-shaped ABoxes in (ii). The same holds when CQs are replaced with rCQs and (b) is dropped from (ii).

Theorem 33 can be proved by an unraveling argument. It is closely related to the notion of unraveling tolerance from [82]. As explained above, Theorem 33 allows us to prove that concept inseparability and query inseparability are the same notion. Here, we state this result only for $\mathcal{EL}$ [52]. Let Σ-$\mathcal{EL}^u$-concept entailment between $\mathcal{EL}$ TBoxes be defined like Σ-concept entailment between $\mathcal{EL}$ TBoxes, except that in $\mathrm{cDiff}_\Sigma(\mathcal{T}_1, \mathcal{T}_2)$ we now admit concept inclusions $C \sqsubseteq D$ where C is an $\mathcal{EL}$-concept and D an $\mathcal{EL}^u$-concept.

Theorem 34. *Let $\mathcal{T}_1$ and $\mathcal{T}_2$ be $\mathcal{EL}$ TBoxes and $\Theta = (\Sigma, \Sigma)$. Then*

(i) $\mathcal{T}_1$ Σ-concept entails $\mathcal{T}_2$ iff $\mathcal{T}_1$ Θ-rCQ entails $\mathcal{T}_2$;
(ii) $\mathcal{T}_1$ Σ-$\mathcal{EL}^u$-concept entails $\mathcal{T}_2$ iff $\mathcal{T}_1$ Θ-CQ entails $\mathcal{T}_2$.

7.2 Deciding Query Inseparability of TBoxes

We now study the decidability and computational complexity of query inseparability. Some results can be obtained by transferring results from Sect. 6 on query
inseparability for KBs (in the $\mathcal{ALC}$ and *DL-Lite* case) or results from Sect. 4 on
concept inseparability of TBoxes (in the $\mathcal{EL}$ case). In other cases, though, this
does not seem possible. To obtain results for Horn-$\mathcal{ALC}$, in particular, we need
new technical machinery; as before, we proceed by first giving model-theoretic
characterizations and then using tree automata.

In *DL-Lite$_{core}$* and *DL-Lite$_{core}^{\mathcal{H}}$*, there is a straightforward reduction of query
inseparability of TBoxes to query inseparability of KBs. Informally, such a reduction is possible since DL-Lite TBoxes are so restricted that they can only perform
deductions from a single ABox assertion, but not from multiple ones together.

Theorem 35. *For $\mathcal{Q} \in \{CQ, rCQ\}$, Θ-$\mathcal{Q}$ entailment and Θ-$\mathcal{Q}$ inseparability of
TBoxes are*

(i) in PTIME *for DL-Lite$_{core}$;*
(ii) EXPTIME*-complete for DL-Lite$_{core}^{\mathcal{H}}$.*

Proof. Let $\Theta = (\Sigma_1, \Sigma_2)$. Using the fact that every CI in a *DL-Lite$_{core}^{\mathcal{H}}$* TBox
has only a single concept of the form A or $\exists r.\top$ on the left-hand side, one can
show that if $\mathcal{T}_1$ does not Θ-$\mathcal{Q}$ entail $\mathcal{T}_2$, then there exists a singleton Σ_1-ABox
(containing either a single assertion of the form $A(c)$ or $r(a, b)$) such that $\mathcal{K}_1$ does
not Σ_2-$\mathcal{Q}$ entail $\mathcal{K}_2$ for $\mathcal{K}_i = (\mathcal{T}_i, \mathcal{A})$ for $i = 1, 2$. Now the upper bounds follow
from Theorem 30. The lower bound proof is a variation of the one establishing
Theorem 30.   □

The undecidability proof for CQ (and rCQ) entailment and inseparability of
$\mathcal{ALC}$ KBs (Theorem 26) can also be lifted to the TBox case; see [85] for details.

Theorem 36. *Let $\mathcal{Q} \in \{CQ, rCQ\}$.*

(i) Θ-$\mathcal{Q}$ entailment of an $\mathcal{ALC}$ TBox by an $\mathcal{EL}$ TBox is undecidable.
(ii) Θ-$\mathcal{Q}$ inseparability of an $\mathcal{ALC}$ and an $\mathcal{EL}$ TBoxes is undecidable.

In contrast to the KB case, decidability of UCQ entailment and inseparability of $\mathcal{ALC}$ TBoxes remains open, as well as for the rUCQ versions. Note that,
for the extension $\mathcal{ALCF}$ of $\mathcal{ALC}$ with functional roles, undecidability of Θ-$\mathcal{Q}$
inseparability can be proved for any class $\mathcal{Q}$ of queries contained in UCQ and
containing an atomic query of the form $A(x)$ or $\exists x A(x)$. The proof is by reduction to predicate and query emptiness problems that are shown to be undecidable
in [90]. Consider, for example, the class of all CQs. It is undecidable whether for
an $\mathcal{ALCF}$ TBox $\mathcal{T}$, a signature Σ, and a concept name $A \notin \Sigma$, there exists a
Σ-ABox $\mathcal{A}$ such that $(\mathcal{T}, \mathcal{A})$ is satisfiable and $(\mathcal{T}, \mathcal{A}) \models \exists x A(x)$ [90]. One can
easily modify the TBoxes $\mathcal{T}$ constructed in [90] to prove that this problem is
still undecidable if Σ-ABoxes $\mathcal{A}$ such that the KB $(\mathcal{T}, \mathcal{A})$ is not satisfiable are
admitted. Now observe that there exists a Σ-ABox $\mathcal{A}$ with $(\mathcal{T}, \mathcal{A}) \models \exists x A(x)$
iff $\mathcal{T}$ is not Θ-$\mathcal{CQ}$ inseparable from the empty TBox for $\Theta = (\Sigma, \{A\})$.

We now consider CQ inseparability in $\mathcal{EL}$. From Theorem 34 and Theorem 9, we obtain ExpTime-completeness of Θ-rCQ inseparability when Θ is of the form (Σ, Σ). ExpTime-completeness of Θ-CQ inseparability in this special case was established in [52]. Both results actually generalize to unrestricted signatures Θ.

Theorem 37. *Let $Q \in \{CQ, rCQ\}$. In $\mathcal{EL}$, Θ-Q-entailment and inseparability of TBoxes are* ExpTime-*complete.*

Theorem 37 has not been formulated in this generality in the literature, so we briefly discuss proofs. In the rooted case, the ExpTime upper bound follows from the same bound for Horn-$\mathcal{ALC}$ which we discuss below. In the non-rooted case, the ExpTime upper bound for the case $\Theta = (\Sigma, \Sigma)$ in [52] is based on Theorem 34 and a direct algorithm for deciding Σ-$\mathcal{EL}^u$-entailment. It is not difficult to extend this algorithm to the general case. Alternatively, one can obtain the same bound by extending the model-theoretic characterization of Σ-concept entailment in $\mathcal{EL}$ given in Theorem 8 to 'Θ-$\mathcal{EL}^u$-concept entailment', where the concept inclusions in $\mathsf{cDiff}_\Sigma(\mathcal{T}_1, \mathcal{T}_2)$ are of the form $C \sqsubseteq D$ with C an $\mathcal{EL}$ concept in signature Σ_1 and D an $\mathcal{EL}^u$ concept in signature Σ_2. Based on such a characterization, one can then modify the automata construction from the proof of Theorem 9 to obtain an ExpTime upper bound.

We note that, for acyclic $\mathcal{EL}$ TBoxes (and their extensions with role inclusions and domain and range restrictions), Θ-CQ entailment can be decided in polynomial time. This can be proved by a straightforward generalization of the results in [11] where it is assumed that $\Theta = (\Sigma_1, \Sigma_2)$ with $\Sigma_1 = \Sigma_2$. The proof extends the approach sketched in Sect. 4 for deciding concept inseparability for acyclic $\mathcal{EL}$ TBoxes. A prototype system deciding Θ-CQ inseparability and computing a representation of the logical difference for query inseparability is presented in [91].

We now consider query entailment and inseparability in Horn-$\mathcal{ALC}$, which requires more effort than the cases discussed so far. We will concentrate on CQs and rCQs. To start with, it is convenient to break down our most basic problem, query entailment, into two subproblems:

1. Θ-Q entailment *over satisfiable ABoxes* is defined in the same way as Θ-Q entailment except that only ABoxes satisfiable with both $\mathcal{T}_1$ and $\mathcal{T}_2$ can witness inseparability; see the remark after Definition 13.
2. A TBox $\mathcal{T}_1$ Σ-*ABox entails* a TBox $\mathcal{T}_2$, for a signature Σ, if for every Σ-ABox $\mathcal{A}$, unsatisfiability of $(\mathcal{T}_2, \mathcal{A})$ implies unsatisfiability of $(\mathcal{T}_1, \mathcal{A})$.

It is easy to see that a TBox $\mathcal{T}_2$ is Θ-Q-entailed by a TBox $\mathcal{T}_1$, $\Theta = (\Sigma_1, \Sigma_2)$, if $\mathcal{T}_2$ is Θ-Q-entailed by $\mathcal{T}_1$ over satisfiable ABoxes and $\mathcal{T}_2$ is Σ_1-ABox entailed by $\mathcal{T}_1$. For proving decidability and upper complexity bounds, we can thus concentrate on problems 1 and 2 above. ABox entailment, in fact, is reducible in polynomial time and in a straightforward way to the containment problem of ontology-mediated queries with CQs of the form $\exists\, x\, A(x)$, which is ExpTime-complete in Horn-$\mathcal{ALC}$ [92]. For deciding query entailment over satisfiable ABoxes, we can find a transparent model-theoretic characterization. The following result from [85] is essentially a consequence of Theorem 24 (i), Theorem 25, and Theorem 33, but additionally establishes a bound on the branching degree of witness ABoxes.

Theorem 38. *Let T_1 and T_2 be Horn-$\mathcal{ALC}$ TBoxes and $\Theta = (\Sigma_1, \Sigma_2)$. Then*

(i) *T_1 Θ-CQ entails T_2 over satisfiable ABoxes iff, for all utree-shaped Σ_1-ABoxes $\mathcal{A}$ of outdegree $\leq |T_2|$ and consistent with T_1 and T_2, $\mathcal{C}_{(T_2,\mathcal{A})}$ is Σ_2-homomorphically embeddable into $\mathcal{C}_{(T_1,\mathcal{A})}$;*
(ii) *T_1 Θ-rCQ entails T_2 over satisfiable ABoxes iff, for all utree-shaped Σ_1-ABoxes $\mathcal{A}$ of outdegree $\leq |T_2|$ and consistent with T_1 and T_2, $\mathcal{C}_{(T_2,\mathcal{A})}$ is con-Σ_2-homomorphically embeddable into $\mathcal{C}_{(T_1,\mathcal{A})}$.*

Based on Theorem 38, we can derive upper bounds for query inseparability in *Horn-$\mathcal{ALC}$* using tree automata techniques.

Theorem 39. *In Horn-$\mathcal{ALC}$,*

(i) *Θ-rCQ entailment and inseparability of TBoxes is* EXPTIME-*complete;*
(ii) *Θ-CQ entailment and inseparability of TBoxes is* 2EXPTIME-*complete.*

The automaton constructions are more sophisticated than those used for proving Theorem 27 because the ABox is not fixed. The construction in [85] uses traditional tree automata whose inputs encode a tree-shaped ABox together with (parts of) its tree-shaped canonical models for the TBoxes T_1 and T_2. It is actually convenient to first replace Theorem 27 with a more fine-grained characterization that uses simulations instead of homomorphisms and is more operational. Achieving the upper bounds stated in Theorem 39 requires a careful automaton construction using appropriate bookkeeping in the input and mixing alternating with non-deterministic automata. The lower bound is based on an ATM reduction.

Interestingly, the results presented above for query inseparability between *DL-Lite* TBoxes have recently been applied to analyse containment and inseparability for TBoxes with declarative mappings that relate the signature of the data one wants to query to the signature of the TBox that provides the interface for formulating queries [93]. We conjecture that the results we presented for $\mathcal{EL}$ and Horn-$\mathcal{ALC}$ can also be lifted to the extension by declarative mappings.

We note that query inseparability between TBoxes is closely related to program expressiveness [94] and to CQ-equivalence of schema mappings [95,96]. The latter is concerned with declarative mappings from a source signature Σ_1 to a target signature Σ_2. Such mappings $\mathcal{M}_1$ and $\mathcal{M}_2$ are CQ-equivalent if, for any data instance in Σ_1, the certain answers to CQs in the signature Σ_2 under $\mathcal{M}_1$ and $\mathcal{M}_2$ coincide. The computational complexity of deciding CQ-equivalence of schema mappings has been investigated in detail [95,96]. Regarding the former, translated into the language of DL the program expressive power of a TBox T is the set of all triples $(\mathcal{A}, q, a)$ such that $\mathcal{A}$ is an ABox, q is a CQ, and a is a tuple in ind($\mathcal{A}$) such that $T, \mathcal{A} \models q(a)$. It follows that two TBoxes T_1 and T_2 are Θ-CQ inseparable for a pair $\Theta = (\Sigma_1, \Sigma_2)$ iff T_1 and T_2 have the same program expressive power.

8   Discussion

In this chapter, we have discussed a few inseparability relations between description logic TBoxes and KBs, focussing on model-theoretic characterizations and deciding inseparability. In this section, we briefly survey three other important topics that were not covered in the main text. (1) We observe that many inseparability relations considered above (in particular, concept inseparability) fail to satisfy natural robustness conditions such as robustness under replacement, and discuss how to overcome this. (2) Since inseparability tends to be of high computational complexity or even undecidable, it is interesting to develop approximation algorithms; we present a brief overview of the state of the art. (3) One is often not only interested in deciding inseparability, but also in computing useful members of an equivalence class of inseparable ontologies such as uniform interpolants and the result of forgetting irrelevant symbols from an ontology. We briefly survey results in this area as well.

Inseparability and Robustness. We have seen that robustness under replacement is a central property in applications of model inseparability to ontology reuse and module extraction. In principle, one can of course also use other inseparability relations for these tasks. The corresponding notion of robustness under replacement can be defined in a straightforward way [10,23].

Definition 14. Let $\mathcal{L}$ be a DL and $\equiv_\Sigma$ an inseparability relation. Then $\mathcal{L}$ is *robust under replacement* for $\equiv_\Sigma$ if $\mathcal{T}_1 \equiv_\Sigma \mathcal{T}_2$ implies that $\mathcal{T}_1 \cup \mathcal{T} \equiv_\Sigma \mathcal{T}_2 \cup \mathcal{T}$ for all $\mathcal{L}$ TBoxes $\mathcal{T}_1, \mathcal{T}_2$ and $\mathcal{T}$ such that $\mathrm{sig}(\mathcal{T}) \cap \mathrm{sig}(\mathcal{T}_1 \cup \mathcal{T}_2) \subseteq \Sigma$.[8]

Thus, robustness under replacement ensures that Σ-inseparable TBoxes can be equivalently replaced by each other even if a new TBox that shares with $\mathcal{T}_1$ and $\mathcal{T}_2$ only Σ-symbols is added to both. This seems a useful requirement not only for TBox re-use and module extraction, but also for versioning and forgetting. Unfortunately, with the exception of model inseparability, none of the inseparability relations considered in the main part of this survey is robust under replacement for the DLs in question. The following counterexample is a variant of examples given in [23,97].

Example 33. Suppose $\mathcal{T}_1 = \varnothing$, $\mathcal{T}_2 = \{A \sqsubseteq \exists r.B, E \sqcap B \sqsubseteq \bot\}$, and $\Sigma = \{A, E\}$. Then $\mathcal{T}_1$ and $\mathcal{T}_2$ are Σ-concept inseparable in expressive DLs such as $\mathcal{ALC}$ and they are Θ-CQ inseparable for $\Theta = (\Sigma, \Sigma)$. However, for $\mathcal{T} = \{\top \sqsubseteq E\}$ the TBoxes $\mathcal{T}_1 \cup \mathcal{T}$ and $\mathcal{T}_2 \cup \mathcal{T}$ are neither Σ-concept inseparable nor Θ-CQ inseparable.

The only DLs for which concept inseparability is robust under replacement are certain extensions of $\mathcal{ALC}$ with the universal role. Indeed, recall that by $\mathcal{L}^u$ we denote the extension of a DL $\mathcal{L}$ with the universal role u. Assume that

---

[8] Robustness under replacement can be defined for KBs as well and is equally important in that case. In this short discussion, however, we only consider TBox inseparability.

T_1 and T_2 are Σ-concept inseparable in $\mathcal{L}^u$ and let T be an $\mathcal{L}^u$ TBox with $\text{sig}(T) \cap \text{sig}(T_1 \cup T_2) \subseteq \Sigma$. As $\mathcal{L}$ extends $\mathcal{ALC}$, it is known that T is logically equivalent to a TBox of the form $\{C \equiv \top\}$, where C is an $\mathcal{L}^u$ concept. Let $D_0 \sqsubseteq D_1$ be a Σ-CI in $\mathcal{L}^u$. Then

$$T_1 \cup T \models D_0 \sqsubseteq D_1 \quad \text{iff} \quad T_1 \models D_0 \sqcap \forall u.C \sqsubseteq D_1$$
$$\text{iff} \quad T_2 \models D_0 \sqcap \forall u.C \sqsubseteq D_1$$
$$\text{iff} \quad T_2 \cup T \models D_0 \sqsubseteq D_1,$$

where the second equivalence holds by Σ-concept inseparability of T_1 and T_2 *if we assume that* $\text{sig}(T) \subseteq \Sigma$ (and so $\text{sig}(C) \subseteq \Sigma$). Recall that, in the definition of robustness under replacement, we only require $\text{sig}(T) \cap \text{sig}(T_1 \cup T_2) \subseteq \Sigma$, and so an additional step is needed for the argument to go through. This step is captured by the following definition.

Definition 15. Let $\mathcal{L}$ be a DL and $\equiv_\Sigma$ an inseparability relation. Then $\mathcal{L}$ is *robust under vocabulary extensions* for $\equiv_\Sigma$ if $T_1 \equiv_\Sigma T_2$ implies that $T_1 \equiv_{\Sigma'} T_2$ for all $\Sigma' \supseteq \Sigma$ with $\text{sig}(T_1 \cup T_2) \cap \Sigma' \subseteq \Sigma$.

Let us return to the argument above. Clearly, if $\mathcal{L}^u$ is robust under vocabulary extensions for concept inseparability, then the second equivalence is justified and we can conclude that $\mathcal{L}^u$ is robust under replacement for concept inseparability. In [10], robustness under vocabulary extensions is investigated for many standard DLs and inseparability relations. In particular, the following is shown:

Theorem 40. *The DLs $\mathcal{ALC}^u$, $\mathcal{ALCI}^u$, $\mathcal{ALCQ}^u$, and $\mathcal{ALCQI}^u$ are robust under vocabulary extensions for concept inseparability, and thus also robust under replacement.*

Because of Theorem 40, it would be interesting to investigate concept inseparability for DLs with the universal role and establish, for example, the computational complexity of concept inseparability. We conjecture that the techniques used to prove the 2ExpTime upper bounds without the universal role can be used to obtain 2ExpTime upper bounds here as well.

We now consider robustness under replacement for DLs without the universal role. To simplify the discussion, we consider *weak robustness under replacement*, which preserves inseparability only if TBoxes T with $\text{sig}(T) \subseteq \Sigma$ are added to T_1 and T_2, respectively. It is then a separate task to lift weak robustness under replacement to full robustness under replacement using, for example, robustness under vocabulary extensions. It is, of course, straightforward to extend the inseparability relations studied in this survey in a minimal way so that weak robustness under replacement is achieved. For example, say that two $\mathcal{L}$ TBoxes T_1 and T_2 are *strongly Σ-concept inseparable in $\mathcal{L}$* if, for all $\mathcal{L}$ TBoxes T with $\text{sig}(T) \subseteq \Sigma$, we have that $T_1 \cup T$ and $T_2 \cup T$ are Σ-concept inseparable. Similarly, say that two $\mathcal{L}$ TBoxes T_1 and T_2 are *strongly Θ-$\mathcal{Q}$-inseparable* if, for all $\mathcal{L}$ TBoxes T with $\text{sig}(T) \subseteq \Sigma_1 \cap \Sigma_2$, we have that $T_1 \cup T$ and $T_2 \cup T$ are Θ-$\mathcal{Q}$ inseparable (we assume $\Theta = (\Sigma_1, \Sigma_2)$). Unfortunately, with the exception of results for the *DL-Lite* family, nothing is known about the properties of the resulting inseparability relations. It is proved in [97] that strong Θ-CQ inseparability is still in ExpTime for $DL\text{-}Lite_{core}^{\mathcal{H}}$ if $\Sigma_1 = \Sigma_2$. We conjecture that this result still holds for arbitrary Θ. A variant of strong Θ-CQ inseparability is also discussed in [23] and analyzed for $DL\text{-}Lite_{core}$ extended with (some) Boolean operators and unqualified number restrictions. However, the authors of [23] do not consider CQ-inseparability as defined in this survey but inseparability with respect to generalized CQs that use atoms $C(x)$ with C a Σ-concept in $\mathcal{L}$ instead of a concept name in Σ. This results in a stronger notion of inseparability that is preserved under definitorial extensions and has, for the *DL-Lite* dialects considered, many of the robustness properties introduced above. It would be of interest to extend this notion of query inseparability to DLs such as $\mathcal{ALC}$. Regarding strong concept and query inseparability, it would be interesting to investigate its algorithmic properties for $\mathcal{EL}$ and Horn-$\mathcal{ALC}$.

Approximation. We have argued throughout this survey that inseparability relations and conservative extensions can play an important role in a variety of applications including ontology versioning, ontology refinement, ontology reuse, ontology modularization, ontology mapping, knowledge base exchange and forgetting. One cannot help noticing, though, another common theme: the high computational complexity of the corresponding reasoning tasks, which can hinder the *practical use* of these notions or even make it infeasible. We now give a brief overview of methods that approximate the notions introduced in the previous sections while incurring lower computational costs. We will focus on modularization and logical difference.

Locality-based approximations have already been discussed in Sect. 5.3, where we showed how the extraction of depleting modules can be reduced to standard ontology reasoning. Notice that $\varnothing$-locality, in turn, can be approximated with a simple syntactic check. Following [72], let Σ be a signature. Define two sets of $\mathcal{ALCQI}$ concepts $\mathcal{C}_\Sigma^\perp$ and $\mathcal{C}_\Sigma^\top$ as follows:

$$\mathcal{C}_\Sigma^\perp ::= A^\perp \mid \neg C^\top \mid C \sqcap C^\perp \mid \exists r^\perp.C \mid \exists r.C^\perp \mid \geq n r^\perp.C \mid \geq n r.C^\perp,$$

$$\mathcal{C}_\Sigma^\top ::= \neg C^\perp \mid C_1^\top \sqcap C_2^\top,$$

where $A^\perp \notin \Sigma$ is an atomic concept, r is a role (a role name or an inverse role) and C is a concept, $C^\perp \in \mathcal{C}_\Sigma^\perp$, $C_i^\top \in \mathcal{C}_\Sigma^\top$, $i = 1, 2$, and $r^\perp$ is r or r^-, for $r \in \mathsf{N_R} \setminus \Sigma$. A CI α is *syntactically $\perp$-local* w.r.t. Σ if it is of the form $C^\perp \sqsubseteq C$ or $C \sqsubseteq C^\top$. A TBox $\mathcal{T}$ is *$\perp$-local* if all CIs in $\mathcal{T}$ are $\perp$-local. Then every TBox $\mathcal{T}$ that is syntactically $\perp$-local w.r.t. a signature Σ is $\varnothing$-local w.r.t. Σ, as shown in [72]. Notice that checking whether a CI is syntactically $\perp$-local can be done in linear time. A dual notion of syntactic $\top$-locality has been introduced in [20]. Both notions can be used to define $\perp$-local and $\top$-local modules; $\top$- and $\perp$-locality module extraction can be iterated leading to smaller modules [75].

A comprehensive study of different locality flavours [98] identified that there is no statistically significant difference in the sizes of semantic and syntactic locality modules. In contrast, [63] found that the difference in size between minimal

modules (only available for acyclic $\mathcal{EL}$ TBoxes) and locality-based approximations can be large. In a separate line of research, [99] showed that intractable
depleting module approximations for unrestricted OWL ontologies based on
reductions to QBF can also be significantly smaller, indicating a possibility for
better tractable approximations. Reachability-based approximations [100,101]
refine syntactic locality modules. While they are typically smaller, self-contained
and justification preserving, reachability modules are only Σ-concept inseparable from the original TBox but not Σ-model inseparable. A variety of tractable
approximations based on notions of inseparability ranging form classification
inseparability to model inseparability can be computed by reduction to Datalog
reasoning [78].

Syntactic restrictions on elements of $\mathsf{cDiff}_\Sigma(\mathcal{T}_1, \mathcal{T}_2)$ lead to approximations
of concept inseparability. In [16], the authors consider counterexamples of the
form $A \sqsubseteq B$, $A \sqsubseteq \neg B$, $A \sqsubseteq \exists r.B$, $A \sqsubseteq \forall r.B$ and $r \sqsubseteq s$ only, where A, B
are Σ-concept names and r, s are Σ-roles, and use standard reasoners to check
for entailment. This approach has been extended in [102] to allow inclusions
between Σ-concepts to be constructed in accordance with some grammar rules.
In [97], CQ-inseparability for $DL\text{-}Lite_{core}^{\mathcal{H}}$ is approximated by reduction to a
tractable simulation check between the canonical models. An experimental evaluation showed that this approach is incomplete in a very small number of cases
on real-world ontologies.

Computing Representatives. Inseparability relations are equivalence relations on classes of TBoxes. One is often interested not only in deciding inseparability, but also in computing useful members of an equivalence class of inseparable
ontologies such as uniform interpolants (or, equivalently, the result of forgetting
irrelevant symbols from an ontology). Recall from Sect. 3 that an ontology $\mathcal{O}_{\mathsf{forget}}$
is the result of forgetting a signature Γ in $\mathcal{O}$ for an inseparability relation $\equiv$ if
$\mathcal{O}$ uses only symbols in $\Sigma = \mathsf{sig}(\mathcal{O}) \setminus \Gamma$ and $\mathcal{O}$ and $\mathcal{O}_{\mathsf{forget}}$ are Σ-inseparable for
$\equiv$. Clearly, $\mathcal{O}_{\mathsf{forget}}$ can be regarded as a representation of its equivalence class
under $\equiv_\Sigma$. For model-inseparability, this representation is unique up to logical
equivalence while this need not be the case for other inseparability relations.

Forgetting has been studied extensively for various inseparability relations.
A main problems addressed in the literature is that, for most inseparability
relations and ontology languages, the result of forgetting is not guaranteed to
be expressible in the language of the original ontology. For example, for the
TBox $\mathcal{T} = \{A \sqsubseteq \exists r.B, B \sqsubseteq \exists r.B\}$, there is no $\mathcal{ALC}$ TBox using only symbols
from $\Sigma = \{A, r\}$ that is Σ-concept inseparable from $\mathcal{T}$. This problem gives
rise to three interesting research problems: given an ontology $\mathcal{O}$ and signature
Σ, can we decide whether the result of forgetting exists in the language of $\mathcal{O}$
and, if so, compute it? If not, can we approximate it in a principled way? Or
can we express it in a more powerful ontology language? The existence and
computation of uniform interpolants for TBoxes under concept inseparability has
been studied in [38] for acyclic $\mathcal{EL}$ TBoxes, in [43,53] for arbitrary $\mathcal{EL}$ TBoxes,
and for $\mathcal{ALC}$ and more expressive DLs in [40,42]. The generalization to KBs has
been studied in [39,41,44]. Approximations of uniform interpolants obtained by

putting a bound on the role depth of relevant concept inclusions are studied in [41,103–105]. In [42,44], (a weak form of) uniform interpolants that do not exist in the original DL are captured using fix-point operators. The relationship between deciding concept inseparability and deciding the existence of uniform interpolants is investigated in [40]. Forgetting under model inseparability has been studied extensively in logic [106] and more recently for DLs [107]. Note that the computation of universal CQ solutions in knowledge exchange [29] is identical to forgetting the signature of the original KB under Σ-CQ-inseparability.

Uniform interpolants are not the only useful representatives of equivalence classes of inseparable ontologies. In the KB case, for example, it is natural to ask whether for a given KB $\mathcal{K}$ and signature Σ there exists a KB $\mathcal{K}'$ with empty TBox that is Σ-query inseparable from $\mathcal{K}$. In this case, answering a Σ-query in $\mathcal{K}$ could be reduced to evaluating the query in an ABox. Another example is TBox rewriting, which asks whether for a given TBox $\mathcal{T}$ in an expressive DL there exists a TBox $\mathcal{T}'$ that is $\mathsf{sig}(\mathcal{T})$-inseparable from $\mathcal{T}$ in a less expressive DL. In this case tools that are only available for the less expressive DL but not for the expressive DL would become applicable to the rewritten TBox. First results regarding this question have been obtained in [49].

Acknowledgments. Elena Botoeva was supported by EU IP project Optique, grant n. FP7-318338. Carsten Lutz was supported by ERC grant 647289. Boris Konev, Frank Wolter and Michael Zakharyaschev were supported by the UK EPSRC grants EP/M012646, EP/M012670, EP/H043594, and EP/H05099X.

References

1. Baader, F., Calvanese, D., McGuinness, D., Nardi, D., Patel-Schneider, P.F., et al. (eds.): The Description Logic Handbook Theory Implementation and Applications. Cambridge University Press, Cambridge (2003)
2. Baader, F., Brandt, S., Lutz, C.: Pushing the EL envelope. In: Proceedings of the 19th International Joint Conference on Artificial Intelligence (IJCAI 2005), pp. 364–369 (2005)
3. Calvanese, D., De Giacomo, G., Lembo, D., Lenzerini, M., Rosati, R.: Tractable reasoning and efficient query answering in description logics: the DL-Lite family. J. Autom. Reason. **39**(3), 385–429 (2007)
4. Artale, A., Calvanese, D., Kontchakov, R., Zakharyaschev, M.: The DL-Lite family and relations. J. Artif. Intell. Res. (JAIR) **36**, 1–69 (2009)
5. Hustadt, U., Motik, B., Sattler, U.: Reasoning in description logics by a reduction to disjunctive Datalog. J. Autom. Reason. **39**(3), 351–384 (2007)
6. Kazakov, Y.: Consequence-driven reasoning for Horn-SHIQ ontologies. In: Proceedings of the 21st International Joint Conference on Artificial Intelligence (IJCAI 2009), pp. 2040–2045 (2009)
7. Glimm, B., Lutz, C., Horrocks, I., Sattler, U.: Answering conjunctive queries in the $\mathcal{SHIQ}$ description logic. J. Artif. Intell. Res. (JAIR) **31**, 150–197 (2008)
8. Calvanese, D., De Giacomo, G., Lembo, D., Lenzerini, M., Rosati, R.: Data complexity of query answering in description logics. In: Proceedings of the 10th International Conference on the Principles of Knowledge Representation and Reasoning (KR 2006), pp. 260–270 (2006)

9. Calvanese, D., Eiter, T., Ortiz, M.: Answering regular path queries in expressive description logics: an automata-theoretic approach. In: Proceedings of the 22nd National Conference on Artificial Intelligence (AAAI 2007), pp. 391–396 (2007)

10. Konev, B., Lutz, C., Walther, D., Wolter, F.: Formal properties of modularisation. In: Stuckenschmidt et al. [21], pp. 25–66. http://dx.doi.org/10.1007/978-3-642-01907-4_3

11. Konev, B., Ludwig, M., Walther, D., Wolter, F.: The logical difference for the lightweight description logic EL. J. Artif. Intell. Res. (JAIR) **44**, 633–708 (2012)

12. Conradi, R., Westfechtel, B.: Version models for software configuration management. ACM Comput. Surv. (CSUR) **30**(2), 232–282 (1998)

13. Noy, N.F., Musen, M.A.: PromptDiff: a fixed-point algorithm for comparing ontology versions. In: Proceedings of the 18th National Conference on Artificial Intelligence (AAAI 2002), pp. 744–750. AAAI Press, Menlo Park (2002)

14. Klein, M., Fensel, D., Kiryakov, A., Ognyanov, D.: Ontology versioning and change detection on the web. In: Gómez-Pérez, A., Benjamins, V.R. (eds.) EKAW 2002. LNCS (LNAI), vol. 2473, pp. 197–212. Springer, Heidelberg (2002). doi:10.1007/3-540-45810-7_20

15. Redmond, T., Smith, M., Drummond, N., Tudorache, T.: Managing change: an ontology version control system. In: Proceedings of the 5th International Workshop on OWL: Experiences and Directions (OWLED 2008). CEUR Workshop Proceedings, vol. 432, CEUR-WS.org (2008)

16. Jimenez-Ruiz, E., Cuenca Grau, B., Horrocks, I., Llavori, R.B.: Supporting concurrent ontology development: framework, algorithms and tool. Data Knowl. Eng. **70**(1), 146–164 (2011)

17. Konev, B., Walther, D., Wolter, F.: The logical difference problem for description logic terminologies. In: Armando, A., Baumgartner, P., Dowek, G. (eds.) IJCAR 2008. LNCS (LNAI), vol. 5195, pp. 259–274. Springer, Heidelberg (2008). doi:10.1007/978-3-540-71070-7_21

18. Kontchakov, R., Wolter, F., Zakharyaschev, M.: Can you tell the difference between DL-Lite ontologies. In: Proceedings of the 11th International Conference on the Principles of Knowledge Representation and Reasoning (KR 2008), pp. 285–295 (2008)

19. Ghilardi, S., Lutz, C., Wolter, F.: Did I damage my ontology? A case for conservative extensions in description logic. In: Proceedings of the 10th International Conference on the Principles of Knowledge Representation and Reasoning (KR 2006), pp. 187–197. AAAI Press (2006)

20. Cuenca-Grau, B., Horrocks, I., Kazakov, Y., Sattler, U.: Modular reuse of ontologies: theory and practice. J. Artif. Intell. Res. (JAIR) **31**, 273–318 (2008)

21. Stuckenschmidt, H., Parent, C., Spaccapietra, S. (eds.): Modular Ontologies: Concepts, Theories and Techniques for Knowledge Modularization. LNCS, vol. 5445. Springer, Heidelberg (2009)

22. Kutz, O., Mossakowski, T., Lücke, D.: Carnap, Goguen, and the hyperontologies: logical pluralism and heterogeneous structuring in ontology design. Log. Univers. **4**(2), 255–333 (2010). doi:10.1007/s11787-010-0020-3

23. Kontchakov, R., Wolter, F., Zakharyaschev, M.: Logic-based ontology comparison and module extraction, with an application to DL-Lite. Artif. Intell. **174**, 1093–1141 (2010)

24. Shvaiko, P., Euzenat, J.: Ontology matching: state of the art and future challenges. IEEE Trans. Knowl. Data Eng. **25**(1), 158–176 (2013). doi:10.1109/TKDE.2011.253

25. Solimando, A., Jiménez-Ruiz, E., Guerrini, G.: Detecting and correcting conservativity principle violations in ontology-to-ontology mappings. In: Mika, P., et al. (eds.) ISWC 2014. LNCS, vol. 8797, pp. 1–16. Springer, Heidelberg (2014). doi:10.1007/978-3-319-11915-1_1
26. Kharlamov, E., et al.: Ontology based access to exploration data at Statoil. In: Arenas, M., et al. (eds.) ISWC 2015. LNCS, vol. 9367, pp. 93–112. Springer, Heidelberg (2015). doi:10.1007/978-3-319-25010-6_6
27. Jiménez-Ruiz, E., Payne, T.R., Solimando, A., Tamma, V.A.M.: Limiting logical violations in ontology alignnment through negotiation. In: Proceedings of the 15th International Conference on the Principles of Knowledge Representation and Reasoning (KR 2016), pp. 217–226 (2016)
28. Fagin, R., Kolaitis, P.G., Miller, R.J., Popa, L.: Data exchange: semantics and query answering. Theor. Comput. Sci. **336**(1), 89–124 (2005). doi:10.1016/j.tcs.2004.10.033
29. Arenas, M., Botoeva, E., Calvanese, D., Ryzhikov, V.: Knowledge base exchange: the case of OWL 2 QL. Artif. Intell. **238**, 11–62 (2016). ISSN 0004-3702
30. Cuenca-Grau, B., Motik, B.: Reasoning over ontologies with hidden content: the import-by-query approach. J. Artif. Intell. Res. (JAIR) **45**, 197–255 (2012)
31. Reiter, R., Lin, F.: Forget it! In: Proceedings of AAAI Fall Symposium on Relevance, pp. 154–159 (1994)
32. Pitts, A.: On an interpretation of second-order quantification in first-order intuitionistic propositional logic. J. Symb. Logic **57**, 33–52 (1992)
33. D'Agostino, G., Hollenberg, M.: Uniform interpolation, automata, and the modal μ-calculus. In: Advances in Modal Logic, vol. 1 (1998)
34. Visser, A.: Uniform interpolation and layered bisimulation. In: Hájek, P. (ed.) Gödel '96 (Brno, 1996). Lecture Notes Logic, vol. 6. Springer, Berlin (1996)
35. Ghilardi, S., Zawadowski, M.: Undefinability of propositional quantifiers in the modal system S4. Stud. Logica **55**, 259–271 (1995)
36. French, T.: Bisimulation quantifiers for modal logics. Ph.D. thesis, University of Western Australia (2006)
37. Su, K., Sattar, A., Lv, G., Zhang, Y.: Variable forgetting in reasoning about knowledge. J. Artif. Intell. Res. (JAIR) **35**, 677–716 (2009). doi:10.1613/jair.2750
38. Konev, B., Walther, D., Wolter, F.: Forgetting and uniform interpolation in large-scale description logic terminologies. In: Proceedings of the 21st International Joint Conference on Artificial Intelligence (IJCAI 2009), pp. 830–835 (2009). http://ijcai.org/papers09/Papers/IJCAI09-142.pdf
39. Wang, Z., Wang, K., Topor, R.W., Pan, J.Z.: Forgetting for knowledge bases in DL-Lite. Ann. Math. Artif. Intell. **58**(1–2), 117–151 (2010). doi:10.1007/s10472-010-9187-9
40. Lutz, C., Wolter, F.: Foundations for uniform interpolation and forgetting in expressive description logics. In: Proceedings of the 22nd International Joint Conference on Artificial Intelligence (IJCAI 2011), pp. 989–995. IJCAI/AAAI (2011)
41. Wang, K., Wang, Z., Topor, R.W., Pan, J.Z., Antoniou, G.: Eliminating concepts and roles from ontologies in expressive descriptive logics. Comput. Intell. **30**(2), 205–232 (2014). doi:10.1111/j.1467-8640.2012.00442.x
42. Koopmann, P., Schmidt, R.A.: Count and forget: uniform interpolation of $\mathcal{SHQ}$-ontologies. In: Demri, S., Kapur, D., Weidenbach, C. (eds.) IJCAR 2014. LNCS (LNAI), vol. 8562, pp. 434–448. Springer, Heidelberg (2014). doi:10.1007/978-3-319-08587-6_34

43. Nikitina, N., Rudolph, S.: (Non-)succinctness of uniform interpolants of general terminologies in the description logic EL. Artif. Intell. **215**, 120–140 (2014). doi:10.1016/j.artint.2014.06.005
44. Koopmann, P., Schmidt, R.A.: Uniform interpolation and forgetting for $\mathcal{ALC}$ ontologies with ABoxes. In: Proceedings of the 29th National Conference on Artificial Intelligence (AAAI 2015), pp. 175–181 (2015). http://www.aaai.org/ocs/index.php/AAAI/AAAI15/paper/view/9981
45. Goranko, V., Otto, M.: Model theory of modal logic. In: Blackburn, P., van Benthem, J., Wolter, F. (eds.) Handbook of Modal Logic, pp. 249–330. Elsevier, Amsterdam (2006)
46. Lutz, C., Piro, R., Wolter, F.: Description logic TBoxes: model-theoretic characterizations and rewritability. In: Proceedings of the 22nd International Joint Conference on Artificial Intelligence (IJCAI 2011), pp. 983–988. AAAI Press, Menlo Park (2011)
47. Lutz, C., Walther, D., Wolter, F.: Conservative extensions in expressive description logics. In: Proceedings of the 20th Internatioanl Joint Conference on Artificial Intelligence (IJCAI), pp. 453–458. AAAI Press, Menlo Park (2007)
48. Wilke, T.: Alternating tree automata, parity games, and modal μ-calculus. Bull. Belgian Math. Soc. **8**(2), 359–391 (2001)
49. Konev, B., Lutz, C., Wolter, F., Zakharyaschev, M.: Conservative rewritability of description logic TBoxes. In: Proceedings of the 25th International Joint Conference on Artificial Intelligence (IJCAI 2016) (2016)
50. Horrocks, I., Sattler, U.: A description logic with transitive and inverse roles and role hierarchies. J. Log. Comput. **9**(3), 385–410 (1999). doi:10.1093/logcom/9.3.385
51. Ghilardi, S., Lutz, C., Wolter, F., Zakharyaschev, M.: Conservative extensions in modal logics. In: Proceedings of the AiML, vol. 6, pp. 187–207 (2006)
52. Lutz, C., Wolter, F.: Deciding inseparability and conservative extensions in the description logic EL. J. Symb. Comput. **45**(2), 194–228 (2010)
53. Lutz, C., Seylan, I., Wolter, F.: An automata-theoretic approach to uniform interpolation and approximation in the description logic EL. In: Proceedings of the 13th International Conference on the Principles of Knowledge Representation and Reasoning (KR 2012), pp. 286–296. AAAI Press (2012)
54. Nikitina, N., Rudolph, S.: ExpExpExplosion: uniform interpolation in general EL terminologies. In: Proceedings of the 20th European Conference on Artificial Intelligence (ECAI 2012), pp. 618–623 (2012). doi:10.3233/978-1-61499-098-7-618
55. Clarke, E., Schlingloff, H.: Model checking. In: Handbook of Automated Reasoning, vol. II, chap. 24, pp. 1635–1790. Elsevier (2001)
56. Ludwig, M., Walther, D.: The logical difference for $\mathcal{ELH}^r$-terminologies using hypergraphs. In: Proceedings of the 21st European Conference on Artificial Intelligence (ECAI 2014). Frontiers in Artificial Intelligence and Applications, vol. 263, pp. 555–560. IOS Press (2014)
57. Feng, S., Ludwig, M., Walther, D.: Foundations for the logical difference of EL-TBoxes. In: Global Conference on Artificial Intelligence (GCAI 2015). EPiC Series in Computing, vol. 36, pp. 93–112. EasyChair (2015). ISSN 2040-557X
58. Byers, P., Pitt, D.: Conservative extensions: a cautionary note. EATCS-Bull. **41**, 196–201 (1990)
59. Veloso, P.: Yet another cautionary note on conservative extensions: a simple case with a computing flavour. EATCS-Bull. **46**, 188–193 (1992)
60. Veloso, P., Veloso, S.: Some remarks on conservative extensions. A socratic dialog. EATCS-Bull. **43**, 189–198 (1991)

61. Diaconescu, J.G.R., Stefaneas, P.: Logical support for modularisation. In: Huet, G., Plotkin, G. (eds.) Logical Environments (1993)
62. Maibaum, T.S.E.: Conservative extensions, interpretations between theories and all that!. In: Bidoit, M., Dauchet, M. (eds.) CAAP 1997. LNCS, vol. 1214, pp. 40–66. Springer, Heidelberg (1997). doi:10.1007/BFb0030588
63. Konev, B., Lutz, C., Walther, D., Wolter, F.: Model-theoretic inseparability and modularity of description logic ontologies. Artif. Intell. **203**, 66–103 (2013)
64. Berger, R.: The Undecidability of the Domino Problem. Memoirs of the AMS, issue 66. American Mathematical Society, Providence (1966)
65. Robinson, R.: Undecidability and nonperiodicity for tilings of the plane. Inventiones Math. **12**, 177–209 (1971)
66. Börger, E., Grädel, E., Gurevich, Y.: The Classical Decision Problem. Perspectives in Mathematical Logic. Springer, Heidelberg (1997)
67. Feferman, S., Vaught, R.L.: The first-order properties of algebraic systems. Fundamenta Math. **47**, 57–103 (1959)
68. Bonatti, P., Faella, M., Lutz, C., Sauro, L., Wolter, F.: Decidability of circumscribed description logics revisited. In: Eiter, T., Strass, H., Truszczyński, M., Woltran, S. (eds.) Advances in Knowledge Representation, Logic Programming, and Abstract Argumentation. LNCS (LNAI), vol. 9060, pp. 112–124. Springer, Heidelberg (2015). doi:10.1007/978-3-319-14726-0_8
69. Maher, M.J.: Equivalences of logic programs. In: Foundations of Deductive Databases and Logic Programming, pp. 627–658. Morgan Kaufmann (1988)
70. Lifschitz, V., Pearce, D., Valverde, A.: Strongly equivalent logic programs. ACM Trans. Comput. Logic **2**(4), 526–541 (2001). doi:10.1145/502166.502170
71. Eiter, T., Fink, M.: Uniform equivalence of logic programs under the stable model semantics. In: Palamidessi, C. (ed.) ICLP 2003. LNCS, vol. 2916, pp. 224–238. Springer, Heidelberg (2003). doi:10.1007/978-3-540-24599-5_16
72. Cuenca Grau, B., Horrocks, I., Kazakov, Y., Sattler, U.: Extracting modules from ontologies: a logic-based approach. In: Stuckenschmidt, H., Parent, C., Spaccapietra, S. (eds.) Modular Ontologies. LNCS, vol. 5445, pp. 159–186. Springer, Heidelberg (2009). doi:10.1007/978-3-642-01907-4_8
73. Grau, B.C., Horrocks, I., Kazakov, Y., Sattler, U.: Just the right amount: extracting modules from ontologies. In: Proceedings of the 16th International World Wide Web Conference (WWW 2007), pp. 717–726. ACM (2007)
74. Grau, B.C., Horrocks, I., Kazakov, Y., Sattler, U.: A logical framework for modularity of ontologies. In: Proceedings of the 20th International Joint Conference on Artificial Intelligence (IJCAI 2007), pp. 298–303 (2007)
75. Sattler, U., Schneider, T., Zakharyaschev, M.: Which kind of module should I extract? In: Proceedings of the 22th International Workshop on Description Logics (DL 2009). CEUR Workshop Proceedings, vol. 477. CEUR-WS.org (2009)
76. Romero, A.A., Grau, B.C., Horrocks, I., Jiménez-Ruiz, E.: MORe: a modular OWL reasoner for ontology classification. In: ORE. CEUR Workshop Proceedings, vol. 1015, pp. 61–67. CEUR-WS.org (2013)
77. Romero, A.A., Grau, B.C., Horrocks, I.: Modular combination of reasoners for ontology classification. In: Proceedings of the 25th International Workshop on Description Logics (DL 2012). CEUR Workshop Proceedings, vol. 846. CEUR-WS.org (2012)
78. Romero, A.A., Kaminski, M., Grau, B.C., Horrocks, I.: Module extraction in expressive ontology languages via Datalog reasoning. J. Artif. Intell. Res. (JAIR) **55**, 499–564 (2016)

79. Bienvenu, M., ten Cate, B., Lutz, C., Wolter, F.: Ontology-based data access a study through Disjunctive Datalog, CSP, and MMSNP. ACM Trans. Database Syst. **39**(4), 33:1–33:44 (2014)

80. Poggi, A., Lembo, D., Calvanese, D., De Giacomo, G., Lenzerini, M., Rosati, R.: Linking data to ontologies. J. Data Semant. **10**, 133–173 (2008)

81. Botoeva, E., Kontchakov, R., Ryzhikov, V., Wolter, F., Zakharyaschev, M.: Games for query inseparability of description logic knowledge bases. Artif. Intell. **234**, 78–119 (2016). doi:10.1016/j.artint.2016.01.010. http://www.sciencedirect.com/science/article/pii/S0004370216300017, ISSN 0004-3702

82. Lutz, C., Wolter, F.: Non-uniform data complexity of query answering in description logics. In: Proceedings of KR. AAAI Press (2012)

83. Chang, C.C., Keisler, H.J.: Model Theory. Studies in Logic and the Foundations of Mathematics, vol. 73. Elsevier, Amsterdam (1990)

84. Botoeva, E., Lutz, C., Ryzhikov, V., Wolter, F., Zakharyaschev, M.: Query-based entailment and inseparability for ALC ontologies (Full Version). CoRR Technical report abs/1604.04164, arXiv.org e-Print archive (2016). http://arxiv.org/abs/1604.04164

85. Botoeva, E., Lutz, C., Ryzhikov, V., Wolter, F., Zakharyaschev, M.: Query-based entailment and inseparability for ALC ontologies. In: Proceedings of the 25th International Joint Conference on Artificial Intelligence (IJCAI 2016), pp. 1001–1007 (2016)

86. Schaerf, A.: Query answering in concept-based knowledge representation systems: algorithms, complexity, and semantic issues. Ph.D. thesis, Dipartimento di Informatica e Sistemistica, Università di Roma La Sapienza (1994)

87. Abiteboul, S., Hull, R., Vianu, V.: Foundations of Databases. Addison-Wesley, Boston (1995)

88. Mazala, R.: Infinite games. In: Grädel, E., Thomas, W., Wilke, T. (eds.) Automata Logics, and Infinite Games. LNCS, vol. 2500, pp. 23–38. Springer, Heidelberg (2002). doi:10.1007/3-540-36387-4_2

89. Chatterjee, K., Henzinger, M.: An $O(n^2)$ time algorithm for alternating Büchi games. In: Proceedings of the 23rd Annual ACM-SIAM Symposium on Discrete Algorithms (SODA), pp. 1386–1399. SIAM (2012)

90. Baader, F., Bienvenu, M., Lutz, C., Wolter, F.: Query and predicate emptiness in ontology-based data access. J. Artif. Intell. Res. (JAIR) **56**, 1–59 (2016)

91. Konev, B., Ludwig, M., Wolter, F.: Logical Difference Computation with CEX2.5. In: Gramlich, B., Miller, D., Sattler, U. (eds.) IJCAR 2012. LNCS (LNAI), vol. 7364, pp. 371–377. Springer, Heidelberg (2012). doi:10.1007/978-3-642-31365-3_29

92. Bienvenu, M., Hansen, P., Lutz, C., Wolter, F.: First order-rewritability and containment of conjunctive queries in Horn description logics. In: Proceedings of the 25th International Joint Conference on Artificial Intelligence (IJCAI 2016), pp. 965–971 (2016)

93. Bienvenu, M., Rosati, R.: Query-based comparison of mappings in ontology-based data access. In: Proceedings of the 15th International Conference on the Principles of Knowledge Representation and Reasoning (KR 2016), pp. 197–206 (2016). http://www.aaai.org/ocs/index.php/KR/KR16/paper/view/12902

94. Arenas, M., Gottlob, G., Pieris, A.: Expressive languages for querying the semantic web. In: Proceedings of the 33rd ACM SIGMOD-SIGACT-SIGART Symposium on Principles of Database Systems, PODS 2014, Snowbird, UT, USA, 22–27 June 2014, pp. 14–26 (2014). doi:10.1145/2594538.2594555

95. Fagin, R., Kolaitis, P.G., Nash, A., Popa, L.: Towards a theory of schema-mapping optimization. In: Proceedings of the 27th ACM SIGACT SIGMOD SIGART Symp. on Principles of Database Systems (PODS 2008), pp. 33–42 (2008). doi:10. 1145/1376916.1376922

96. Pichler, R., Sallinger, E., Savenkov, V.: Relaxed notions of schema mapping equivalence revisited. Theory Comput. Syst. **52**(3), 483–541 (2013). doi:10.1007/ s00224-012-9397-0

97. Konev, B., Kontchakov, R., Ludwig, M., Schneider, T., Wolter, F., Zakharyaschev, M.: Conjunctive query inseparability of OWL 2 QL TBoxes. In: Proceedings of the 25th National Conference on Artificial Intelligence (AAAI 2011), pp. 221–226. AAAI Press (2011)

98. Vescovo, C., Klinov, P., Parsia, B., Sattler, U., Schneider, T., Tsarkov, D.: Empirical study of logic-based modules: cheap is cheerful. In: Alani, H., et al. (eds.) ISWC 2013. LNCS, vol. 8218, pp. 84–100. Springer, Heidelberg (2013). doi:10. 1007/978-3-642-41335-3_6

99. Gatens, W., Konev, B., Wolter, F.: Lower and upper approximations for depleting modules of description logic ontologies. In: Proceedings of the 21st European Conference on Artificial Intelligence (ECAI 2014), vol. 263, pp. 345–350. IOS Press (2014)

100. Nortje, R., Britz, A., Meyer, T.: Module-theoretic properties of reachability modules for SRIQ. In: Proceedings of the 26th International Workshop on Description Logics (DL 2013). CEUR Workshop Proceedings, vol. 1014, pp. 868–884. CEUR-WS.org (2013)

101. Nortje, R., Britz, K., Meyer, T.: Reachability modules for the description logic $\mathcal{SRIQ}$. In: McMillan, K., Middeldorp, A., Voronkov, A. (eds.) LPAR 2013. LNCS, vol. 8312, pp. 636–652. Springer, Heidelberg (2013). doi:10.1007/ 978-3-642-45221-5_42

102. Gonçalves, R.S., Parsia, B., Sattler, U.: Concept-based semantic difference in expressive description logics. In: Cudré-Mauroux, P., et al. (eds.) ISWC 2012. LNCS, vol. 7649, pp. 99–115. Springer, Heidelberg (2012). doi:10.1007/ 978-3-642-35176-1_7

103. Wang, K., Wang, Z., Topor, R., Pan, J.Z., Antoniou, G.: Concept and role forgetting in $\mathcal{ALC}$ ontologies. In: Bernstein, A., Karger, D.R., Heath, T., Feigenbaum, L., Maynard, D., Motta, E., Thirunarayan, K. (eds.) ISWC 2009. LNCS, vol. 5823, pp. 666–681. Springer, Heidelberg (2009). doi:10.1007/978-3-642-04930-9_42

104. Zhou, Y., Zhang, Y.: Bounded forgetting. In: Burgard, W., Roth, D. (eds.) Proceedings of the 25th National Conference on Artificial Intelligence (AAAI 2011). AAAI Press (2011)

105. Ludwig, M., Konev, B.: Practical uniform interpolation and forgetting for $\mathcal{ALC}$ tboxes with applications to logical difference. In: Proceedings of the 14th International Conference on the Principles of Knowledge Representation and Reasoning (KR 2014). AAAI Press (2014)

106. Gabbay, D.M., Schmidt, R.A., Szałas, A.: Second-Order Quantifier Elimination: Foundations, Computational Aspects and Applications. Studies in Logic: Mathematical Logic and Foundations, vol. 12. College Publications (2008). ISBN 978-1-904987-56-7

107. Zhao, Y., Schmidt, R.A.: Concept forgetting in $\mathcal{ALCOI}$-ontologies using an ackermann approach. In: Arenas, M., et al. (eds.) ISWC 2015. LNCS, vol. 9366, pp. 587–602. Springer, Heidelberg (2015). doi:10.1007/978-3-319-25007-6_34

Navigational and Rule-Based Languages for Graph Databases

Juan L. Reutter[1,2]([✉]) and Domagoj Vrgoč[1,2]

[1] Escuela de Ingeniería, Pontificia Universidad Católica de Chile, Santiago, Chile
{jreutter,dvrgoc}@ing.puc.cl
[2] Center for Semantic Web Research, Santiago, Chile

Abstract. One of the key differences between graph and relational databases is that on graphs we are much more interested in navigational queries. As a consequence, graph database systems are specifically engineered to answer these queries efficiently, and there is a wide body of work on query languages that can express complex navigational patterns.

The most commonly used way to add navigation into graph queries is to start with a basic pattern matching language and augment it with navigational primitives based on regular expressions. For example, the friend-of-a-friend relationship in a social network is expressed via the primitive (friend)+, which looks for paths of nodes connected via the friend relation. This expression can be then added to graph patterns, allowing us to retrieve, for example, all nodes A, B and C that have a common friend-of-a-friend.

But, in order to alleviate some of the drawbacks of isolating navigation in a set of primitives, we have recently witnessed an effort to study languages which integrate navigation and pattern matching in an intrinsic way. A natural candidate to use is Datalog, a well known declarative query language that extends first order logic with recursion, and where pattern matching and recursion can be arbitrarily nested to provide much more expressive navigational queries.

In this chapter we review the most common navigational primitives for graphs, and explain how these primitives can be embedded into Datalog. We then show current efforts to restrict Datalog in order to obtain a query language that is both expressive enough to express all these primitives, but at the same time feasible to use in practice. We illustrate how this works both over the base graph model and over the more general RDF format underlying the semantic web.

1 Introduction

Graph structured data is quickly becoming one of the more popular data paradigms for storing data in computer applications. Social networks, bioinformatics, astronomic databases, digital libraries, Semantic Web, and linked government data, are only a few examples of applications in which structuring data as graphs is essential. There is a growing body of literature on the subject and there are now several vendors of graph database systems [1–3]. See also [4,5] for surveys of the area. The simplest model of a graph database is that of edge-labelled

© Springer International Publishing AG 2017
J.Z. Pan et al. (Eds.): Reasoning Web 2016, LNCS 9885, pp. 90–123, 2017.
DOI: 10.1007/978-3-319-49493-7_3

directed graphs, where the nodes of the graph represent objects and the labelled edges represent relationships between these objects.

A standard way of querying graph data is to use a pattern matching language to look for small subgraphs of interest. For example, in a social network one can match the following pattern to look for a clique of three individual that are all friends with each other:

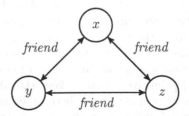

However, in several graph database applications one also needs to look for more complex conditions between nodes that are not necessarily adjacent to each other, which are known as navigational queries. The most commonly used way to add navigation into graph queries is to start with a basic pattern matching language and augment it with navigational primitives based on regular expressions. For example, the friend-of-a-friend relationship in a social network is expressed via the primitive $friend^+$, which looks for paths of nodes connected via an indirect chain of friends. We can then add this primitive into the following pattern, to look for pairs of persons with an indirect friend p_1 in common:

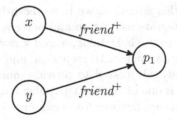

Navigational queries showcase one of the key differences between graph and relational databases, because these queries use a form of recursion that traditional relational languages and engines are not designed to deal with. As a consequence, graph database systems are specifically engineered to answer navigational queries efficiently, and there is a wide body of work on query languages that can express complex navigational patterns [5].

The first and most common navigational primitive that was proposed is that of *Regular Path Queries*, or RPQs [6], which corresponds precisely to regular expressions. But nowadays there exists a wide range of extensions of RPQs that add numerous other features such as the ability to traverse edges backwards, existential test operators such as the one used in XPath [7], negation over paths [8], etc. Graph systems also implement these types of navigational primitives; for example the graph system Neo4j [1] implements a subset of RPQs, and SPARQL [9], the standard query language for RDF graphs [10], features *Property Paths*:

another extension of RPQs. The combination of graph patterns and RPQs is usually modelled as a language where RPQs can be added to standard *conjunctive queries*, resulting in *Conjunctive Regular Path Queries*, or CRPQs [11]. This basic navigational pattern matching language are nowadays well understood, and there are also several extensions that have been proposed or are now implemented (see [5]).

But isolating navigation in a set of primitives has drawbacks for both systems and users. First, the algorithmic challenges needed to support efficient navigation are different from those needed to support efficient pattern matching. Thus, one usually ends up implementing two separate engines to compute the answers of a navigational query: one for pattern matching and one for dealing with the navigational primitives. This makes other database problems such as query planning and query optimization substantially more difficult, since now one has to implement techniques that work across both engines. This issue is also closely related to the fact that the navigational queries are generally not an *algebraically closed* language, in the sense that one cannot re-apply the same operators used for navigation to a graph pattern (for example, there is no notion of applying the transitive Kleene star of regular expressions to the navigational pattern of indirect friends shown above). Thus, posing queries in these language can also be uncomfortable for users familiar with algebraically closed languages such as the relational algebra, or SQL in general. But, additionally, by focusing on primitives designed to deal with paths we leave out the possibility of expressing other complex navigational relationships that cannot be reduced to a set of path operations.

In order to alleviate this situation, we have recently witnessed an effort to study languages which integrate navigation and pattern matching in an intrinsic way. A natural candidate to use is Datalog, a well known declarative query language that extends first order logic with recursion, and where pattern matching and recursion can be arbitrarily nested to provide much more expressive navigational queries. Datalog is one of the most popular languages in databases and has been used in numerous applications (one example is information extraction on the Web [12]).

The first attempt to define a Datalog-like language for graph databases is GraphLog [13], a language specifically designed to be closed, in the sense that GraphLog queries not only mix navigation and pattern matching directly, but also one can design queries where the Kleene star is directly applied to patterns in order to form a different graph database. However, dealing with Datalog has its own challenges, since it makes standard problems such as query evaluation and query containment substantially more difficult than they are for navigational queries based on pattern matching and primitives based on regular expressions. Thus, in the last years there have been numerous proposals to restrict GraphLog and similar languages so that they enjoy these other good algorithmic properties [14–17,19,42]. A good example of such a restriction are Regular Queries [42], a subset of Datalog that has almost the same algorithmic properties as CRPQs but whose expressive power is a good approximation of what GraphLog can do.

In this chapter we review the most common navigational primitives for graphs, and explain how these primitives can be embedded into Datalog. We then show current efforts to restrict Datalog in order to obtain a query language that is both expressive enough to subsume all these primitives, but at the same time feasible to use in practice. To show how these concepts can be used in a specific graph application, we then move to RDF databases, the graph format underlying the Semantic Web. We review Property Paths and Nested Regular Expressions, the choices of navigational primitives for RDF, and then show the specific problems we encounter when trying to design an algebraically closed language for RDF.

About the Languages Included in this Survey. We would like to stress that this is not intended to be a complete survey of graph querying features and languages. The objective of this article is to provide a good overview of query features that separate graph databases from the traditional relational format, and to do this we will focus on navigational aspects of graph querying. In particular, we will place a strong emphasis on languages based on regular expressions, and also show how more declarative formalisms (such as e.g. Datalog) can be used to capture interesting properties over graphs. Note that we do not consider several other important features of graph query languages such as e.g. attribute values and how these mix with navigational queries [20].

Organization. We introduce the formal model of graph databases and review relational queries in Sect. 2. Basic graph query languages based on regular expression are described in Sect. 3. In Sect. 4 we talk about how Datalog can be used to capture navigation over graphs, and in Sect. 5 we illustrate what problems we face when applying graph query languages over the RDF format underlying the Semantic Web. We conclude in Sect. 6.

2   Notation

Graph Databases. Let Σ be a finite alphabet. A *graph database* G over Σ is a pair (V, E), where V is a finite set of nodes and $E \subseteq V \times \Sigma \times V$ is a set of edges. That is, we view each edge as a triple $(v, a, v') \in V \times \Sigma \times V$, whose interpretation is an a-labelled edge from v to v' in G. When Σ is clear from the context, we shall simply speak of a graph database. Figure 1 shows an example of a graph database that stores information about a social network: here the nodes represent individuals that can be connected by relation *knows*, indicating that a person knows another person, or by *helps*, indicating that a person has helped another person in the past. Unless we specify otherwise, the size $|G|$ of G is simply the number of nodes in V plus the number of tuples in E.

We define the finite alphabet $\Sigma^{\pm} = \Sigma \cup \{a^- \mid a \in \Sigma\}$, that is, $\Sigma^{\pm}$ is the extension of Σ with the *inverse* of each symbol. The *completion* $G^{\pm}$ of a graph database G over Σ, is a graph database over $\Sigma^{\pm}$ that is obtained from G by adding the edge (v', a^-, v) for each edge (v, a, v') in G.

A *path* ρ from v_0 to v_m in a graph $G = (V, E)$ is a sequence (v_0, a_0, v_1), $(v_1, a_1, v_2), \cdots, (v_{m-1}, a_{m-1}, v_m)$, for some $m \geq 0$, where each (v_i, a_i, v_{i+1}), for

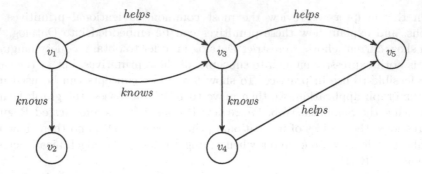

Fig. 1. A graph database over alphabet $\{knows, helps\}$, in which nodes are v_i, $1 \leq i \leq 5$.

$i < m$, is an edge in E. In particular, all the v_i's are nodes in V and all the a_j's are letters in Σ. The *label* of ρ, denoted by $\lambda(\rho)$, is the word $a_0 \cdots a_{m-1} \in \Sigma^*$. We also define the empty path as (v, ϵ, v) for each $v \in N$; the label of such path is the empty word ϵ.

Relational Queries: CQs and UCQs. A relational schema is a set $\sigma = \{R_1, \ldots, R_n\}$ of relation symbols, with each R_i having a fixed arity. Let **D** be a countably infinite domain. An instance I of σ assigns to each relation R in σ of arity n a finite relation $R^I \subseteq \mathbf{D}^n$. We denote by $\mathrm{dom}(I)$ the set of all elements from **D** that appear in any of the relations in I.

A *conjunctive query* (CQ) over a relational schema σ is a formula $Q(\bar{x}) = \exists \bar{y} \varphi(\bar{x}, \bar{y})$, where $\bar{x}$ and $\bar{y}$ are tuples of variables and $\varphi(\bar{x}, \bar{y})$ is a conjunction of relational atoms from σ that use variables from $\bar{x}$ and $\bar{y}$. We say that $\bar{x}$ are the free variables of the query Q. Conjunctive queries with no free variables are called boolean CQs; if Q is a boolean CQ, we identify the answer **false** with the empty relation, and **true** with the relation containing the 0-ary tuple.

We want to use CQs for querying graph databases over a finite alphabet Σ. In order to do this, given an alphabet Σ, we define the schema $\sigma(\Sigma)$ that consists of one binary predicate symbol E_a, for each symbol $a \in \Sigma$. For readability purposes, we identify E_a with a, for each symbol $a \in \Sigma$. Each graph database $G = (V, E)$ over Σ can be represented as a relational instance $\mathbf{D}(G)$ over the schema $\sigma(\Sigma)$: The database $\mathbf{D}(G)$ consists of all facts of the form $E_a(v, v')$ such that (v, a, v') is an edge in G (for this we assume that **D** includes all the nodes in V).

A conjunctive query over Σ is simply a conjunctive query over $\sigma(\Sigma^{\pm})$. The answer $Q(G)$ of a CQ Q over G is $Q(\mathbf{D}(G^{\pm}))$. A union of CQs (UCQ) Q over Σ is a disjunction $\theta_1(\bar{x}) \vee \cdots \vee \theta_k(\bar{x})$ of CQs over Σ with the same free variables. The answer $Q(G)$ is $\bigcup_{1 \leq j \leq k} \theta_j(G)$, for each graph database G.

Example 1. Consider a social network over alphabet $\{knows, helps\}$ such as the one from Fig. 1. The CQ

$$Q(x) = \exists y \exists z \big(knows(x, y) \wedge helps(y, z)\big)$$

retrieves all people that know a helper (a person that helps someone else). In the graph G from Fig. 1, we have that $v_1 \in Q(G)$, since v_1 knows v_3, and v_3 helps v_5. Similarly, $v_3 \in Q(G)$.

On the other hand, if we want to retrieves all people that either help someone, or that know someone who is a helper, we can use the following UCQ

$$Q'(x) = \exists p\ helps(x, p) \vee \exists y \exists z\ (knows(x, y) \wedge helps(y, z)).$$

3   Navigational Languages for Graph Databases

In this section we review the most widely used graph navigational primitives, and introduce query languages that are obtained when we combine these primitives in order to build more complex graph patterns. For each of these languages we study their expressive power, as well as the complexity of some computational tasks associated with them.

3.1   Path Queries

The most simple navigational querying mechanism for graph databases is provided by means of regular expressions, which are commonly known as *regular path queries*, or *RPQs* [6,21,22]. Formally, an RPQ Q over Σ is a regular language $L \subseteq \Sigma^*$, and it is specified using a regular expression R. The idea behind regular path queries is to select all pairs of nodes whose labels belong to the language of the defining regular expression. That is, given a graph database $G = (V, E)$ and an RPQ R, both over Σ, the evaluation $[\![R]\!]_G$ of R over G is the set of all pairs $(v, v') \in V$ such that there is path ρ between v and v', and the label $\lambda(\rho)$ of this path is a word in the language of R.

Example 2. Consider again the social network with relations *knows* and *helps* from Fig. 1. We can use RPQs to extract basic navigational information about this graph. For example, the query $knows^+$ retrieves all pairs of persons that are connected by a path $v_1, \ldots, v_n$ of individuals, where each v_i knows v_{i+1}. Furthermore, $knows \cdot helps$ can be used to retrieve all individuals that know a helper, and $knows + helps$ retrieves all nodes connected by a paths of individuals linked by either a *knows* edge or a *helps* edge.

The idea of using regular expressions for querying graph databases has been well established in the literature [11,23], and several extensions have been proposed for RPQs. The most popular is 2RPQs [22], which adds to RPQs the possibility of traversing the edges in a backwards direction. Furthermore, the language of 2RPQs has been subsequently extended to *Nested regular path queries* (NRPQs), with the inclusion of an *existential test* operator, similar to the one in XPath [7]. NRPQs were proposed in [24] for querying Semantic Web data, and as a querying formalism they offer a substantial increase in expressive power in comparison with 2RPQs, while maintaining the same query evaluation properties [24]. For this reason the language of nested regular path queries has received a fair deal of attention in the last few years [15,25,26].

Just as RPQs and 2RPQs, NRPQs specify pairs of node ids in a graph database, subject to the existence of a path satisfying a certain regular condition among them. The syntax of NRPQs over an alphabet Σ is given by the following grammar:

$$R \ := \ \varepsilon \ | \ a \ (a \in \Sigma) \ | \ a^- \ (a \in \Sigma) \ | \ R \cdot R \ | \ R^* \ | \ R + R \ | \ [R]$$

As usual we use R^+ as shorthand for $R \cdot R^*$. Moreover, when it is clear from the context, we omit the concatenation operator. Thus, if r_1 and r_2 are NRPQs, we sometimes just write $r_1 r_2$ instead of $r_1 \cdot r_2$. The size $|R|$ of an NRPQ is the number of characters used to represent R, and the nesting depth of R is the maximum number of nested [] operators in R (that is, the nesting depth of an expression that does not use any [] operator is 0, the nesting depth of $[R]$ is 1 plus the nesting depth of R, and all other operations preserve the maximum nesting depth of its subexpressions).

Although the semantics of an NRPQ R can be defined in terms of paths, it is best to define the binary relation $[\![R]\!]_G$, corresponding to the evaluation of R over a graph database G, in an inductive fashion. We present the definition in Table 1.

Table 1. Semantics of NRPQs. Here a is a symbol in Σ, and R, R_1 and R_2 are arbitrary NRPQs. The symbol ∘ denotes the usual composition of binary relations.

$$
\begin{aligned}
[\![\varepsilon]\!]_G &= \{(u, u) \mid u \text{ is a node id in } G\} \\
[\![a]\!]_G &= \{(u, v) \mid (u, a, v) \in G\} \\
[\![a^-]\!]_G &= \{(u, v) \mid (v, a, u) \in G\} \\
[\![R_1 \cdot R_2]\!]_G &= [\![R_1]\!]_G \circ [\![R_2]\!]_G \\
[\![R_1 + R_2]\!]_G &= [\![R_1]\!]_G \cup [\![R_2]\!]_G \\
[\![R^*]\!]_G &= [\![\varepsilon]\!]_G \cup [\![R]\!]_G \cup [\![R \cdot R]\!]_G \cup [\![R \cdot R \cdot R]\!]_G \cup \cdots \\
[\![[R]]\!]_G &= \{(u, u) \mid \text{there exists } v \text{ s.t. } (u, v) \in [\![R]\!]_G\}.
\end{aligned}
$$

Note that NRPQs subsume RPQs and 2RPQs. In fact, 2RPQs are just NRPQs that do not use the test operator $[R]$, and RPQs are NRPQs that use neither $[R]$ nor the inverse operator $^-$.

Example 3. Let G be the graph database in Fig. 1 that used labels *knows* and *helps*. Recall that in Example 2 we used the RPQ (which is also an NRPQ) $R_1 = knows^+$ to retrieve all pairs of nodes connected by a path in which all the edges are labelled *knows*. In particular, the pair (v_1, v_2) belongs to $[\![R_1]\!]_G$, and so do (v_1, v_3) and (v_1, v_4). If we now consider instead the NRPQ

$$R_2 \ = \ (knows + knows^-)^+,$$

we are now searching for a path of *knows*-labelled edges that may be traversed in either direction. Thus, the pair (v_2, v_4) now belongs to $[\![R_2]\!]_G$, as we can travel from v_2 to v_1 by using $knows^-$, and then to v_4 via two $knows$-labelled edges. This query is not an RPQ, but it is a 2RPQ. The NRPQ

$$R_3 = (knows[helps])^+$$

asks for all nodes x and y that are connected by a path of $knows$'s, but such that from each node in this path, except from x, there is also an outgoing edge labelled $helps$. The pair (v_1, v_4) belongs to $[\![R_3]\!]_G$, but $(v_1, v_2) \notin [\![R_3]\!]_G$, as v_2 has no outgoing $helps$-labelled edge. This query is neither an RPQ nor a 2RPQ.

Query Evaluation. As usual, the query evaluation problem asks, given a query R, a graph database G and a pair (u, v) of nodes from G, whether (u, v) belongs to the evaluation $[\![R]\!]_G$. The problem of evaluating RPQs was first considered in [6], where the resemblance between graph databases and automata was exploited to produce a simple algorithm that is linear in both the size of the graph and the size of the query. The idea is the following. Given a graph database $G = (V, E)$ over Σ, an RPQ R and a pair (u, v) of nodes from V, in order to decide wether (u, v) belongs to $[\![R]\!]_G$ one constructs from G the automaton $A_G(u, v) = (V, \Sigma, u, \{v\}, E)$ and the automaton A_R that accepts the language given by R. Note that $A_G(u, v)$ is obtained by viewing G as an NFA in which the initial state is u, the only final state is v, and the transition function is given by the edge relation E. Then it is not difficult to show that (u, v) belong to $[\![R]\!]_G$ if and only if the language of the product automaton $A_G(u, v) \times A_R$ is nonempty. Since the size of the automata is linear in the size of R, the size of the resulting product automata is $O(|G| \cdot |R|)$, and the reachability test is linear, giving us the desired time bounds. But we can also obtain an NLOGSPACE upper bound by performing the usual on-the-fly reachability test on the said product automaton. Hardness follows by reduction from the connectivity problem.

This result was lifted to 2RPQs in [11], the evaluation algorithm is based on the idea that evaluating a 2RPQ R over Σ on a graph G is the same as evaluating R over the completion $G^\pm$ of G, but now treating R as an RPQ over the extended alphabet $\Sigma^\pm$. Thus, all one needs to do is to obtain $G^\pm$ and then compute $[\![R]\!]_{G^\pm}$ just as explained above.

Proposition 1 [11]. *The query evaluation problem for a 2RPQ R and a graph G is* NLOGSPACE-*complete, and can be solved in* $O(|G| \cdot |R|)$.

It turns out that one can also obtain the same linear bounds even for NRPQs. The idea is to start with the innermost sub-expressions of the form $[R']$ in R, where the nesting depth of R' is 0. We evaluate R' using the algorithm for 2RPQs, and then augment graph G with a self-loop labeled $[R']$ in node u' whenever there is a node v' such that $(u', v') \in [\![R']\!]_G$. We can now repeat the process, treating R as an as an NRPQ with 1 less level of nesting over the extended alphabet that considers $[R]$ as an additional label. However, this time we need to assume that the nesting depth of the expression is fixed in order to obtain an NLOGSPACE upper bound.

Proposition 2 [24]. *The query evaluation problem for an NRPQ R and a graph G can be solved in $O(|G| \cdot |R|)$. The problem is* NLOGSPACE-*complete if the nesting depth of R is assumed to be fixed.*

Query answering has also been studied in the context of description logics, and interestingly, the complexity of the evaluation problem for nested regular path queries is usually higher than that for 2RPQs, even when considering knowledge bases given by simple DL-lite ontologies [26].

Query Containment. Another important problem in the study of query languages is that of containment. Formally, the containment problem asks, given queries R_1 and R_2 over Σ, whether $[\![R_1]\!]_G \subseteq [\![R_2]\!]_G$ on all possible graph databases over Σ. Checking query containment is a fundamental problem in database theory, and is relevant to several database tasks such as data integration [27], query optimisation [28], view definition and maintenance [29], and query answering using views [30].

As an example, query $(aa)^+$ is contained in (a^+), because all nodes connected by a path of even number of as are also connected by a path of a's. The problem becomes more interesting when considering 2RPQs, for example, the query aa is contained in $a(a^-a)^+a$, since in particular every path of two as will be part of the evaluation of the second query.

The containment problem has also received substantial attention in the graph database community. Calvanese et al. showed that the problem can be solved in PSPACE for RPQs and 2RPQs [11]. The proof uses the fact that an RPQ R_1 is contained in an RPQ R_2 over all graph databases if and only if they are contained only over *paths*, which allows us to work instead over strings: one can show that R_1 is contained in R_2 over graphs if an only if the language given by the regular expression R_1 is contained in the language of R_2. This gives us an immediate PSPACE tight bound for the complexity of containment since testing containment of two regular expressions is PSPACE-complete [31]. Likewise, for 2RPQs we can limit the search space for a counterexample to *semipaths*, or paths in which edges may be reversed, enabling us to reason about containment of 2RPQs by a clever rewriting of queries into 2-way automata [11], showing that the containment problem for 2RPQs is still in PSPACE.

For NRPQs the picture is a bit more complicated, since we cannot concentrate anymore on path-like structures. For example, consider the NRPQ $R_1 = a[b]a[b] + c^*$. Since the left disjunct of R_1 is not satisfiable over paths, we have that R_1 is contained in $R_2 = c^*$ over paths. However, R_1 is not contained in R_2 if we consider all possible graphs over $\{a, b, c\}$. Nevertheless, one can still solve the containment of NRPQs in PSPACE. The idea of the algorithm is to transform NRPQs into alternating two-way automata over a special types of trees, which can be subsequently encoded into strings.

Proposition 3 [11,32]. *The containment problem for two NRPQs is* PSPACE-*complete. It is* PSPACE-*hard even when the input are RPQs.*

3.2   Adding Conjunction, Union and Projection

It has been argued (see, e.g., [6,11,13,21]) that analogs of conjunctive queries whose atoms are navigational primitives such as RPQs, 2RPQs or NRPQs are much more useful in practice than the simple binary primitives. This motivated in [33] the study of *conjunctive regular path queries*, or *CRPQs*, and the further definition of *conjunctive two-way regular path queries* (C2RPQs, [11]) and *conjunctive nested regular path queries* (CNRPQs, [26,34]).

In such queries, multiple path queries can be combined, and some variables can be existentially quantified. Formally, a CNRPQ Q over a finite alphabet Σ is an expression of the form:

$$Q(\bar{z}) = \bigwedge_{1 \leq i \leq m} (x_i, L_i, y_i), \tag{1}$$

such that $m > 0$, each L_i is an NRPQ, and $\bar{z}$ is a tuple of variables among $\bar{x}$ and $\bar{y}$. The atom $Q(\bar{z})$ is the *head* of the query, the expression on the right of the equality is its *body*. A query with the head $Q()$ (i.e., no variables in the output) is called a *Boolean* query. CRPQs and C2RPQs are defined analogously, requiring instead the L_is to be RPQs or 2RPQs, respectively.

Intuitively, a query of the form (1) selects tuples $\bar{z}$ for which there exist values of the remaining node variables from $\bar{x}$ and $\bar{y}$ such that each RPQ (respectively, 2RPQ or NRPQ) in the body is satisfied. Formally, given Q of the form (1) and a graph $G = (V, E)$, a valuation is a map $\tau : \bigcup_{1 \leq i \leq m} \{x_i, y_i\} \rightarrow V$. We write $(G, \tau) \models Q$ if $(\tau(x_i), \tau(y_i))$ is in $[\![L_i]\!]_G$. Then the evaluation $Q(G)$ of Q over G is the set of all tuples $\tau(\bar{z})$ such that $(G, \tau) \models Q$. If Q is Boolean, we let $Q(G)$ be true if $(G, \tau) \models Q$ for some τ (that is, as usual, the singleton set with the empty tuple models true, and the empty set models false).

Example 4. Recall the social network from Fig. 1 that connects people via the *helps* and *knows* relationships. The following query looks for two individuals u and v that are connected both by a path of *helps* relations and by a path of *friends* relations:

$$Q(x, y) = (x, helps^+, y) \wedge (x, knows^+, y).$$

Note that the query in the example above has the same structure as the pattern $(x, helps, y) \wedge (x, knows, y)$ we used to compute people connected by both *knows* and *helps* in Example 1. And indeed, there is a tight connection between relational conjunctive queries and the notion of CRPQs, C2RPQs and CNRPQs. Namely, CQs can be seen as queries over a relational representation of a graph, where we can use the edge labels as basic navigational primitives. In CRPQs this is generalised, and we can now use any regular language in place of simple edge labels. Likewise, in C2RPQs we can use regular expressions over $\Sigma^{\pm}$, and in CNRPQs we can use any NRPQ (see [35] for a more detailed study of this type of graph patterns).

Query Evaluation. All three classes of graph queries we consider here contain the class of CQs, so they inherit the NP-hardness bound for query evaluation from CQs [36]. And using the fact that the evaluation of each primitive is in polynomial time, it is not difficult to show that this bound is tight: to check wether a tuple $\bar{a}$ belongs to the answer of a CNRPQ $Q(\bar{z})$ of the form (1) over a graph G, one just need to guess a valuation τ that maps $\bar{z}$ to $\bar{a}$, and then verify, for each conjunct (x_i, L_i, y_i) of Q, that $(\tau(x_i), \tau(y_i))$ is in $[\![L_i]\!]_G$ (in polynomial time since L_i is an NRPQ). In databases it is also customary to study the evaluation problem when the query is considered to be fixed, which is known as the *data complexity* of the evaluation problem. For CQs the data complexity is in AC^0, which is contained in NLOGSPACE, and we can plug-in the NLOGSPACE algorithm to compute the answers of each NRPQ to obtain an NLOGSPACE upper bound for the evaluation of any fixed CNRPQ. Hardness follows directly from Proposition 1, since RPQs are a special case of CNRPQs.

Observation 1. *The query evaluation problem for CNRPQs is NP-complete. It is NLOGSPACE-complete in data complexity (when the query is fixed).*

Query Containment. The containment problem for CRPQs was first studied by Calvanese et al. [11], and from there onwards we have seen a great deal of work devoted to the containment problem for various restrictions and extensions of these languages. The first observation is that, for containment, we only need to focus on boolean queries.

Observation 2. *There is a polynomial time reduction from the containment problem for nested C2RPQs to the containment problem for boolean nested C2RPQs.*

The idea of the reduction is to replace each query $Q(\bar{z})$ of the form (1) over Σ with free variables $\bar{z} = z_1, \ldots, z_n$ with the following boolean query Q^b over an extended alphabet $\Sigma \cup \{\$_1, \ldots, \$_n\}$

$$Q^b = \bigwedge_{1 \le i \le m} (x_i, L_i, y_i) \wedge \bigwedge_{1 \le j \le n} (z_j, \$_j, z_j) \tag{2}$$

It is straightforward to show that a query Q_1 is contained in a query Q_2 iff Q_1^b is contained in Q_2^b.

Let us now explain how to decide the containment problem for boolean CRPQs. Let Q_1 and Q_2 be two boolean CRPQs over Σ. The basic idea in [11] is the following. Given two CRPQs, Q_1 and Q_2, we first construct an NFA $\mathcal{A}_1$, of exponential size, that accepts precisely the "codifications" of the graph databases that satisfy Q_1, and then construct an NFA $\mathcal{A}_2$, of double-exponential size, that accepts precisely the "codifications" of the graph databases that do not satisfy Q_2. Then it is possible to prove that $Q_1 \not\subseteq Q_2$ if and only the language accepted by $\mathcal{A}_1 \cap \mathcal{A}_2$ is nonempty. Since $\mathcal{A}_1$ and $\mathcal{A}_2$ are of exponential size, the latter can be done in EXPSPACE by using a standard "on-the-fly" verification algorithm [37]. The same work also shows that the containment is also hard for EXPSPACE, so this algorithm is essentially the best one can do. Moreover, the same technique is shown to work when both Q_1 and Q_2 are C2RPQs. The containment problem for CNRPQs was studied indirectly in [34] in the context of graph data exchange, and an EXPSPACE upper bound also follows from [16,17].

Proposition 4. *The query containment problem for CNRPQs is* EXPSPACE-*complete. It remains* EXPSPACE-*hard even for CRPQs.*

Adding Unions. Further extensions make a case for considering unions of these queries, obtaining the classes of UCRPQs, UC2RPQs and UCNRPQs (where the capital U stands for union). It is not difficult to show that these queries have actually more expressive power than their union-free counterparts, and one can also show that the same bounds hold for both containment and query evaluation problems.

4 Datalog for Querying Graphs

It is evident that the base navigational languages we introduced in the previous section lack the expressive power to be used as a standalone query language for graph databases. But unfortunately none of extensions we have seen so far (from CRPQs to UCNRPQs) is *algebraically closed*, which is a key disadvantage from both the user and the system point of view. Indeed, algebraic closure has proved to be a prevalent property in several other widely used query languages. To name a few examples, note first that relational algebra is defined as the closure of a set of relational operators [28]. Also, the class of CQs is closed under projection and join, while UCQs are also closed under union [28]. Similarly, the class of 2RPQs is closed under concatenation, union, and transitive closure. In contrast, UC2RPQs and UCNRPQs are not closed under transitive closure, because even the transitive closure of a binary UC2RPQ query is not a UC2RPQ query.

In this section we show how to obtain query languages that are *algebraically closed*. All of these languages are based on Datalog, so we must start by introducing Datalog programs, and showing how they are used to query graphs. The problem, however, is that when using full Datalog as our query language we lose the nice evaluation and containment properties that UCNRPQs enjoy. Is there a navigational graph language that is algebraically closed, but that at the same time enjoys the good properties of UCNRPQs? We answer this positively, with the introduction of Regular Queries.

4.1 Datalog as a Graph Query Language

A Datalog *program* Π consists of a finite set of rules of the form $S(\bar{x}) \leftarrow R_1(\bar{y}_1), \ldots, R_m(\bar{y}_m)$, where $S, R_1, \ldots, R_m$ are predicate symbols and $\bar{x}, \bar{y}_1, \ldots, \bar{y}_m$ are tuples of variables. A predicate that occurs in the head of a rule is called *intensional* predicate. The rest of the predicates are called *extensional* predicate. We assume that each program has a distinguished intensional predicate called *Ans*. Let P be an intensional predicate of a Datalog program Π and I an instance over the schema given by the extensional predicates of Π. For $i \geq 0$, $P_\Pi^i(I)$ denote the collection of facts about the intensional predicate P that can be deduced from I by at most i applications of the rules in Π. Let $P_\Pi^\infty(I)$ be $\bigcup_{i \geq 0} P_\Pi^i(I)$. Then, the *answer* $\Pi(I)$ of Π over I is $Ans_\Pi^\infty(I)$.

A predicate P *depends* on a predicate Q in a Datalog program Π, if Q occurs in the body of a rule ρ of Π and P is the predicate in the head of ρ. The *dependence graph* of Π is a directed graph whose nodes are the predicates of Π and whose edges capture the dependence relation: there is an edge from Q to P if P depends on Q. A program Π is *nonrecursive* if its dependence graph is acyclic, that is, no predicate depends recursively on itself.

We can view Datalog queries as a graph language. In order to do this we proceed just as for CQs: given an alphabet Σ, we use the schema $\sigma(\Sigma)$ that consists of one binary predicate for each symbol $a \in \Sigma$. We can then represent each graph G over Σ as its straightforward relational representation $\mathbf{D}(G)$ over $\sigma(\Sigma)$. A (nonrecursive) Datalog program over a finite alphabet Σ is a (nonrecursive) Datalog program Π whose extensional predicates belong to $\sigma(\Sigma^{\pm})$. The *answer* $\Pi(G)$ of a (nonrecursive) Datalog program Π over a graph database G over Σ is $\Pi(\mathbf{D}(G^{\pm}))$.

The idea of using Datalog as a graph query language comes from Consens and Mendelzon [13], where it was introduced under the name GraphLog, as an alternative to UCRPQs that could express other types of graph properties, in particular those which are not monotone. To keep the complexity low, and to ensure that the intentional predicates in Datalog programs continue to resemble graphs, the arity of all predicates in the programs where restricted to be *binary*. However, GraphLog includes features that we shall not review in this section, such as including negated predicates. Nevertheless, Datalog programs as we have defined here are still enough to express all of the primitives we reviewed in the previous section, as well as conjunctions, unions, and of course more expressive forms of recursion.

Example 5. Consider again the CRPQ in Example 4. It can be expressed via the following Datalog program:

$$Hpath(x, y) \leftarrow helps(x, y).$$
$$Hpath(x, z) \leftarrow Hpath(x, y), helps(y, z).$$
$$Kpath(x, y) \leftarrow knows(x, y).$$
$$Kpath(x, z) \leftarrow Kpath(x, y), knows(y, z).$$
$$Ans(x, y) \leftarrow Hpath(x, y), Kpath(x, y).$$

The program uses three intentional predicates: *Hpath*, whose intention is to store all pairs of nodes that belong to the evaluation of the RPQ *helps*$^+$ (that is, the transitive closure of *helps*), *Kpath*, intended to store the result of *knows*$^+$, and *Ans*, which selects those pairs that appear both in *Hpath* and *Kpath*.

Datalog can also express NRPQs. For example, consider the NRPQ

$$(knows[helps])^+,$$

which intuitively computes those pairs of nodes connected by a path of people that know one another, but requiring as well that each node in the path is a helper. This query is computed by the program

$$N(x, y) \leftarrow knows(x, y), helps(y, z).$$
$$Ans(x, y) \leftarrow N(x, y).$$
$$Ans(x, z) \leftarrow Ans(x, y), N(y, z).$$

4.2 Binary Linear Datalog

The examples in the previous section suggest that Datalog programs subsume all of our navigational primitives, and even all CNRPQs. But we can actually show more: each CNRPQ (and in fact, each UCNRPQ) can be expressed by a fragment of Datalog that is particularly well behaved for query evaluation. We say that a Datalog program Π is *linear* if we can partition its rules into sets $\Pi_1, \ldots, \Pi_n$ such that (1) the predicates in the head of the rules in Π_i do not ocurr in the body of any rule in any set Π_j, with $j < i$; and (2) the body of each rule in Π_i has at most one occurrence of a predicate that occurs in the head of a rule in Π_i[1]. A binary linear Datalog program is just a linear program where all intensional predicates have arity 2, except possibly for *Ans*.

As usual, we say that a language $\mathcal{L}_1$ can be expressed using a language $\mathcal{L}_2$ if for every query in $\mathcal{L}_1$ there is an equivalent query in $\mathcal{L}_2$. If in addition $\mathcal{L}_2$ has a query not expressible in $\mathcal{L}_1$, then $\mathcal{L}_2$ is strictly more expressive than $\mathcal{L}_1$. The languages are equivalent if each can be expressed using the other. They are incomparable if none can be expressed using the other.

Observation 3. *Binary linear Datalog programs are strictly more expressive than UCNRPQs.*

To show that every UCNRPQ can be expressed as a Datalog program we proceed just as in the example above. Unions, conjunctions and concatenations, and the empty string are all straightforwardly expressed in Datalog, and if one has programs that compute expressions R_1 and R_2 into predicates P_{R_1} and P_{R_2}, then the query $R_1[R_2]$ can be computed using the rule $And(x, y) \leftarrow P_{R_1}(x, y), P_{R_2}(y, z)$. Finally, if one has a program to compute R into predicate P_R, then R^+ is given by the predicate P_{R^+}, computed as follows:

$$P_{R^+}(x, y) \leftarrow P_R(x, y).$$
$$P_{R^+}(x, y) \leftarrow P_{R^+}(x, z), P_R(z, y).$$

Moreover, as the following examples show, one can use binary linear Datalog to express a large number of interesting queries that cannot be expressed as UCNRPQs.

Example 6. Let us come back to our graph of relationships with labels *knows* and *helps* from Fig. 1. We say that a person p is a *friend* of a person p' if p knows

---

[1] These programs are sometimes referred to as *stratified linear* programs, or *piecewise linear Programs* [38].

and helps p' at the same time. The following program returns all the *indirect* friends, that is, all pairs of people connected by a chain of friends.

$$F(x, y) \leftarrow knows(x, y), helps(x, y).$$
$$Fchain(x, y) \leftarrow F(x, y).$$
$$Fchain(x, y) \leftarrow Fchain(x, y), F(y, z).$$
$$Ans(x, y) \leftarrow Fchain(x, y).$$

Suppose now that a person p' is an *acquaintance* of p if p knows p' and they have an indirect friend in common. The pairs of people connected by a chain of acquaintances can be expressed by the following RQ.

$$F(x, y) \leftarrow knows(x, y), helps(x, y).$$
$$Fchain(x, y) \leftarrow F(x, y).$$
$$Fchain(x, y) \leftarrow Fchain(x, y), F(y, z).$$
$$A(x, y) \leftarrow knows(x, y), Fchain(x, z), Fchain(y, z).$$
$$Achain(x, y) \leftarrow A(x, y).$$
$$Achain(x, y) \leftarrow Achain(x, y), A(y, z).$$
$$Ans(x, y) \leftarrow Achain(x, y).$$

With a little bit of work one can use the techniques from [15] to show that the two queries above cannot be expressed with UCNRPQs.   □

Thus, it appears that binary linear Datalog programs are a good candidate for querying graph databases: the language is algebraically closed, and it can express all UCNRPQs. But what are the algorithmic properties of this language? The good news is query evaluation, as we can show that binary linear Datalog programs enjoy the same complexity as the conjunctive languages we have reviewed in the previous section.

Proposition 5 [13,39]. *The query evaluation problem for binary linear Datalog programs is NP-complete in combined complexity and* NLogspace-*complete in data complexity.*

But, as it usually happens when working with Datalog programs, the containment problem becomes substantially more difficult when we move from UCNRPQs to binary linear Datalog. The following upper bound follows from [40,41] (there is also a refinement in [17]), while the lower bound follows from lower bounds of slightly less expressive languages in [15].

Proposition 6. *The query containment problem for binary linear Datalog queries is non-elementary.*

As we see, the problem with this language is that we are allowing too much freedom in choosing the way these programs are navigated. Thus, we need to further restrict the language in order to obtain something manageable from the point of view of containment.

4.3 Regular Queries

An *extended Datalog rule* is a rule of the form $S(\bar{x}) \leftarrow R_1(\bar{y}_1), \ldots, R_m(\bar{y}_m)$, where S is a predicate and, for each $1 \leq i \leq m$, R_i is either a predicate or an expression P^+ for a binary predicate P. An *extended* Datalog program is a finite set of extended Datalog rules. For an extended Datalog program, we define its extensional/intensional predicates and its dependence graph in the obvious way. Again we assume that there is a distinguished intensional predicate *Ans*. As expected, a nonrecursive extended Datalog program over an alphabet Σ is an extended Datalog program whose extensional predicates are in $\sigma(\Sigma^\pm)$ and whose dependence graph is acyclic.

Intuitively, extended Datalog rules offer some degree of recursion, but the recursion is limited so that it mimics the transitive closure operator of regular expressions. One can further define the language that results of combining multiple of these rules, which is known as Regular Queries [42]. Formally, A *regular query* (RQ) Ω over a finite alphabet Σ is a nonrecursive extended Datalog program over Σ, where all intensional predicates, except possibly for *Ans*, have arity 2.

The semantics of an extended Datalog rule is defined as in the case of a standard Datalog rule considering the semantics of an atom $P^+(y, y')$ as the pairs (v, v') that are in the transitive closure of the relation P. The semantics of a RQ is then inherited from the semantics of Datalog in the natural way. We denote by $\Omega(G)$ the answer of a RQ Ω over a graph database G.

Example 7. Recall the first query in Example 6, that computed chains of friends over a graph of relationships with labels *knows* and *helps*, and where a person p is a *friend* of a person p' if p knows and helps p' at the same time. The following RQ returns the desired information:

$$F(x, y) \leftarrow knows(x, y), helps(x, y).$$
$$Ans(x, y) \leftarrow F^+(x, y).$$

The second query in Example 6 computed all chains of acquaintances, where a person p' is an *acquaintance* of p if p knows p' and they have an indirect friend in common. This query can be expressed with the following RQ:

$$F(x, y) \leftarrow knows(x, y), helps(x, y).$$
$$A(x, y) \leftarrow knows(x, y), F^+(x, z), F^+(y, z).$$
$$Ans(x, y) \leftarrow A^+(x, y).$$

$\square$

Expressive Power. Note that RQs are also a closed language, since in particular the transitive closure of a binary RQ is always a RQ. This makes RQs a natural graph query language, but what about its expressive power?

The first observation is that every NRPQ can be expressed as a regular query, and in fact RQs subsume UCNPQs. In order to provide a translation from UCN-RPQs to RQs one can do a construction such as the one in Example 5, except that now the + operator in NRPQs is translated directly as an an expression P^+. Furthermore, we have just shown, in Example 7, that both the *chain-of-friends* and the *chain-of-acquaintances* queries in Example 6 can be expressed as RQs, and since these queries are not expressible as UCNRPQs, it follows that regular queries are strictly more expressive than UCNRPQs.

The next observation is RQs are actually contained in binary linear Datalog. To see this, note first that each expression P^+ in an extended Datalog program can be computed by the following linear Datalog rules (assuming now P^+ is a new predicate):

$$P^+(x, y) \leftarrow P(x, y).$$
$$P^+(x, y) \leftarrow P^+(x, z), P(z, y).$$

Thus, every RQ Ω can be translated in polynomial time into a binary linear Datalog program Π_Ω: one just transforms Ω into a regular Datalog program by treating each of the expressions P^+ as a new predicate, and then adds the rules shown above for each such predicate P^+. It is not difficult to see that the resulting program is indeed linear: since Regular Queries are nonrecursive we can use the same ordering on the rules of Ω to derive a partition for the rules in Π_Ω.

Being a subset of binary linear Datalog, the query evaluation problem for regular queries remains NP-complete in combined complexity and NLOGSPACE-complete in data complexity. Moreover, as promised, the complexity of query containment is elementary.

Proposition 7 (from [42]). *The query containment problem for regular queries is 2EXPSPACE-complete.*

Other Fragments. There are several other languages that are either more expressive or incomparable to regular queries. Amongst the list of the most modern ones we have *extended* CRPQs [43], which extends CRPQs with *path variables*, XPath for graph databases [44,45], and algebraic languages such as [8,46]. Although all these languages have interesting evaluation properties, the containment problem for all of them is undecidable. Another body of research has focused on fragments of Datalog with decidable containment problems. In fact, regular queries are also investigated in [16,17] (under the name of *nested positive* 2RPQs). But there are other restrictions that also lead to the decidability of the containment problem. Some of these include Monadic Datalog programs [14,15], programs whose rules are either guarded or frontier-guarded [15,17,19], and first order logic with transitive closure [16]. However, most of these fragments have non-elementary tight bounds for the containment problem, and elementary upper bounds are only obtained when the depth of the programs is fixed. Thus, regular queries seems to be the most expressive fragment of first-order logic with transitive closure that is known to have an elementary containment problem.

5 Moving to RDF

The Semantic Web and its underlying data model, RDF, are usually cited as one of the key applications of graph databases, but there is some mismatch between them. The basic concept of RDF is a *triple* (s, p, o), that consists of the subject s, the predicate p, and the object o, drawn from a domain of internationalised resource identifiers (IRI's). Thus, the middle element need not come from a finite alphabet, and may in addition play the role of a subject or an object in another triple. For instance, $\{(s, p, o), (p, s, o')\}$ is a valid set of RDF triples, but in graph databases, it is impossible to have two such edges.

We take the notion of reachability for granted in graph databases, but what is the corresponding notion for triples, where the middle element can serve as the source and the target of an edge? Then there are multiple possibilities, two of which are illustrated below.

Query Reach$_\rightarrow$ looks for pairs (x, z) connected by paths of the following shape:

and Reach$_f$ looks for the following connection pattern:

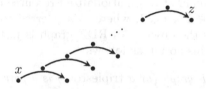

SPARQL, the standard query language for RDF graphs, defines property paths, a navigational query language that resembles 2RPQs, but is more tailored at querying graphs that can draw labels from infinite alphabets (for example, they include an explicit operator $!a$ to specify that two nodes should be connected by an edge labelled with something different from a, which makes little sense when dealing with a finite alphabet). However, property paths do not allow navigation through the middle element in triples, so queries such as Reach$_f$ cannot be expressed with property paths. To alleviate the situation, [24] propose to treat navigation primitives over a different graph encoding of RDF files that uses a finite alphabet of just three labels, and study the addition of NRPQs over this representation to the language of SPARQL, a language they denote as nSPARQL. We describe both property paths and nSPARQL in Sect. 5.2

However, as shown by [46], there are natural reachability patterns for triples, similar to those shown above, that *cannot* be defined in graph encodings of RDF [47] using nested regular path queries, nor in nSPARQL itself. In fact, queries Reach$_\rightarrow$ and Reach$_f$ demonstrate that there is no such thing as *the reachability* for triples. Moreover, we have again the problem of *closure*, as using graph languages for RDF leads us again to non-composable languages that need two different engines. The alternative is to completely redefine the concept of reachability for

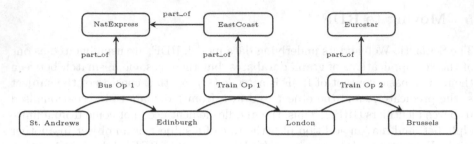

Fig. 2. RDF graph storing information about cities and transport services between them

RDF graphs. One possibility is to take all possible analogs of compositions of tertiary relation, and devise a navigational language built from these operators. We review this approach in Sect. 5.2.

5.1   Preliminaries

RDF Graphs. RDF graphs consist of triples in which, unlike in graph data-bases, the middle component need not come from a fixed set of labels. Let **I** be a countably infinite domain of Internationalized resource identifiers (IRI's). An RDF triple is $(s, p, o) \in \mathbf{I} \times \mathbf{I} \times \mathbf{I}$, where s is referred to as the subject, p as the predicate, and o as the object. An RDF graph is just a collection of RDF triples.[2] We formalise this notion as follows:

Definition 1. *An RDF graph (or a* triplestore*) is a pair $T = (O, E)$, where*

- $O \subseteq \mathbf{I}$ *is a finite set of IRIs;*
- $E \subseteq O \times O \times O$ *is a set of triples; and*
- *for each $o \in O$ there is a triple $t \in E$ such that o appears in T.*

Note that the final condition is used in order to simulate how RDF data is structured in practice, namely that it is presented in terms of sets of triples, so all the objects we are interested in actually appear in the triple relation.

Example 8. The RDF graph T in Fig. 2 contains information about cities, modes of transportation between them, and operators of those services. Each triple is represented by an arrow from the subject to the object, with the arrow itself labeled with the predicate. Examples of triples in T are (`Edinburgh`, `Train Op 1`, `London`) and (`Train Op 1`, `part_of`, `EastCoast`). For simplicity, we assume from now on that we can determine implicitly whether an object is a city or an operator. This can of course be modeled by adding an additional outgoing edge labeled `city` from each city and `operator` from each service operator.

---

[2] To simplify the presentation in this chapter we only consider *ground* RDF graphs [24], i.e. RDF graphs which do not contain any blank nodes nor literals.

5.2   Property Paths

Navigational properties (e.g. reachability) are among the most important functionalities of RDF query languages. In this section we introduce property paths, the W3C standard for querying navigational patterns in RDF, show how they work and how difficult it is to evaluate them, and also discuss some of their shortcomings and some proposals to fix those.

Property paths are a feature of SPARQL, the standard query language for RDF [9], that allow asking navigational queries over RDF graphs. Intuitively, a property path views an RDF document as a labelled graph where the predicate IRI in each triple acts as an edge label. It then extracts each pair of nodes connected by a path such that the word formed by the edge labels along this path belongs to the language of the expression specifying the property path. This idea is of course based on the family of regular path queries for graphs, but as we will see, there are several important differences.

Let us first review the definition of property paths, following the SPARQL 1.1 specification [9]. For consistency we stick to the graph database notation, but note that the standard sometimes uses different symbols for operators; for example, inverse paths e^- and alternative paths $e_1 + e_2$ from our definition are denoted there by $\hat{\ }e$ and $e_1 \mid e_2$, respectively.

Definition 2. Property paths *are defined by the grammar*

$$e := a \mid e^- \mid e_1 \cdot e_2 \mid e_1 + e_2 \mid e^+ \mid e^* \mid e? \mid !\{a_1, \ldots, a_k\} \mid !\{a_1^-, \ldots, a_k^-\},$$

where $a, a_1, \ldots, a_k$ *are IRIs in* $\mathcal{I}$. *Expressions of the last two forms (i.e., starting with* !) *are called* negated property sets.

The definition is based on 2RPQs, with the only difference being negated property sets. When dealing with singleton negated property sets brackets may be omitted: for example, $!a$ is a shortcut for $!\{a\}$. Besides the forms in Definition 2 the SPARQL 1.1 specification includes a third version of the negated property sets $!\{a_1, \ldots, a_k, b_1^-, \ldots, b_\ell^-\}$, which allows for negating both normal and inverted IRIs at the same time. We however do not include this extra form in our formalisation, since it is equivalent to the expression $!\{a_1, \ldots, a_k\} + !\{b_1^-, \ldots, b_\ell^-\}$.

The normative semantics for property paths is given in the following definition.

Definition 3. *The evaluation* $[\![e]\!]_T$ *of a property path* e *over an RDF graph* $T = (O, E)$ *is a set of pairs of IRIs from* O *defined as follows:*

$$[\![a]\!]_T = \{(s, o) \mid (s, a, o) \in E\},$$
$$[\![e^-]\!]_T = \{(s, o) \mid (o, s) \in [\![e]\!]_T\},$$
$$[\![e_1 \cdot e_2]\!]_T = [\![e_1]\!]_T \circ [\![e_2]\!]_T,$$
$$[\![e_1 + e_2]\!]_T = [\![e_1]\!]_T \cup [\![e_2]\!]_T,$$
$$[\![e^+]\!]_T = \bigcup_{i \geq 1} [\![e^i]\!]_T,$$
$$[\![e^*]\!]_T = [\![e^+]\!]_T \cup \{(a, a) \mid a \in O\},$$
$$[\![e?]\!]_T = [\![e]\!]_T \cup \{(a, a) \mid a \in O\},$$
$$[\![!\{a_1, \ldots, a_k\}]\!]_T = \{(s, o) \mid \exists a \text{ with } (s, a, o) \in E \text{ and } a \notin \{a_1, \ldots, a_k\}\},$$
$$[\![!\{a_1^-, \ldots, a_k^-\}]\!]_T = \{(s, o) \mid (o, s) \in [\![!\{a_1, \ldots, a_k\}]\!]_T\},$$

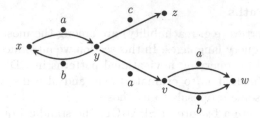

Fig. 3. Illustrating how negated property sets work. Triples in this RDF graph are $(x,a,y),(y,b,x),(y,c,z),(y,a,v),(v,a,w)$ and (v,b,w).

where $\circ$ is the usual composition of binary relations, and e^i is the concatenation $e \cdot \ldots \cdot e$ of i copies of e.

As we can see, for the most part, the semantics is the same as when dealing with 2RPQs. The only real difference comes from the interpretation of negated property sets. Intuitively, two IRIs are connected by a negated property set if they are subject and object of a triple in the graph whose predicate is not mentioned in the set under negation. Note that, according to Definition 3, the expression $!\{a_1^-, \ldots, a_k^-\}$ retrieves the inverse of $!\{a_1, \ldots, a_k\}$, and thus it respects the direction: a negated inverted IRI returns all pairs of nodes connected by some other inverted IRI. To exemplify, consider the RDF graph T from Fig. 3. We have that $[\![!a]\!]_T = \{(y,x),(y,z),(v,w)\}$ as we can find a forward looking predicate different from a for any of these pairs. Note that there is an a-labelled edge between v and w, but since there is also a b-labelled one, the pair (v,w) is in the answer. On the other hand, $[\![!a^-]\!]_T = \{(x,y),(z,y),(w,v)\}$, because we can traverse a backward looking predicate (either b^- or c^-) between these pairs.

Note that $!\{a_1, \ldots, a_k\}$ is not equivalent to $!a_1 + \ldots + !a_k$. To see this consider again the graph T from Fig. 3. We have $[\![!a]\!]_T = \{(y,x),(y,z),(v,w)\}$ and $[\![!b]\!]_T = \{(x,y),(y,z),(y,v),(v,w)\}$, while $[\![!\{a,b\}]\!]_T = \{(y,z)\}$.

Query Evaluation and Query Containment. Syntactically, property paths without negated property sets are nothing more than 2RPQs, with the only minor exception that the empty 2RPQ ε is not expressible as a property path expression. However, negated property sets are a unique feature that we have not reviewed yet. Note that if we were working with graph databases, where predicates come from a finite alphabet Σ, then one could easily replace $!a$ with a disjunction of all other symbols in Σ. But since we are dealing with RDF graphs, which have predicates from the infinite set of IRIs $\mathcal{I}$, we cannot treat this feature in such a naive way. However, one can still show that the query evaluation problem remains in low polynomial time.

Proposition 8 (from [18]). *For every property path e and RDF graph T the problem of deciding whether a pair (a,b) of IRIs belongs to $[\![e]\!]_T$ can be solved in time $O(|T| \cdot |e|)$.*

The idea of the algorithm is to do the usual product of the graph and the automata, now taking into account the negated property sets. We proceed in two steps.

- First we transform $T = (O, E)$ into a graph database $G = (V, E')$ over Σ as follows. Let $Pred(T)$ be the set of all predicates appearing in the triples in T, that is, $Pred(T) = \{p \mid \exists s, o \text{ such that } (s, p, o) \in E\}$. The set Σ of labels is $Pred(T) \cup \{p^- \mid p \in Pred(T)\}$. The set of nodes V is defined as the set of all the objects and subjects appearing in the triples of T. Finally, the set of edges E' contains an edge labelled with p from a node u to a node v if the triple (u, p, v) is in T, and an edge labelled with p^- from u to v if the triple (v, p, u) is in T.
- To do the cross product construction we treat e as an automata over the extended alphabet that includes all negated property sets as additional labels. We can then do the usual cross product construction, except we force a transition labelled with $\{!p_1, \ldots, !p_n\}$ in e to be matched to edges in G that are labelled with anything not in $\{p_1, \ldots, p_n\}$.

When it comes to query containment one can show that property paths behave similarly as 2RPQs. However, the techniques required to show this now differ slightly due to the inclusion of negated property sets. Just as in the case of 2RPQs, we can show that given two property paths e_1 and e_2, we can check in PSPACE if it holds that $[\![e_1]\!]_T \subseteq [\![e_2]\!]_T$ for every RDF graph T. Interestingly, when we allow conjunctions, projections and unions, the bound for UC2RPQs is still preserved, but the basic ideas for CRPQ containment [11] can no longer be used to prove this directly. Overall, we get:

Proposition 9 (from [18]). *The query containment problem for property paths is* PSPACE*-complete. If we allow combining property paths using union, conjunction and projection, the problem becomes* EXPSPACE*-complete.*

Nested Regular Path Queries in the RDF Context. As we have mentioned, the navigational capabilities currently in use in SPARQL are quite limited, in the sense that one cannot define paths that follow through properties such as the one in the query Reach$_f$. To address this limitation, Pérez et al. [24] propose to use NRPQs over a codification which transforms RDF graphs into graph databases.

Formally, given an RDF graph T, we define the transformation $\delta_{\text{NNE}}(T) = (V, E)$ as a graph database over alphabet $\Sigma = \{\text{next}, \text{node}, \text{edge}\}$, where V contains all IRIs from T, and for each triple (s, p, o) in T, the edge relation E contains edges (s, edge, p), (p, node, o) and (s, next, o). An example of coding an RDF graph using this technique is shown in Fig. 4.

Notice that the RDF document from Fig. 4 corresponds to a part of the reachability pattern Reach$_f$ introduced above. As we stated, this type of pattern is not expressible using property paths, but it can be computed as an NPRQ (and in fact, an RPQ) over the translation δ_{NNE} of an RDF graph. To be more precise, one can show that evaluating Reach$_f$ over an arbitrary RDF graph T yields

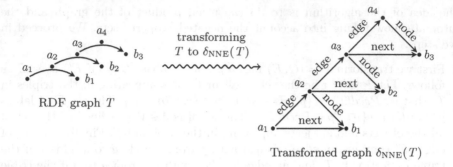

Fig. 4. Transforming an RDF graph into a graph database using δ_{NNE}.

the same answer as evaluating the RPQ $edge^+ node$ over the transformation $\delta_{\text{NNE}}(T)$. Besides allowing to express more complicated navigational queries, this transformation scheme was also used in several important practical RDF applications (e.g. to address the problem of interpreting RDFS features within SPARQL [24]). Since the transformation δ_{NNE} can be computed in linear time, using NRPQs over these codification is basically the same as using them over the original RDF graph, from the point of view of computational complexity. In particular, the bounds from Proposition 2 still hold with respect to the size of the RDF graph.

Although more powerful than property paths, NRPQs are still not capable of expressing some queries which one would naturally ask in the RDF context. To see this, consider the transportation network from Fig. 2. Suppose one wants to answer the following query:

Q: Find pairs of cities (x, y) such that one can travel from x to y
using services operated by the same company.

A query like this is likely to be relevant, for instance, when integrating numerous transport services into a single ticketing interface. For the graph T in Fig. 2, the pair (**Edinburgh**, **London**) belongs to $Q(T)$ (here $Q(T)$ denotes the answer of the query Q over the RDF graph T), and one can also check that (**St. Andrews**, **London**) is in $Q(T)$, since recursively both operators are part of NatExpress (using the transitivity of part_of). However, the pair (**St. Andrews**, **Brussels**) does not belong to $Q(T)$, since we can only travel that route if we change companies, from NatExpress to Eurostar.

If we try to use NRPQs, or even their conjunctive variants, to answer such a query, we immediately run into problem, as we are trying to verify that the company at the end of a chain of **part_of** edges is the same before we proceed to the next city. In essence, to answer the query Q, we would need to find patterns such as the one in Fig. 5 in our RDF graph.

In fact, it was shown in e.g. [17,46] that the query Q above can not be expressed using NRPQs over RDF graphs (or their δ_{NNE}-codification). On the

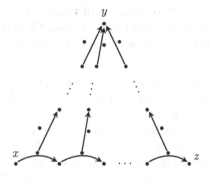

Fig. 5. A pattern required to answer the query Q.

other hand, queries such as Q seem quite natural in the context of RDF, so is there an efficient way of specifying and answering them? We give an answer to this question in the following section, where we introduce a (recursive) algebra for RDF graphs which can ask queries such as Q, and many others.

5.3 Native Navigation in Triples

In previous section we saw that treating RDF graphs as ordinary graph databases might not always allow us to answer the queries we want, and we also showed that the notion of reachability is not as straightforward when dealing with RDF as it is in graph databases. There is also another problem we did not consider so far. Namely, applying graph queries to RDF graphs leaves us with a set of pairs of nodes, while the initial data we started with contained triples. This means that all of these languages violate the *closure* property; that is, they start in one data model (triples), but end up in binary relations. To see why this might be a problem, note that once we obtain an answer to such query over RDF, we can no longer ask another query over this answer. Another way of saying this is that graph queries over RDF data do not *compose*.

So how can we overcome these issues? It is clear that for this we need a language which works directly over triples and which is composable in a sense that it does not leave the initial data model. Natural candidates to start with are of course Datalog, as we have shown in the previous section, but also the relational algebra [28], perhaps the most famous database composable language. We take for now the algebraic approach to language design, and introduce an algebra designed specifically for triples. We start with a plain version and then add recursive primitives that provide the crucial functionality for handling navigational queries. We also show how this algebra can be transformed into a Datalog fragment that resembles the binary linear Datalog programs we defined previously (but of course the binary restriction has to be dropped).

The operations of the usual relational algebra are selection, projection, union, difference, and cartesian product. Our language must remain *closed*, i.e., the result of each operation ought to be a set of triples. This clearly rules out projection. Selection and Boolean operations are fine. Cartesian product, however, would create a relation of arity six, so instead we use *joins* that only keep three positions in the result.

Triple Joins. To better understand what kind of joins we need, let us first look at the *composition* of two relations. For binary relations S and S', their composition $S \circ S'$ has all pairs (x, y) so that $(x, z) \in S$ and $(z, y) \in S'$ for some z. We can now define reachability with respect to relation S by recursively applying composition: $S \cup S \circ S \cup S \circ S \circ S \cup \ldots$. Note that this is how RPQs or property paths define reachability. So we need an analog of composition for triples. To understand how it may look, we can view $S \circ S'$ as the *join* of S and S' on the condition that the 2nd component of S equals the first of S', and the output consist of the remaining components. We can write it as

$$S \overset{1,2'}{\underset{2=1'}{\bowtie}} S'$$

Here we refer to the positions in S as 1 and 2, and to the positions in S' as $1'$ and $2'$, so the join condition is $2 = 1'$ (written below the join symbol), and the output has positions 1 and $2'$. This suggests that our join operations on triples should be of the form $R \bowtie_{\text{cond}}^{i,j,k} R'$, where R and R' are ternary relations, $i, j, k \in \{1, 2, 3, 1', 2', 3'\}$, and cond is a condition (to be defined precisely later).

But what is the most natural analog of relational composition? Note that to keep three indexes among $\{1, 2, 3, 1', 2', 3'\}$, we ought to project away three, meaning that two of them will come from one argument, and one from the other. Any such join operation on triples is bound to be *asymmetric*, and thus cannot be viewed as a full analog of relational composition.

So what do we do? Our solution is to add *all* such join operations. Formally, given two ternary relations R and R', *join* operations are of the form

$$R \overset{i,j,k}{\underset{\theta}{\bowtie}} R',$$

where

- $i, j, k \in \{1, 1', 2, 2', 3, 3'\}$,
- θ is a set of (in)equalities between elements in $\{1, 1', 2, 2', 3, 3'\} \cup \mathbf{I}$.

As before, we use the indices $1, 2, 3$ to denote positions in the relation to the left of the join symbol, and $1', 2', 3'$ for the ones to the right. In θ we allow comparing if the elements in some position are equal or different, and we also allow comparing them to some IRI as e.g. property paths do.

The semantics is defined as follows: (o_i, o_j, o_k) is in the result of the join iff there are triples $(o_1, o_2, o_3) \in R$ and $(o_{1'}, o_{2'}, o_{3'}) \in R'$ such that

- each condition from θ holds; that is, if $l = m$ is in θ, then $o_l = o_m$, and if $l = o$, where o is an IRI, is in θ, then $o_l = o$, and likewise for inequalities.

Using triple joins we can now define the language with the desired properties.

Triple Algebra. We now define the expressions of the *Triple Algebra*, or TriAL for short. It is a restriction of relational algebra that guarantees closure over triples, i.e., the result of each expression is an RDF graph.

- The set E of all the triples in an RDF graph is a TriAL expression.
- If e is a TriAL expression and θ a set of equalities and inequalities over $\{1, 2, 3\} \cup \mathbf{I}$, then $\sigma_\theta(e)$ is a TriAL expression, called a *selection expression*.
- If e_1, e_2 are TriAL expressions, then the following are TriAL expressions:
    - $e_1 \cup e_2$;
    - $e_1 - e_2$;
    - $e_1 \Join_\theta^{i,j,k} e_2$, with i, j, k, θ as in the definition of the join above.

The semantics of the join operation has already been defined. The semantics of the Boolean operations is the usual one. The semantics of the selection is defined in the same way as the semantics of the join (in fact, the operator itself can be defined in terms of joins): one just chooses triples (o_1, o_2, o_3) satisfying θ.

Given an RDF graph T, we write $e(T)$ for the result of evaluating the expression e over T. Note that $e(T)$ is again an RDF graph, and thus TriAL defines closed operations on triplestores.

Example 9. To get some intuition about the Triple Algebra consider the following TriAL expression:

$$R = E \overset{1,3',3}{\underset{2=1'}{\Join}} E$$

Indexes $(1, 2, 3)$ refer to positions of the first triple, and indexes $(1', 2', 3')$ to positions of the second triple in the join. Thus, for two triples (x_1, x_2, x_3) and $(x_{1'}, x_{2'}, x_{3'})$, such that $x_2 = x_{1'}$, expression R outputs the triple $(x_1, x_{3'}, x_3)$. E.g., in the triplestore of Fig. 2, (London, Train Op 2, Brussels) is joined with (Train Op 2, part_of, Eurostar), producing (London, Eurostar, Brussels); the full result is the following set of triples

St. Andrews	NatExpress	Edinburgh
Edinburgh	EastCoast	London
London	Eurostar	Brussels

When interpreted over the RDF document from Fig. 2, R gives us pairs of European cities together with companies one can use to travel from the first city to the second one. Note that this expression fails to take into account that EastCoast is a part of NatExpress. To add such information to query results (and produce triples such as (Edinburgh, NatExpress, London)), we use $R' = R \cup (R \Join_{2=1'}^{1,3',3} E)$.

Adding Recursion. One problem with Example 9 above is that it does not include triples (city$_1$, service, city$_2$) so that relation R contains a triple (city$_1$, service$_0$, city$_2$), and there is a chain, of some length, indicating that service$_0$ is a part of service. The second expression in Example 9 only accounted for such paths of length 1. To deal with paths of arbitrary length, we need reachability, which relational algebra is well known to be incapable of expressing. Thus, we need to add recursion to our language.

To do so, we expand TriAL with *right* and *left Kleene closure* of any triple join $\bowtie_\theta^{i,j,k}$ over an expression R, denoted as $(R \bowtie_\theta^{i,j,k})^*$ for right, and $(\bowtie_\theta^{i,j,k} R)^*$ for left. These are defined as

$$(R \bowtie)^* = \emptyset \cup R \cup R \bowtie R \cup (R \bowtie R) \bowtie R \cup \ldots,$$
$$(\bowtie R)^* = \emptyset \cup R \cup R \bowtie R \cup R \bowtie (R \bowtie R) \cup \ldots$$

We refer to the resulting algebra as *Triple Algebra with Recursion* and denote it by TriAL*.

When dealing with binary relations we do not have to distinguish between left and right Kleene closures, since the composition operation for binary relations is associative. However, as the following example shows, joins over triples are not necessarily associative, which explains the need to make this distinction.

Example 10. Consider an RDF graph $T = (O, E)$, with $E = \{(a, b, c), (c, d, e), (d, e, f)\}$. The expression

$$R_1 = (E \underset{3=1'}{\overset{1,2,2'}{\bowtie}})^*$$

computes $R_1(T) = E \cup \{(a, b, d), (a, b, e)\}$, while

$$R_2 = (\underset{3=1'}{\overset{1,2,2'}{\bowtie}} E)^*$$

computes $R_2(T) = E \cup \{(a, b, d)\}$.

Now we show how to formulate the queries mentioned earlier in this section using Triple Algebra.

Example 11. We started Sect. 5 by saying how there are different types of reachability over RDF graphs, and presented two queries, Reach$_\rightarrow$ and Reach$_\nearrow$, which illustrate this claim. It is easy to see that Reach$_\rightarrow$ and Reach$_\nearrow$ can be expressed using the TriAL* expressions E_1 and E_2 below:

$$E_1 = (E \underset{3=1'}{\overset{1,2,3'}{\bowtie}})^* \quad \text{and} \quad E_2 = (\underset{1=2'}{\overset{1',2',3}{\bowtie}} E)^*.$$

Next consider the query Q from Sect. 5.2. Recall that in this query we are asking for all pair of cities such that one can travel from the first city to the second one using services provided by the same company in a travel network such as the one presented in Fig. 2. Abstracting away from a particular RDF graph, this query asks us to find patterns of the form illustrated in Fig. 5. Although not expressible using property paths, or NRPQs (over codifications of RDF graphs), we can express this query using the following TriAL* expression:

$$((E \underset{2=1'}{\overset{1,3',3}{\bowtie}})^* \underset{3=1',2=2'}{\overset{1,2,3'}{\bowtie}})^*.$$

Note that the interior join $(E \underset{2=1'}{\overset{1,3',3}{\bowtie}})^*$ computes all triples (x, y, z), such that $E(x, w, z)$ holds for some w, and y is reachable from w using some E-path. The outer join now simply computes the transitive closure of this relation, taking into account that the service that witnesses the connection between the cities is the same.

Datalog for RDF. Triple Algebra and its recursive versions are in their essence *procedural* languages. It would therefore be nice to see a more declarative option for specifying TriAL queries. We have already seen a good candidate for capturing algebraic and recursive properties of a language in Sect. 4, that is: Datalog. So it seems natural to look for Datalog fragments that capture TriAL and its recursive version.

Since Datalog works over relational vocabularies, once again we need to explain how to interpret an RDF graph $T = (O, E)$ as a relational structure. However, this is rather straightforward: our relation schema will consist of a single ternary relation E, and in an instance I_T of this schema the interpretation of the relation E is equal to the relation E which stores all the triples in T. Using this we can now describe a Datalog fragment called TripleDatalog, which captures TriAL.

A TripleDatalog rule is of the form

$$S(\overline{x}) \leftarrow S_1(\overline{x_1}), S_2(\overline{x_2}), u_1 = v_1, \ldots, u_m = v_m \tag{3}$$

where

1. S, S_1 and S_2 are (not necessarily distinct) predicate symbols of arity 3;
2. $\overline{x}$, $\overline{x_1}$ and $\overline{x_2}$ are variables;
3. u_is and v_is are either variables or IRIs from **I**;
4. all variables in $\overline{x}$ and all variables in u_j, v_j are contained in $\overline{x_1} \cup \overline{x_2}$.

A TripleDatalog$^\neg$ rule is like the rule (3) but all equalities and predicates, except the head predicate S, can appear negated. A TripleDatalog$^\neg$ *program* Π is a finite set of TripleDatalog$^\neg$ rules. Such a program Π is *non-recursive* if there is an ordering $r_1, \ldots, r_k$ of the rules of Π so that the relation in the head of r_i does not occur in the body of any of the rules r_j, with $j \leq i$.

As is common with non-recursive programs, the semantics of nonrecursive TripleDatalog$^\neg$ programs is given by evaluating each of the rules of Π, according to the order $r_1, \ldots, r_k$ of its rules, and taking unions whenever two rules have the same relation in their head (see [28] for the precise definition). We are now ready to present the first capturing result.

Proposition 10. TriAL *is equivalent to nonrecursive* TripleDatalog⁻ *programs.*

Of course, here we are more interested in expressing navigational properties, so we now turn to TriAL*, the recursive variant of Triple Algebra. To capture it, we of course add recursion to Datalog rules, and impose a restriction that was previously used in [13]. A ReachTripleDatalog⁻ *program* is a (potentially recursive) TripleDatalog⁻ program in which each recursive predicate S is the head of exactly two rules of the form:

$$
\begin{aligned}
S(\overline{x}) &\leftarrow R(\overline{x}) \\
S(\overline{x}) &\leftarrow S(\overline{x}_1), R(\overline{x}_2), V(y_1, z_1), \ldots, V(y_k, z_k)
\end{aligned}
\tag{4}
$$

where each $V(y_i, z_i)$ is one of the following: $y_i = z_i$, or $y_i \neq z_i$, and R is a nonrecursive predicate of arity 3, or a recursive predicate defined by a rule of the form 4 that appears before S. These rules essentially mimic the standard linear reachability rules (for binary relations) in Datalog, and in addition one can impose equality and inequality constraints, along the paths.

Note that the negation in ReachTripleDatalog⁻ programs is *stratified*. The semantics of these programs is the standard least-fixpoint semantics [28]. The language GraphLog corresponds to almost the same syntactic class, except it is defined for graph databases, rather than triplestores. Interestingly, one can show that these classes capture the expressive power of FO with the transitive closure operator [13]. In our case, we have a capturing result for TriAL*.

Theorem 4. *The expressive power of* TriAL* *and* ReachTripleDatalog⁻ *programs is the same.*

We now give an example of a simple Datalog program computing the query Q from Sect. 5.2 and Example 11.

Example 12. The following ReachTripleDatalog⁻ *program is equivalent to query Q from Sect. 5.2. Note that the answer is computed in the predicate Ans.*

$$
\begin{aligned}
S(x_1, x_2, x_3) &\leftarrow E(x_1, x_2, x_3) \\
S(x_1, x_3', x_3) &\leftarrow S(x_1, x_2, x_3), E(x_2, x_2', x_3') \\
\mathrm{Ans}(x_1, x_2, x_3) &\leftarrow S(x_1, x_2, x_3) \\
\mathrm{Ans}(x_1, x_2, x_3') &\leftarrow \mathrm{Ans}(x_1, x_2, x_3), S(x_3, x_2, x_3')
\end{aligned}
$$

Recall that this query can be written in TriAL* *as* $Q = ((E \bowtie_{2=1'}^{1,3',3})^* \bowtie_{3=1',2=2'}^{1,2,3'})^*$. *The predicate S in the program computes the inner Kleene closure of the query, while the predicate* Ans *computes the outer closure.*

Query Evaluation and Query Containment over RDF Graphs. We have seen that TriAL* is a powerful language capable of expressing a wide range of queries over RDF graphs. The question is then, can we evaluate these queries efficiently? As before, to answer this question, we will look at the query evaluation problem, which asks, given an RDF graph T, TriAL* expression e, and a triple t, if it is true that $t \in e(T)$.

From previous sections we know that many graph query languages (RPQs, NRPQs) have a PTIME upper bound for the evaluation problem, and the data complexity (i.e. when e is assumed to be fixed) is generally NLOGSPACE (which can not be improved since basic reachability is already NLOGSPACE-hard). It can be shown that similar bounds hold for Triple algebra, and even its recursive variant.

Proposition 11 (from [46]). *The query evaluation problem for* TriAL* *queries is* PTIME-*complete, and it is* NLOGSPACE-*complete when the algebra expression is fixed.*

Of course, the high expressive power of TriAL* has to come with price, and this is reflected when we consider the query containment problem. Combining the fact that TriAL* subsumes a variant of XPath over graphs [46], and the fact that the latter has undecidable query containment [48], we obtain the following:

Proposition 12. *The query containment problem for* TriAL* *expressions is undecidable.*

5.4 More Expressive Languages

In previous sections we introduced several popular languages for extracting basic navigational patterns from RDF graphs, and showed that they can be evaluated efficiently. However, there are still many interesting properties of RDF graphs that cannot be expressed using these languages, so several more expressive formalisms for extracting information from RDF data have been proposed in the literature. The first one we would like to mention is TriQ [49], which is a Datalog-based language for expressing RDF queries, and which subsumes property paths, NRPQs and TriAL*. In its essence TriQ can be viewed as an extension of the Datalog characterisation of TriAL which does not place such severe restriction on the shape of the recursion as ReachTripleDatalog⁻ does, so it allows us to express more powerful queries while still keeping good query evaluation properties. On a more practical side there has been quite a bit of work on efficiently implementing property paths and their extensions, starting from engines exclusively dedicated to fine-tune the performance of reachability queries over RDF [50,51], to systems supporting fully recursive queries with SPARQL as their base [52,53].

Finally, we would like to mention that all of the previously mentioned languages work under the classic assumption that we have all of the data available locally and can process and rewrite it as needed. However, when we consider the native setting where RDF data is used, namely, the Web, this assumption is no longer valid, since it is no longer feasible to keep all the data locally, and it is often not possible to run usual query evaluation algorithms which compute the entire set of answers, but we need to limit ourselves to a subset, or an approximation of the answers which would be available if we were able to have the data locally. In particular, several basic principles of navigational query languages have to be refined to work in this context, and there has been some recent work [54–56] proposing how to do this over the data published under the Linked Data initiative [57], whose aim is to encourage the publishing of RDF data in a way that allows connecting datasets residing on different servers into one big dataset by enabling them to reference each other.

6   Conclusion

In this chapter, we have shown that the problem of defining and evaluating navigational graph queries has been the subject of numerous studies in recent years, and we now have several options of languages available for use, that offer a wide range of querying possibilities under a relatively low computational cost.

However, there is still much work to be done. For example, there are several other options for obtaining closed languages that we have not reviewed, and that so far have received far less attention than the Datalog variants. One such option is to directly define iterations and transitive closures of any graph pattern [58], and there are many other possibilities to explore, each of which having their own algorithmic challenges.

The second important problem that need to be dealt with is that of queries capable of returning the complete paths (see for example [43]). This feature is widely used and needed in practice, but yet the theoretical studies are just starting to appear, and there is still no consensus on what is a good way of integrating navigational queries with the ability to return paths.

Acknowledgments. We would like to thank Pablo Barceló, Jorge Pérez, Miguel Romero, Moshe Vardi, Egor Kostlyev and Leonid Libkin for helpful discussions on the topics of this chapter, as several of the results presented here where published in papers co-authored with them. The authors are supported by the Millennium Nucleus Center for Semantic Web Research Grant NC120004.

References

1. Robinson, I., Webber, J., Eifrem, E., Databases, G.: New Opportunities for Connected Data. O'Reilly Media, Inc., Sebastopol (2015)
2. Martinez-Bazan, N., Gomez-Villamor, S., Escale-Claveras, F.: DEX: a high-performance graph database management system. In: 2011 IEEE 27th International Conference on Data Engineering Workshops (ICDEW), pp. 124–127 (2011)
3. Angles, R.: A comparison of current graph database models. In: 2012 IEEE 28th International Conference on Data Engineering Workshops (ICDEW), pp. 171–177. IEEE (2012)
4. Angles, R., Gutierrez, C.: Survey of graph database models. ACM Comput. Surv. (CSUR) **40**(1), 1 (2008)
5. Barceló Baeza, P.: Querying graph databases. In: PODS, pp. 175–188. ACM (2013)
6. Cruz, I., Mendelzon, A., Wood, P.: A graphical query language supporting recursion. In: ACM Special Interest Group on Management of Data 1987 Annual Conference (SIGMOD), pp. 323–330 (1987)

7. Gottlob, G., Koch, C., Pichler, R.: Efficient algorithms for processing XPath queries. ACM Trans. Database Syst. **30**(2), 444–491 (2005)
8. Fletcher, G.H., Gyssens, M., Leinders, D., Surinx, D., Van den Bussche, J., Van Gucht, D., Vansummeren, S., Wu, Y.: Relative expressive power of navigational querying on graphs. Inf. Sci. **298**, 390–406 (2015)
9. Harris, S., Seaborne, A., Prud'hommeaux, E.: SPARQL 1.1 Query Language. W3C Recommendation (2013)
10. Klyne, G., Carroll, J.J.: Resource description framework (RDF): concepts and abstract syntax (2006)
11. Calvanese, D., De Giacomo, G., Lenzerini, M., Vardi, M.: Containment of conjunctive regular path queries with inverse. In: 7th International Conference on Principles of Knowledge Representation and Reasoning (KR), pp. 176–185 (2000)
12. Gottlob, G., Koch, C.: Logic-based web information extraction. SIGMOD Rec. **33**(2), 87–94 (2004)
13. Consens, M., Mendelzon, A.: GraphLog: a visual formalism for real life recursion. In: 9th ACM Symposium on Principles of Database Systems (PODS), pp. 404–416 (1990)
14. Rudolph, S., Krötzsch, M.: Flag & check: data access with monadically defined queries. In: Proceedings of the 32nd Symposium on Principles of Database Systems, pp. 151–162. ACM (2013)
15. Bourhis, P., Krötzsch, M., Rudolph, S.: Query containment for highly expressive datalog fragments. arXiv preprint arXiv:1406.7801 (2014)
16. Bourhis, P., Krötzsch, M., Rudolph, S.: How to best nest regular path queries. In: Description Logics (2014)
17. Bourhis, P., Krötzsch, M., Rudolph, S.: Reasonable highly expressive query languages. In: IJCAI, pp. 2826–2832 (2015)
18. Kostylev, E.V., Reutter, J.L., Romero, M., Vrgoč, D.: SPARQL with property paths. In: Arenas, M., Corcho, O., Simperl, E., Strohmaier, M., d'Aquin, M., Srinivas, K., Groth, P., Dumontier, M., Heflin, J., Thirunarayan, K., Staab, S. (eds.) ISWC 2015. LNCS, vol. 9366, pp. 3–18. Springer, Heidelberg (2015). doi:10.1007/978-3-319-25007-6_1
19. Bienvenu, M., Ortiz, M., Simkus, M.: Navigational queries based on frontier-guarded datalog: preliminary results. In: Alberto Mendelzon International Workshop on Foundations of Data Management, p. 162 (2015)
20. Libkin, L., Martens, W., Vrgoč, D.: Querying graphs with data. J. ACM **63**(2), 14 (2016)
21. Abiteboul, S., Buneman, P., Suciu, D.: Data on the Web: From Relations to Semi-structured Data and XML. Morgan Kauffman, Burlington (1999)
22. Calvanese, D., De Giacomo, G., Lenzerini, M., Vardi, M.: Rewriting of regular expressions and regular path queries. J. Comput. Syst. Sci. **64**(3), 443–465 (2002)
23. Mendelzon, A., Wood, P.: Finding regular simple paths in graph databases. SIAM J. Comput. **24**(6), 1235–1258 (1995)
24. Pérez, J., Arenas, M., Gutierrez, C.: nSPARQL: a navigational language for RDF. J. Web Semant. **8**(4), 255–270 (2010)
25. Barceló, P., Pérez, J., Reutter, J.L.: Relative expressiveness of nested regular expressions. In: AMW, pp. 180–195 (2012)
26. Bienvenu, M., Calvanese, D., Ortiz, M., Simkus, M.: Nested regular path queries in description logics. In: KR (2014)
27. Lenzerini, M.: Data integration: a theoretical perspective. In: PODS, pp. 233–246 (2002)

28. Abiteboul, S., Hull, R., Vianu, V.: Foundations of Databases. Addison-Wesley, Boston (1995)
29. Gupta, A., Mumick, I.S., et al.: Maintenance of materialized views: problems, techniques, and applications. IEEE Data Eng. Bull. **18**(2), 3–18 (1995)
30. Calvanese, D., Giacomo, G., Lenzerini, M., Vardi, M.Y.: View-based query answering and query containment over semistructured data. In: Ghelli, G., Grahne, G. (eds.) DBPL 2001. LNCS, vol. 2397, pp. 40–61. Springer, Heidelberg (2002). doi:10. 1007/3-540-46093-4_3
31. Meyer, A.R., Stockmeyer, L.J.: The equivalence problem for regular expressions with squaring requires exponential space. In: 13th Annual Symposium on Switching and Automata Theory, College Park, Maryland, USA, 25–27 October 1972, pp. 125–129 (1972)
32. Reutter, J.L.: Containment of nested regular expressions. arXiv preprint arXiv:1304.2637 (2013)
33. Florescu, D., Levy, A., Suciu, D.: Query containment for conjunctive queries with regular expressions. In: 17th ACM Symposium on Principles of Database Systems (PODS), pp. 139–148 (1998)
34. Barceló, P., Pérez, J., Reutter, J.L.: Schema mappings and data exchange for graph databases. In: ICDT, TBD (2013)
35. Barceló, P., Libkin, L., Reutter, J.L.: Querying regular graph patterns. J. ACM (JACM) **61**(1), 8 (2014)
36. Chandra, A.K., Merlin, P.M.: Optimal implementation of conjunctive queries in relational data bases. In: Proceedings of the Ninth Annual ACM Symposium on Theory of Computing, pp. 77–90. ACM (1977)
37. Vardi, M.Y.: An automata-theoretic approach to linear temporal logic. In: Moller, F., Birtwistle, G. (eds.) Logics for Concurrency. LNCS, vol. 1043, pp. 238–266. Springer, Heidelberg (1996). doi:10.1007/3-540-60915-6_6
38. Ullman, J.D.: Principles of Database and Knowledge-Base Systems. Computer Science Press, Rockville (1989)
39. Chaudhuri, S., Vardi, M.Y.: On the equivalence of recursive and nonrecursive datalog programs. In: Proceedings of the Eleventh ACM SIGACT-SIGMOD-SIGART Symposium on Principles of Database Systems, pp. 55–66. ACM (1992)
40. Courcelle, B.: The monadic second-order logic of graphs. I. Recognizable sets of finite graphs. Inf. Comput. **85**(1), 12–75 (1990)
41. Courcelle, B.: Recursive queries and context-free graph grammars. Theoret. Comput. Sci. **78**(1), 217–244 (1991)
42. Reutter, J.L., Romero, M., Vardi, M.Y.: Regular queries on graph databases. In: 18th International Conference on Database Theory (ICDT 2015), vol. 31, pp. 177–194 (2015)
43. Barcelo, P., Libkin, L., Lin, A.W., Wood, P.T.: Expressive languages for path queries over graph-structured data. ACM TODS **37**(4), 31 (2012)
44. Libkin, L., Martens, W., Vrgoč, D.: Querying graph databases with XPath. In: Proceedings of the 16th International Conference on Database Theory, pp. 129–140. ACM (2013)
45. Kostylev, E.V., Reutter, J.L., Vrgoč, D.: Containment of data graph queries. In: ICDT, pp. 131–142 (2014)
46. Libkin, L., Reutter, J., Vrgoč, D.: Trial for RDF: adapting graph query languages for RDF data. In: PODS, pp. 201–212. ACM (2013)

47. Arenas, M., Gutierrez, C., Pérez, J.: Foundations of RDF databases. In: Tessaris, S., Franconi, E., Eiter, T., Gutierrez, C., Handschuh, S., Rousset, M.-C., Schmidt, R.A. (eds.) Reasoning Web 2009. LNCS, vol. 5689, pp. 158–204. Springer, Heidelberg (2009). doi:10.1007/978-3-642-03754-2_4
48. Kostylev, E.V., Reutter, J.L., Vrgoč, D.: Static analysis of navigational XPath over graph databases. Inf. Process. Lett. **116**(7), 467–474 (2016)
49. Arenas, M., Gottlob, G., Pieris, A.: Expressive languages for querying the semantic web. In: Proceedings of the 33rd ACM SIGMOD-SIGACT-SIGART Symposium on Principles of Database Systems, PODS 2014, Snowbird, UT, USA, 22–27 June 2014, pp. 14–26 (2014). http://doi.acm.org/10.1145/2594538.2594555
50. Gubichev, A., Bedathur, S.J., Seufert, S.: Sparqling kleene: fast property paths in RDF-3X. In: First International Workshop on Graph Data Management Experiences and Systems, GRADES 2013, Co-located with SIGMOD/PODS 2013, New York, NY, USA, 24 June 2013, p. 14 (2013)
51. Yakovets, N., Godfrey, P., Gryz, J.: Query planning for evaluating SPARQL property paths. In: Proceedings of the 2016 International Conference on Management of Data, SIGMOD Conference 2016, San Francisco, CA, USA, 26 June–01 July 2016, pp. 1875–1889 (2016)
52. Atzori, M.: Computing recursive SPARQL queries. In: 2014 IEEE International Conference on Semantic Computing, Newport Beach, CA, USA, 16–18 June 2014, pp. 258–259 (2014)
53. Reutter, J.L., Soto, A., Vrgoč, D.: Recursion in SPARQL. In: Arenas, M., et al. (eds.) ISWC 2015. LNCS, vol. 9366, pp. 19–35. Springer, Heidelberg (2015). doi:10.1007/978-3-319-25007-6_2
54. Hartig, O., Pérez, J.: LDQL: a query language for the web of linked data. In: Arenas, M., et al. (eds.) ISWC 2015. LNCS, vol. 9366, pp. 73–91. Springer, Heidelberg (2015). doi:10.1007/978-3-319-25007-6_5
55. Hartig, O., Pirrò, G.: A context-based semantics for SPARQL property paths over the web. In: Gandon, F., Sabou, M., Sack, H., d'Amato, C., Cudré-Mauroux, P., Zimmermann, A. (eds.) ESWC 2015. LNCS, vol. 9088, pp. 71–87. Springer, Heidelberg (2015). doi:10.1007/978-3-319-18818-8_5
56. Fionda, V., Pirrò, G., Gutierrez, C.: NautiLOD: a formal language for the web of data graph. TWEB **9**(1), 5:1–5:43 (2015)
57. Berners-Lee, T.: Linked data (2006). http://www.w3.org/DesignIssues/LinkedData.html
58. He, H., Singh, A.K.: Graphs-at-a-time: query language and access methods for graph databases. In: Proceedings of the 2008 ACM SIGMOD International Conference on Management of Data, pp. 405–418. ACM (2008)

LOD Lab: Scalable Linked Data Processing

Wouter Beek, Laurens Rietveld, Filip Ilievski,
and Stefan Schlobach$^{(\boxtimes)}$

Department of Computer Science, VU University Amsterdam,
Amsterdam, Netherlands
{w.g.j.beek,laurens.rietveld,f.ilievski,k.s.schlobach}@vu.nl

Abstract. With tens if not hundreds of billions of logical statements, the Linked Open Data (LOD) is one of the biggest knowledge bases ever built. As such it is a gigantic source of information for applications in various domains, but also given its size an ideal test-bed for knowledge representation and reasoning, heterogeneous nature, and complexity.

However, making use of this unique resource has proven next to impossible in the past due to a number of problems, including data collection, quality, accessibility, scalability, availability and findability. The LOD Laundromat and LOD Lab are recent infrastructures that addresses these problems in a systematic way, by automatically crawling, cleaning, indexing, analysing and republishing data in a unified way. Given a family of simple tools, LOD Lab allows researchers to query, access, analyse and manipulate hundreds of thousands of data documents seamlessly, e.g. facilitating experiments (e.g. for reasoning) over hundreds of thousands of (possibly integrated) datasets based on content and meta-data.

This chapter provides the theoretical basis and practical skills required for making ideal use of this large scale experimental platform. First we study the problems that make it so hard to work with Semantic Web data in its current form. We'll also propose generic solutions and introduce the tools the reader needs to get started with their own experiments on the LOD Cloud.

1 Introduction

When we look at empirical data about the rudimentary infrastructure of the Semantic Web today, we see multiple problems: Millions of data documents exist that potentially contain information that is relevant for intelligent agents. However, only a tiny percentage of these data documents can be straightforwardly used by software agents. Typically, online data sources cannot be consistently queried over a prolonged period of time. As a consequence no commercial Web Service would dare to depend on general query endpoint availability and consistency. In practice, Semantic Web applications run locally on self-deployed and centralized triple stores housing data that has been integrated and cleaned for a specific application or purpose. Meanwhile, the universally accessible and automatically navigable online Linked Open Data (LOD) Cloud remains structurally

© Springer International Publishing AG 2017
J.Z. Pan et al. (Eds.): Reasoning Web 2016, LNCS 9885, pp. 124–155, 2017.
DOI: 10.1007/978-3-319-49493-7_4

disjointed, unreliable, and — as a result — largely unused for building the next generation of large-scale Web solutions.

It is widely accepted that proper data publishing is difficult. Much data that is published today contains imperfections of various kinds. In the realm of Linked Open Data it is particularly important for data to be published correctly. Data is intended to be processed by machines that cannot always come up with fallback procedures for data imperfections.

Uptake of Linked Open Data has seen a tremendous growth over the past fifteen years. Due to the inherently heterogeneous nature of interlinked datasets that come from different sources, LOD is not only a fertile environment for innovative data reuse, but also for mistakes and incompatibilities [1,2]. Such mistakes include character encoding issues, socket errors, protocol errors, syntax errors, corrupted archive headers and authentication problems.

In [3] the LOD Laundromat was presented; an attempt to clean a large number of RDF documents using a single, uniform and standards-compliant format. The LOD Laundromat has now (re)published a wealth of LOD in a format that can be processed by machines without having to pass through a dataset-specific and cumbersome data cleaning stage. At the time of writing, the LOD Laundromat disseminates over 650,000 data documents containing over 38,000,000,000 triples.

In [4] the LOD Laundromat, which had been serving clean datadumps until that point, was combined with the Linked Data Fragments (LDF) paradigm [5] in order to offer live query access to its entire collection of cleaned datasets.

LOD Laundromat differs from existing initiatives in the following ways:

- The **scale** on which clean data is made available: LOD Laundromat comprises hundreds of thousands of RDF documents, comprising tens of billions of RDF statements.
- The **speed** at which data is cleaned and made available: LOD Laundromat cleans about one billion triples a day and makes them immediately available online for others to download and query.
- The **level of automation**. LOD Laundromat automates the entire data processing pipeline, from dataset discovery to serialization in a standards-compliant canonical format that enables easy reuse.
- The **ease of use** it brings by making Open Web APIs and command-line tools available that allow the data to be accessed.

These characteristics of the LOD Laundromat framework have been illustrated in several applications that were built on top of it, particularly showcasing the properties of scale, automation and ease of use. These applications illustrate that the infrastructure supports many use cases that were difficult to implement before.

One use case is evaluating Semantic Web algorithms on large-scale, heterogeneous and real-world data. This has resulted in the LOD Lab application. Another use case is large-scale Semantic Search, which has resulted in the LOTUS full-text search framework.

In the next section of this chapter, we analyze the problems underlying current LOD deployments. Section 3 presents theoretical solutions to each of these problems. In Sect. 3 the implementation of these solutions into the LOD Laundromat infrastructure is explained. A tutorial-like overview of how to use client tools that interact with this infrastructure is given in Sect. 4. In the final section we enumerate some of the use cases that are supported by the infrastructure and client tools. We also look a bit into the future to see what other use cases are within reach for the community to work on.

2   Analyzing the Problem

In this section we give an overview of the problems that are exhibited in current Linked Data deployments. Even though some of these problems interact, it is possible to formulate a limited number of main obstacles that are currently blocking progress in the field.

2.1   Data Collection

Data collection is a difficult task on the LOD Cloud. With online sources appearing and disappearing all the time it is difficult to claim completeness. There is also the issue of no longer being up-to-date once a data source changes. There are alternative approaches to collect large volumes of LOD, each with its own incompleteness issues. Here we discuss the three main approaches:

Resources. Resource crawlers use the dereferenceability of IRIs to find LOD about that IRI. This approach has the following deficiencies:
- Datasets that do not contain dereferenceable IRIs are ignored. In a comprehensive empirical study [1] 7.2% of the crawled IRIs could not be dereferenced.
- For IRIs that can be dereferenced, back-links are often not included [2]. As a consequence, even datasets that contain only dereferenceable IRIs still have parts that cannot be reached by a crawler.
- Even for datasets that have only dereferenceable IRIs and that include back-links, the crawler can never be certain that the entire dataset has been crawled due to the possibility of there being disconnected graph components.

Endpoints. Querying endpoints provides another way of collecting large volumes of LOD. The disadvantages of this approach are:
- Datasets that do not have a query endpoint are ignored. While hundreds of SPARQL endpoints are known to exist today, there are hundreds of thousands of RDF documents.
- Datasets that have a custom API and/or that require an API key in order to pose questions, are not generally accessible and require either appropriation to a specific API or the creation of an account in order to receive a custom key.

- For practical reasons, otherwise standards-compliant SPARQL endpoints put restrictions on either the number of results that can be retrieved or the number of rows that can be involved in a sorting operation that is needed for paginated retrieval.[1]
- Existing LOD Observatories show that SPARQL endpoint availability is quite low.[2] This may be a result of the fact that keeping a SPARQL endpoint up and running requires considerably more resources than hosting a datadump.

Datadump. Downloading datadumps is a third approach to collecting large volumes of LOD. Its disadvantages are the following:

- Datasets that are not available as datadump are ignored.
- Since the same datadump can often be found on different Web locations, it is not always clear which RDF documents belong to the same dataset and who was the original publisher of the data.
- Datadumps do not expose dynamic data, since they require some time to be generated.

Problem 1 (Data collecting). There is no single approach for collecting LOD that guarantees completeness.

2.2   Data Quality

Many data dumps that are available online are not fully standards-compliant [2, 3] (Problem 2). To give a concrete example, one of the authors of this chapter has tried to run an evaluation on Freebase, one of the most popular Linked Datasets out there. A human agent can assess that the Freebase dereference of the 'monkey' resource[3] consists of approximately 600 RDF statement. However, state-of-the-art RDF parsers such as Rapper[4] are only able to retrieve 32 triples, slightly more than 5% of the actual monkey info. Such results are not uncommon, even among often-used, high-impact datasets.

Problem 2 (Data quality). In practice, most RDF documents are not fully standards compliant. Because of this they cannot be (directly) used by machine processors.

Data quality is also a problem on the original Web. However, on the original Web data quality is less of an issue because of the following three reasons:

1. On the original Web, the typical data consumer is a human being. On the Semantic Web the typical data consumer is a client application whose intelligence is significantly below that of an average human being. Human users can

---

[1] For example, Virtuoso by default limits both the result set size and the number of rows within a sorting operation.

[2] See http://sparqles.okfn.org/.

[3] See http://rdf.freebase.com/ns/m.08pbxl.

[4] Version 2.0.14, retrieved from http://librdf.org/raptor/rapper.html.

deal with data quality issues significantly better than machine users do. For instance, if a human has to download a document whose link is ill-formatted, the human can often come up with fallback mechanisms such as copy/pasting the link from plain text or looking up the same document on a different site. A machine agent is often lost when even a single bracket is pointing in the wrong direction.

2. The original Web exhibited what Tim Berners-Lee called "direct gratification": if you change a few tags in your HTML document, your Web site immediately looks more appealing. On the Semantic Web, most data cleaning operations do not have an immediately noticeable result (especially not when data cannot be directly interacted with).

3. Client applications for the original Web have been optimized for decades to deal with common mistakes made by data publishers. For instance, Web browsers are able to display most Web sites even though few of them will fully follow the HTML specification.

Another reason why data quality is more of an issue on the Semantic Web is that by virtue of linking, the quality of your data is affected by the quality of someone else's data. In other words; *your* data becomes more/less valuable due to changes *someone else* makes. Therefore, on the Semantic Web there is a level of incentive to clean data that goes beyond that of the original dataset creators.

The problem of idiosyncrasies within one dataset is worsened by the fact that *different* datasets exhibit *different* deviations from RDF and Web standards. This means that a custom script that is able to process one RDF document may fail to perform the same job for another, requiring ad hoc and therefore human-supervised operations to be performed.

2.3 Accessibility

In addition to data quality issues, there are also problems with how Linked Data is currently being deployed. The goal of Linked Data is to be able to ask a question to a Semantic Web service and receive an correct answer within an acceptable time-frame. A large hurdle towards Web-scale live querying is that Semantic Web datasets cannot all be queried in the same, uniform way (Problem 3).

Problem 3 (Accessibility). In practice, there is no single, uniform way in which the LOD can be queried.

We describe the main problems that currently make it difficult for a machine agent to query the LOD Cloud:

- Most Semantic Web datasets that are available online are data dumps [2, 6], which implies that they cannot be live queried at all. In order to perform structured queries on such datasets, one has to download and deploy them locally.

- Not all datasets that can be queried use a standardized query language (such as SPARQL). Some require a data consumer to formulate a query in a dedicated query language or to use a custom API.
- Most custom APIs are not self-describing, making it difficult for a machine processor to create such queries on the fly.
- Most online datasets that can be queried live and that are using standardized query languages such as SPARQL are imposing restrictions on queries that can be expressed and results that can be returned [7,8] (see also Sect. 2.1).
- Different SPARQL endpoints impose *different* restrictions [8]. This makes it difficult for a data consumer to predict whether, and if so how, a query will be answered.

2.4   Scalability and Availability

After the first 14 years of Semantic Web deployment there are at least millions of data documents [9,10] but only 260 live query endpoints [8]. Even though the number of endpoints is growing over time [7,8], at the current growth rate, the gap between the number of data dumps and the number of live queryable endpoints will only increase [8]. Figure 1 shows the number of SPARQL endpoints that are monitored by Sparqles over time, as well as their availability statistics. The reason for the growing discrepancy between datadumps and SPARQL endpoint is that the former are easier and cheaper to make available than the latter (Problem 4).

This problems manifests itself not only in the relatively low number of query endpoints, but also in the fact that the few endpoints that do exist are either (1) used very little, or (2) are enforcing restrictions (see Sect. 2.1), or (3) have low availability.

Problem 4 (Scalability and availability). Existing Linked Data deployments cannot keep up with the increasing scale at which LOD is being produced and published. The query endpoints that currently exist are either small-scale, have low availability or are too expensive to maintain.

The following causes contribute to Problem 4:

- It is difficult to deploy Semantic Web data, since this currently requires a complicated stack of software products.
- Existing query endpoints perform most calculations on the server-side, resulting in a relatively high cost and thus a negative incentive for the data publisher to provide high quality data that many clients will consume.
- Some clients prefer datadumps over queryable endpoints, even if the latter is freely available. This may be because of restrictions enforced by contemporary SPARQL endpoints or because the client does not understand how to interface with a query endpoint (and downloading a datadump is easy).
- Even though there is no requirement to do so within the SPARQL standards, all existing SPARQL endpoints return all query results at once. Incremental

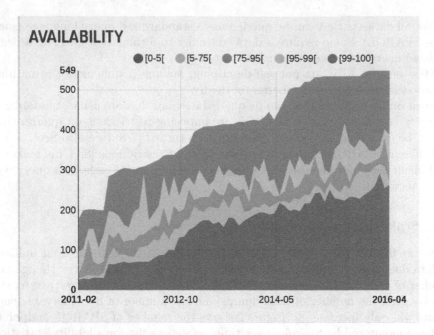

Fig. 1. The number of SPARQL endpoints as monitored by Sparqles over time, as well as their availability.

result retrieval is not always possible, for instance when aggregations are performed over sub-results. However, the ability to incrementally receive results in cases where such constructs are not used seems crucial to ensure a responsive client-side experience.

- The SPARQL standard does not support pagination of results. In practice many clients use a combination of the LIMIT and OFFSET operators in order to mimic pagination. Unfortunately, there is no guarantee that this will return all results without duplicates, especially not when the dataset is dynamic and adding/removing statements in-between requests may alter the internal indexes.

2.5   Findability

In Sect. 2.1 we already saw that locating data sources is difficult. There is also the wider problem that data resources, possibly described in multiple data sources, cannot be searched for, based on textual search terms (Problem 5).

Problem 5. With the current LOD dissemination approaches, it is inherently difficult to find unknown resources as well as statements about known resources.

We give an overview of the ability to find resources under the current LOD publication paradigms:

Datadumps. Datadumps implement a rather simple way of finding resource-denoting terms: one must know the exact Web address of a datadump in order to download and extract resource-denoting IRIs. This means that search is neither text-based nor resilient to typos and spelling variations. The client has to extract the (often relatively small) resource description from the (sometimes very large) datadump. This results in low serviceability. Datadumps do not link explicitly to assertions about the same resource that are published by other sources. In fact, there is no support for finding (assertions about) resources across RDF documents. Finally, it is not trivial to find the locations where datadumps are published in the first place (see Sect. 2.1).

IRI dereferencing. When implemented correctly, an RDF IRI dereferences to a set of statements where that IRI appears in the subject and/or object position. Which statements belong to the dereference result set is decided by the authority of that IRI, i.e., the person or organization that pays for the domain that appears in the IRI's authority component. There is no support for finding non-authoritative assertions about a known IRI. In fact, non-authoritative statements can only be found accidentally by navigating the interconnected graph of dereferencing IRIs. Since blank nodes do not dereference, significant parts of the graph cannot be traversed. This is not merely a theoretical problem, since 7% of all RDF terms are blank nodes [2]. Since RDF literals, which often contain textual content, cannot be dereferenced either, text-based search is not possible at all through dereferencing. Because the IRI is both the name and location of a resource, IRI dereferencing makes it trivial to locate authoritative data about a resource once its name is known (datadumps and SPARQL do not have this feature).

SPARQL. With SPARQL it is possible to search for text strings that (partially) match RDF literals. More advanced matching and ranking approaches such as string similarity or fuzzy matching are generally not available. As for the findability of non-authoritative statements, SPARQL has largely the same problems as datadumps and IRI dereferencing: only statements that appear in the same endpoint can be found. As with the locations of datadumps, the locations of SPARQL endpoints are generally not easy to find.

Resource findability can be additionally hampered when two LOD publication approaches interact. For instance, many servers implement IRI dereferencing on top of a SPARQL back-end. In such a case IRI dereferencing inherits the deficiencies of SPARQL querying. To give a concrete example, at the time of writing dereferencing DBpedia's resource for the city of London (`dbr:London`) returns exactly 10,000 triples. Further inspection shows that there are over 30,000 triples in DBpedia that describe the city of London, but due to SPARQL restrictions over 20,000 of them are not returned. Because IRI dereferencing and SPARQL both lack 'controls' (see Sect. 3.3) it is not even possible to communicate to the client that the description of London is incomplete.

2.6 Metadata

Metadata about datasets has many use cases. For instance, metadata may describe which RDF documents belong together. It may give metrics about data size and distribution of the data. Such metrics can be used to optimize data handling and to choose the right algorithm for the job. Metrics are also useful for query rewriting in the case of federated querying (Sect. 2.7). Unfortunately, many data publishers do not publish a dataset description that can be found by automated means. Dataset descriptions that can be found do not always contain all (formally or *de facto*) standardized metadata (Problem 6).

Problem 6 (Metadata). Metadata descriptions of Linked Datasets are often missing, incomplete or incorrect.

Since not all metadata is created in exactly the same way, metadata values are generally not comparable between datasets. For instance, it is not generally the case that a dataset with a higher value for the void:triples property will also contain more triples. This may either be because the value is outdated or because it has been calculated incorrectly. Sometimes metrics are multiples of 10^n for some n, which may indicate that values are estimates and have been entered by hand.

Another problem is caused by the way in which existing metadata vocabularies are defined. For instance, the most commonly used metadata vocabulary VoID [11] explicitly allows values of its properties to be estimates. Different data publishers will have different ideas about what is still a good estimate.

Definitions of metadata properties are often ambiguous or incorrect. For example, the specification of the VoID property void:properties conflates the semantic notion of an RDF property with the syntactic notion of an RDF predicate term. That these are quite different things is apparent from the following example which contains 4 distinct predicate terms and 9 distinct RDF terms that denote an RDF property.

```
ex:p1 rdf:type rdf:Property;
ex:p2 rdfs:subPropertyOf ex:p3.
ex:p4 rdfs:domain ex:c1.
ex:p5 rdfs:range ex:c2.
```

Because of such incompatibilities between existing dataset descriptions, it is difficult to reliably analyze and compare datasets on a large scale. It is also difficult to optimize algorithms based on those metrics.

2.7 Federation

In federated querying, sub-queries are evaluated by different query endpoints and the sub-results are integrated. For example, one may be interested in who happens to know a given person by querying a collection of HTML files that contain FOAF profiles in RDFa. Or one may have a question without knowing in advance who might be able to answer it. At present, querying multiple endpoints is problematic [12] (Problem 7).

Problem 7. Web-scale federation, performing a query against hundreds of thousands of endpoints, is currently not supported by the standards nor by the implementations.

These are the reasons why federated querying in problematic:

- The problem that the locations of query endpoints are difficult to find in the first place (Problem 1).
- The cumulative effect of slow and/or (temporarily) unavailable individual endpoints (Problem 4). In federation the slowest endpoint determines the response time of the entire query. This also interacts with the fact that existing SPARQL endpoints do not return results incrementally, so the client has to wait for the full result set to be computed.
- The cumulative effect of the heterogeneity of interfaces to LOD (Problem 3).
- The standardized way for SPARQL Federation [13] requires every endpoint to be entered explicitly. This makes it cumbersome to query against hundreds of thousands of query endpoints. Specifically, in many cases the query will exceed the allowed length.

Some of the deficiencies of federated querying can be circumvented by deciding algorithmically which endpoints to send which sub-query to. Such advanced query planning techniques require dataset summaries and other metadata descriptions (VoID, VoID-ext, Bio2RDF metrics) about each endpoint. However, such metadata descriptions are often not available and existing metadata descriptions do not always contain enough metadata in order to make efficient query federation possible.

2.8   Generalizability

This last problem is slightly circular: it is caused by the problems exhibited by existing LOD deployments; but it is also poses a risk when it comes to improving these deployments in the future. It is the problem that existing Semantic Web research evaluations do not currently take the variety of Linked Data into account. In Fig. 2 we count the number of datasets that are used in accepted submissions for the ISWC 2014 full research paper track.

The figure shows that in 20 evaluations only 17 datasets are used in total. The number of datasets per article varies between 1 and 6 and is 2 on average. This implies that many papers evaluate against the same datasets, most often DBpedia. This means that it is generally unclear to what extent published results will transfer to other datasets, specifically those that are only very rarely evaluated against. This is the problem of the generalizability of Semantic Web research results (Problem 8).

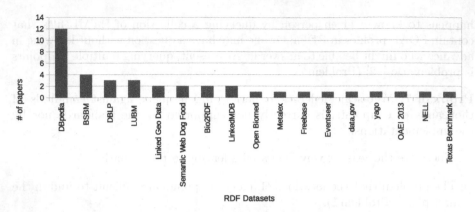

Fig. 2. Overview of datasets used in evaluations of papers accepted in the ISWC 2014 research track. For each dataset the number of articles that use it is shown.

Problem 8 (Generalizability). With the current LOD dissemination approaches it is inherently difficult to find unknown resources as well as statements about known resource.

Most evaluations use only a handful of, often the same, datasets because it is simply impractical to run a Semantic Web algorithm against hundreds of thousands of datasets. Notice that the challenge here is not scalability per se, as most datasets on the Semantic Web are actually quite small (much smaller than DBpedia, for instance). The problem seems to be with the heterogeneity of data formats and idiosyncrasies.

Given the issues with data quality described in Sect. 2.2, we can only guess at what it actually means to run an evaluation against a dataset like Freebase: does it mean that the evaluation was run against its <5% syntactically correct triples?

3   Infrastructure Solutions (LOD Laundromat)

In this section we provide a solution for each of the problems identified in the previous section. Each solution is implemented within the LOD Laundromat framework. The LOD Laundromat [3] is available at http://lodlaundromat.org. The collection of datasets that it comprises is continuously being extended. Figure 3 shows the LOD Laundromat (re)publishing framework which consists of the following components:

LOD Basket. A list of initial pointers to online available RDF documents, plus the ability for human and machine users to add new pointers.
LOD Washing Machine. A full automated cleaning mechanism for RDF documents, implementing a wealth of standards and best practices.

LOD Wardrobe. A centralized store of a large subset of the LOD Cloud. Each clean RDF document is made available in several representations and through several Web APIs.

LOD Laundromat metadataset. Stores metadata about each RDF document's cleaning process, as well as structural metrics about the data graph, together with provenance data about how these metrics are calculated.

Index. A very large-scale key/value store that maps IRIs and namespaces to documents in which those IRIs and namespaces appear.

LOTUS. A full-text search index on all textual RDF literals that allows resources, statements about resources, to be found through a configurable combination of matchers and rankers.

LOD Lab. An approach to reproducing Linked Data research which uses the various components in an intergrated way.

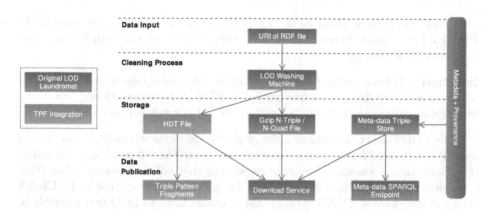

Fig. 3. The LOD Laundromat (re)publishing workflow.

3.1 LOD Basket

The overview in Sect. 2.1 shows that all forms of data publishing exhibit some form of incompleteness. There is, therefore, no perfect solution for Problem 1. Combining multiple approaches for data publishing is also non-trivial since the way in which data that is published through different channels should be combined is itself problematic.

When we look at the degree of incompleteness for each of the three main LOD publication approaches, we see that resource- and endpoint-based data collection cannot even guarantee completeness at the *document* level, while datadump-based collecting at least guarantees that an RDF document can be fully downloaded. This is why the LOD Laundromat mainly uses datadumps for data collection. We say 'mainly', because some SPARQL queries that are encoded inside URLs are also (links to) RDF documents.

In practice, many datasets that are exposed through a SPARQL endpoint are also made available as datadumps. Since the LOD Laundromat does not deal with dynamic data, the deficiencies of datadumps are also less of a problem. At the moment, the search for initial links to datadumps cannot be fully automated. There are four reasons for this:

- Catalogs that collect metadata descriptions must be accessed by Web site-specific APIs.
- Standards-compliant metadata descriptions are stored at multiple locations and cannot always be found by Web search operations that can be automated.
- Metadata descriptions of datasets, whether standards-compliant or catalog-specific, are often outdated (e.g., pointing to an old server) or incomplete (e.g., including the schema but not the data or linkset).
- Many datasets are not described anywhere and require someone to know the server location where the data is currently stored.

Given the current state of the LOD Cloud there is no perfect solution to Problem 1. We must therefore use a 'best effort' approach that can be improved over time (Solution 1).

Solution 1 (Data collecting). The initial data collection should consist of a fair number of links to online available RDF documents. Means should be implemented that allow the initial collection to be extended over time.

The LOD Basket[5] maintains a list of online locations where RDF documents can be downloaded from. It implements Solution 1 in the following way: the initial data collection is assembled by creating scripts that collect links to online RDF documents from existing data catalogs. An good example of this is the CKAN API[6]. A well-known CKAN catalog that contains links to LOD is the Datahub, which includes the datasets in the original LOD Cloud. This approach implies that URLs that are not included in a LOD catalog or portal are less likely to be washed by the LOD Washing Machine. In addition, links to commonly used RDF documents have been added manually.

The second part of Solution 1, i.e., improving the data collection over time, is implemented in two ways. Firstly, users can add new datasets to the LOD Basket by entering a URL that points to an online available datadump or by uploading a local file through the Dropbox plugin. Secondly, if the LOD Washing Machine (Sect. 3.2) comes across VoID metadata that describes a datadump location, these URLs are automatically added to the LOD Basket.

Some URL strings that appear in catalogs or that are entered by users do not parse according to the RFC 3986[7] grammar. Some URL strings are parsed as IRIs but not as URLs, mostly because of unescaped spaces. Some URL strings parse per RFC 3986, but have no IANA-registered scheme[8], or the `file` scheme

---

[5] See http://lodlaundromat.org/basket.

[6] See http://ckan.org/.

[7] See http://tools.ietf.org/html/rfc3986.

[8] See http://www.iana.org/assignments/uri-schemes/uri-schemes.xhtml.

which is host-specific and cannot be used for downloading. Only URLs that parse per RFC 3986 (after IRI-to-URL conversion) and that have an IANA-registered scheme that is not host-specific are passed to the LOD Washing Machine.

3.2   LOD Washing Machine

Before the LOD Laundromat existed, the problem of non-standards compliant data (Problem 2) was already widely recognized. Many efforts had already been made to improve the quality of LOD. These efforts include creating standards, formulating guidelines, and building tools. In addition, Semantic Web educators have informed data producers to follow those guidelines and use those tools. The problem with these existing solutions is that they are targeted towards human data creators, who can (and do) choose not to use them.

Since current approaches have not been able to sufficiently fix the data quality issue, we propose a complementary Solution 2. The idea behind this solution is that coordinating a data cleaning effort with a large number of people is inherently difficult, costly and time consuming. Performing data cleaning operations in a fully automated way is much cheaper and also very fast. The data-oriented approach of LOD Laundromat is complementary to existing efforts, since it is preferable that someday the original dataset is cleaned by its original maintainers. There are many aspects of data quality that cannot be automated.

Solution 2 (Data cleaning). Automatically extract the subset of the LOD Cloud that can be made compliant with a wide array of existing Semantic Web standards, specifications and best practices.

The LOD Washing Machine[9] implements Solution 2. It takes 'dirty' RDF documents from the LOD Basket and tries to download them. Potential TCP/IP and HTTP errors are stored as part of the data document's metadata description that is generated by the LOD Washing Machine. Data documents that occur in archives are recursively unpacked. Archive and compression errors are also recorded in the metadata. Once fully unpacked, the RDF serialization format of the data document is determined heuristically based on the file extension (if any), the value of the Content-Type HTTP header (if present), and a lenient parse of the first chunk of the data file. Most standardized RDF serialization formats are supported: N-Quads, N-Triples, RDFa, RDF/XML, TRiG, and Turtle. N3 and JSON-LD are currently missing. Once the serialization format has been guessed, the document is parsed. Since many data documents contain syntax errors, only compliant triples are retained. Every warning is stored as metadata for the resultant dataset in order to make the cleaning process transparent.

The LOD Washing Machine takes immediate action by targeting the data directly, not its maintainers. Because all data follows the same set of standards and best practices, the LOD Washing Machine greatly improves the chance of data actually being reused.

---

[9] See https://github.com/LOD-Laundromat/LOD-Washing-Machine.

In addition to extracting the standards-compliant subset of the LOD Cloud, LOD Laundromat implements existing standards in such a way that the resultant RDF documents are specifically geared towards easy reuse by other tools. Specifically, data that is cleaned by the LOD Washing Machine has the following benefits:

Easy grammar. N-Quads is use to serialize each RDF statement on one line. For this newlines in RDF literals are escaped. Even non-RDF tools such as Pig, grep, sed, and the like are able to parse these statements/lines with common, well-known techniques such as regular expressions.

Speed. The LOD Washing Machine solves the major parsing inefficiencies for LOD. Some inefficiencies are caused by the use of serialization formats such as RDF/XML and RDFa that contain a complicated tree-based encoding of graph structure. Other inefficiencies are caused by syntax errors, which necessitate the parser to come up with fallback options that may be expensive, are ensured to be absent.

Quantity. The LOD Washing Machine is able to clean large quantities of data in a streamed way, using almost no memory. This allows it to clean large documents as well as small ones. It is able to clean approximately one billion statements a day.

Flexibility. The clean data format produced by the LOD Washing Machine allows data documents to be easily combined. Splitting a document into multiple ones often results in corrupted data since it is difficult to find a splitting point that respects well-formedness. Since the LOD Washing Machine serializes exactly one RDF statement per line, data can be split at every newline character while remaining standards compliant. In addition, appending multiple documents into a single one normally requires the costly process of blank node renaming. Data delivered by the LOD Laundromat can be freely appended, and blank nodes are replaced in a standards-compliant way by well-known IRIs.

Streaming. The LOD Washing Machine supports streamed processing of RDF statements since it guarantees that no duplicate triples occur in the data stream. This means that the streamed processor does not have to perform additional bookkeeping in order to check for statements that were observed earlier.

Completeness. The LOD Washing Machine guarantees that cleaned data is always a complete representation of the input dataset to the extent at which the original dataset is standards-compliant.

3.3   LOD Wardrobe

Cleaned datasets are disseminated in the LOD Wardrobe (http://lodlaundro mat.org/wardrobe). The LOD Wardrobe disseminates the data documents that are cleaned by the LOD Washing Machine.

As with the issue of data quality (Sect. 3.2), standardization alone cannot solve the social problem that accessibility standards, in order to be effective, have to reach a large number of people who all have to act in accordance with the standards. We therefore propose a similar, machine-oriented solution for improving the accessibility of LOD (Solution 3).

Solution 3 (Accessibility). Allow all RDF documents to be queried through a uniform interface that is standards-compatible and self-descriptive.

Solution 3 is implemented by the LOW Wardrobe which publishes hundreds of thousands of RDF documents in exactly the same way. It also allows each document to be queried live in a uniform and machine-accessible way without artificial restrictions.

In Sect. 2.4 we showed that there are issues with the scalability and availability of existing LOD endpoints. Solution 4 addresses this problem by shifting part of the effort from the server to the client side.

Solution 4 (Scalability and availability). Strike a balance between server- and client-side processing, and automatically deploy all LOD as live queryable endpoints.

Linked Data Fragments (LDF) [5] has been designed to minimize server-side processing, while at the same time enabling efficient live querying on the client side. LDF allows data to be queried at the level of Simple Graph Patterns and allows results to be retrieved incrementally through pagination. LDF is able to answer requests efficiently by using the Header Dictionary Triples[10] [14] (HDT) back-end. HDT is a binary, compressed and indexed serialization format that facilitates efficient browsing and low-level querying.

Each page (or 'fragment') of LDF results contains an estimate of the total number of matches for the queried pattern. This allows the client to perform query planning. Pages also contain controls that can be used by an machine processor to interact with the endpoint, e.g., to retrieve the next page of results. This makes each fragment self-describing by presenting the options available to the machine processor. At the same time, the client cannot rely on the server to perform all complicated query operations. Specifically, the client has to calculate the overall query result by joining the results of sub-queries locally.

LDF allows Solution 4 to be implemented. Indeed, query endpoints that use the LDF approach are shown to have significantly higher availability than traditional endpoints [5]. The LOD Wardrobe implements Solution 4 by generating an LDF endpoint for each clean data document.

The LOD Wardrobe republishes a very large subset of the LOD Cloud. For each clean data document it provides the following:

Datadump. The output of the cleaning process is stored in a canonical and easy to process data format and is compressed using gzip. The serialization format is either N-Triples or N-Quads (depending on whether or not at least one quadruple is present in the data file). The statements in the file are sorted

---

[10] See http://www.rdfhdt.org/.

lexicographically and duplicates are removed. This means that two lines are guaranteed to denote the same statement iff they compare identical on a character-by-character basis. Because of these properties it is easy to process clean data files in a uniform way, e.g., by streaming through a dataset knowing that the next triple ends with the next newline character.

HDT binary. The output of the cleaning process is also stored in an Header Dictionary Triples [14] (HDT) file. HDT files are compressed but still allow Triple Pattern Fragment queries to be performed. These files can be downloaded for querying the data locally.

LDF endpoint. For online querying over HTTP a Linked Data Fragments [5] (LDF) endpoint is created for each non-empty data document. The TPF API uses HDT as its back-end and allows TPF queries to be performed. Results are returned in various standardized RDF serialization formats and pagination is used when there are more results.

Origin. A link to the original location. The original data file is currently not stored because this would take too much disk space. Oftentimes the datadump is no longer available after a few months or the datadump has been changed.

Metadata. The metadata about the cleaning process as well as metrics about the structural properties of the data.

3.4 LOD Laundomat Metadataset

The LOD Laundromat metadataset includes metadata for each clean data document. The metadata is published online and can be queried through a SPARQL endpoint (http://lodlaundromat.org/sparql).

As with the cleaning process (Solution 2), metadata is generated automatically. Metadata properties that are ambiguous or that rely on human interpretation are left out (Solution 5).

Solution 5 (Metadata). Automatically generate metadata in a uniform way for all RDF documents.

The LOD Laundromat metadataset implements Solution 5 by automatically calculating metadata for each clean data document in the LOD Wardrobe. Values are calculated algorithmically in exactly the same way, which allows values to be compared across documents.

The LOD Laundromat metadataset reuses the following official and de-facto metadata standards in order to be compatible with other dataset descriptions and to promote reuse:

- VoID [11]
- VoID-ext [15]
- DCAT (http://www.w3.org/TR/vocab-dcat/)
- Bio2RDF [16]

The metadata contains the following types of information:

- Generic metadata, e.g., the homepage of a dataset.
- Access metadata, e.g., which protocols are available.
- Process data, describing the cleaning process.
- Metrics about the structural properties of the data (e.g., average literal length, maximum degree).
- Provenance annotations that explain how and when the metadata was calculated.

Because of the scale at which the LOD Laundromat metadataset describes datasets, we have discovered many a-typical datasets that give unanticipated values for metadata properties. For instance, there are RDF documents whose number of unique predicate terms is close to the total number of predicate terms. For such data documents metadata properties that rely on enumeration or partitions can become impractically large (larger than the original dataset).

Figure 4 the average out-degree of all LOD Laundromat documents. This shows another atypical value: one document contains 10,004 statements, all about one subject, thereby strongly skewing the overall distribution. Such a-typical metric values can be used to explain deviating evaluation results for specific documents (see Sect. 3.7).

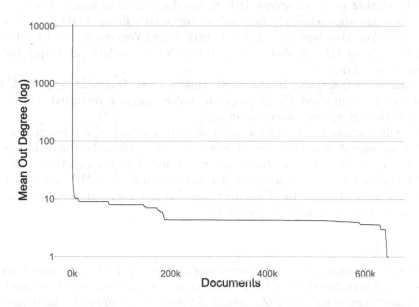

Fig. 4. Average out degree distribution of LOD Laundromat documents

Most metadata metrics can be calculated within a stream, using very little memory. Others, e.g., the distinct number of IRIs, require some (in-memory) datastructure in order to be calculated. Even in such cases the computational resources can be constrained, for instance by using an efficient in-memory IRI dictionary such as RDF Vault [17].

3.5   Index

As we saw in Sect. 2.7 it is currently not possible to evaluate a query against a very large number of endpoints. In SPARQL querying one data source is the default; SPARQL federation is then used to widen the set of data sources on a one-by-one basis. LOD Laundromat takes the opposite route: by default a query is evaluated against all data sources (Solution 6). If a client wants to evaluate a query against a subset of data sources then these source have to be explicitly specified.

Solution 6 (Federation). By default, evaluate queries against all datasets. A query can still be evaluated against a subset of data sources by specifying those explicitly.

The types of queries that can be performed in LOD Laundromat are of the form (?s,?p,?o) with optionally instantiated subject, predicate and object terms. If neither of the terms is instantiated the query is trivially 'evaluated' by streaming all clean data documents. The list of all data documents is stored in the LOD Laundromat metadataset and is retrieved with a SPARQL query.

If one or more terms are instantiated the situation is not so trivial: since a triple pattern often appears in a limited number of data sources it is inefficient to send a query to all 650K LDF endpoints. This problem is solved by adding a key-value index that maps all unique IRIs to the documents in which they appear. Since there are approximately two billion unique IRIs in the LOD Laundromat data collection, this key-value index is very large. We use RocksDB (http://rocksdb.org/) for this. A Web API to this index is available at http://index.lodlaundromat.org.

Whenever a (partially) instantiated triple pattern fragment is queried, the list of documents in which the given ground terms appear is retrieved. The query is only evaluated against those documents.

If a client wants to evaluate a query against a subset of all documents the LOD Laundromat provides several ways in which this subset can be specified. The client can use the IRI-to-document index in combination with the LOD Laundromat metadataset (Sect. 3.4) in order to filter on IRIs and/or structural properties of the data. The client can also use a second index that maps namespaces to documents (see Sect. 4.3).

3.6   LOTUS

LOTUS relates unstructured to structured data using RDF as a paradigm to express such structured data. LOTUS is integrated with a central architecture that exposes a large collection of resource-denoting terms and structured descriptions of those terms, all formulated in RDF. It indexes natural text literals that appear in the object position of RDF statements and allows the denoted resources to be findable based on approximate matching (Solution 7). LOTUS currently includes four different matching algorithms and eight ranking algorithms, which leverage both textual features and relational information from the RDF graph.

Solution 7 (Findability). Allow resources, statements about resource, and documents containing (statements about those) resources to be found by fuzzy string matching on RDF literals.

Approximate matching of a query to RDF literals can be performed at different granularity levels: entire phrases, sets of tokens, or sets of characters. LOTUS implements four matching functions to cope with this diversity):

Phrase matching. Matches a phrase in an RDF literal. Multiple query terms must occur consecutively in the target string.

Disjunctive matching. Disjunctive token matching (OR) of at least one of the tokens specified with at least one of the tokens that appears in a target literal. The order in which the disjunctive tokens are specified does not matter.

Conjunctive matching. Conjunctive token matching (AND) of all query tokens in specified with some of the tokens that appear in a target literal. The order in which the conjunctive tokens are specified does not matter.

Conjunctive matching with character edit distance. As conjunctive term matching (AND), but where a small Levenshtein-based edit distance on a character level is permitted. This allows typos and spelling variations to be matched to the intended results.

LOTUS implements the following three content-based [18] filtering functions:

Character length normalization. A score that is inverse proportional to the number of characters that appear in the lexical form of each RDF literal.

Practical scoring function.[11] A score that is a product of the following token-based IR metrics: term frequency (TF), inverse-document frequency (IDF) and length normalization.

Phrase proximity. A score that is inverse proportional to its edit distance with respect to the query.

LOTUS implements the following five content-based [18] filtering functions:

Terminological richness. A score that is proportional to the presence of controlled vocabularies. This is defined as the ratio of statements expressing schema information about classes and properties rather than instances.

Semantic richness. A score that is proportional to the mean graph connectedness degree of the data document.

Recency ranking. A score that is proportional to the time since the original data document was last modified. Statements from recently updated documents have higher score.

Degree popularity. A score for RDF literals that is proportional to the weighted sum of the overall degree (in- plus out-degree) of subject terms that it appear together with those RDF literals.

---

[11] The default ranking function in ElasticSearch. See https://www.elastic.co/guide/en/elasticsearch/guide/current/scoring-theory.html.

Appearance popularity. A score for RDF literals that is proportional to the cumulative sum of the number of documents in which a subject that appears together with those RDF literals appears.

LOTUS is not a Semantic Search engine for end users, but a framework for Semantic Web researchers that wish to build and evaluate Semantic Search engines. The architecture of LOTUS is displayed in Fig. 5.

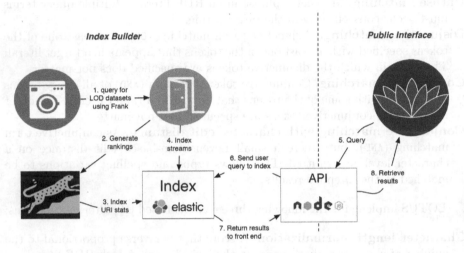

Fig. 5. Overview of the LOTUS architecture. Data is retrieved from the LOD Laundromat (left blue icon) data cleaning and republishing framework. IRIs are their associated textual literals are queried, and their results streamed, through the convenient *Frank* tool (green icon). Rankings of IRIs based on their associated textual literals are stored in RocksDB (yellow icon). IRIs are stored together with their textual literals in records in ElasticSearch. A Node.js wrapper forms the Web-facing API on top of ElasticSearch. The Web-facing API can be publicly accessed by anyone through the LOTUS website. (Color figure online)

3.7   LOD Lab

As we saw in Problem 8, Semantic Web evaluations are often performed on a small number of datasets. Since Linked Data is very heterogeneous, results of current Semantic Web evaluations are not as generalizable as they could be. This is largely due to the almost linear cost of adding one more dataset to an evaluation: the dataset has to be collected, cleaned, index, etc. Until recently performing each of these steps required some human effort. Now that we have the LOD Laundromat framework, scaling in the number of data documents no longer incurs a linear cost. We expect that this change in the 'cost function' for Semantic Web evaluations will result in more large-scale evaluations that take the true variety of LOD into account.

LOD Lab [19] (http://lodlaundromat.org/lodlab) implements Solution 8 by allowing evaluations to be run on the LOD Laundromat data collection. The viability of the LOD Lab approach was shown by rerunning evaluations from three recent Semantic Web research papers.

Solution 8 (Generalizability). Make it easy to run new, and rerun existing, experiments over a large and heterogeneous data collection.

One of these evaluations is taken from the ISWC 2014 paper "Adoption of the Linked Data Best Practices in Different Topical Domains" [20]. In this paper the authors analyze conformance to Linked Data best practices by crawling the LOD Cloud using LDSpider [21]. This crawler takes its seed points from the Billion Triple Challenge as well as from data catalogs and datasets advertised on public LOD mailing lists. The crawl includes results for 900,129 dereferenced IRIs and 8,038,396 unique IRIs. Documents are grouped in 1,014 RDF datasets. Datasets are heuristically identified by using manually supplied metadata from catalogs as well as the Pay-Level-Domain components of IRIs. The paper presents a large and diverse set of statistics about LOD best practice conformance.

Table 1 shows one of these statistics: the 10 most often occurring RDF namespaces. The table confirms this: the OWL Time (`time`) vocabulary is used in 68% of LOD Laundromat documents, but it does not appear in the original top 10. The same is true of the RDF Data Cube vocabulary (`cube`).

It is not surprising that the results of the original publication and our rerun are very different. The crawling approach of LDSpider and LOD Laundromat are also very different (Sect. 3.1). It does not make sense to claim that one study is more 'right' than the other. These results do show that the LOD Lab approach makes it easy to run such large-scale experiments.

Table 1. The 10 most frequently occurring RDF namespaces according to the original ISWC 2014 publication [20] (left) and according to LOD Lab (right).

Original [20]			LOD Lab		
Prefix	#datasets	% datasets	Prefix	#docs	% docs
rdf	996	98.22%	rdf	639,575	98.40%
rdfs	736	72.58%	time	443,222	68.19%
foaf	701	69.13%	cube	155,460	23.92%
dcterm	568	56.01%	sdmxdim	154,940	23.84%
owl	370	36.49%	worldbank	147,362	22.67%
wgs84	254	25.05%	interval	69,270	10.66%
sioc	179	17.65%	rdfs	30,422	4.68%
admin	157	15.48%	dcterms	26,368	4.06%
skos	143	14.11%	foaf	20,468	3.15%
void	137	13.51%	dc	14,423	2.22%

4   Client Tooling

While the LOD Lab base infrastructure is quite complex, using the LOD Lab to scale your Semantic Web evaluation is relatively easy. LOD Lab relies on the following Web APIs the LOD Laundromat makes available:

- **LOTUS.** Search for IRIs and triples based on free text matching and filtering.
- **LDF.** Search for triples based on TPF queries.
- **Namespace index.** Search for documents wherein a given namespace occurs.
- **IRI index.** Search for documents in which a given IRI occurs.
- **SPARQL.** Search for metadata about the crawling process and (structural) properties of the data to filter documents.
- **Datadump.** Bulk access to full RDF documents.

For some use cases, e.g. downloading an RDF document or querying a small number of documents, the Web UIs works well. For other use cases the Web APIs can still be used, but they have to be combined in non-trivial ways. For instance, to query the entire LOD Laundromat data collection someone needs to first request the list of all documents from the SPARQL endpoint and then pose an LDF request for each document. For use cases like these we have created *Frank*, a Command-Line Interface (CLI) that allows the most common operations to be performed more easily. *Frank* is implemented as a single-file Bash script. It is installed in the following way:

```
$ git clone https://github.com/LOD-Laundromat/Frank
$ cd Frank
$ ./frank --help
```

As with other Unix-inspired tools, *Frank* can be easily combined with other commands through Bash pipes. This makes it relatively easy to run large-scale evaluations over the LOD Laundromat data collection. In fact, the three evaluations that were (re)run in our original LOD Lab paper [19] all consisted of only a couple of calls to *Frank*. Figure 6 shows the relationships between the three *Frank* modes ('statements', 'documents' and 'meta') and the various LOD Laundromat Web Services.

We now discuss the four main tasks that can be performed with the LOD Laundromat Web APIs and with *Frank*. Where possible/useful we perform the same task with both the Web APIs and *Frank*. These are the four main LOD Lab tasks:

1. Find IRIs.
2. Find statements.
3. Find documents.
4. Combine finding IRIs, statements and documents with other tools and custom algorithms.

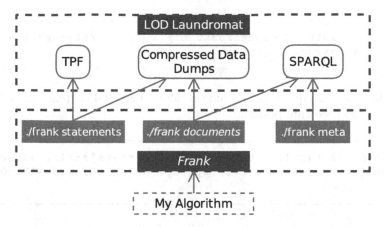

Fig. 6. The implementation architecture for *Frank* and its dependence on the LOD Laundromat Web Services.

4.1 Find IRIs

IRIs can be found by LOTUS. Usually, we can come up with a few terms that describe our interest, e.g. "circus monkeys". It is however not always trivial to find the right IRIs that denote circuses, monkeys and circus monkeys.

The LOTUS Web UI can be found at http://lotus.lodlaundromat.org. A query is issued by filling in the respective HTML forms. Different matchers and rankers can be chosen from dropdown menus. The same queries that are issued through the Web UI can also be performed algorithmically. The URI-encoded query is shown inside the Web UI. For instance, the following searches for RDF terms and statements about monkeys:

```
http://lotus.lodlaundromat.org/retrieve?string=monkey
```

For each option that is changed in the Web UI, the URI changes as well. For instance, the following excludes results where the subject term is not an IRI (i.e., excluding blank nodes).

```
http://lotus.lodlaundromat.org/retrieve?string=monkey&\
noblank=true
```

LOTUS is not limited to searching textual RDF literals, descriptive text often appears in IRIs as well. The following query searches for monkeys that are defined in OpenCyc:

```
http://lotus.lodlaundromat.org/retrieve?string=monkey&\
noblank=true&subject=sw.opencyc.org
```

In the following query we exclude results where 'label' (e.g., rdfs:label) appears in the predicate position:

```
http://lotus.lodlaundromat.org/retrieve?string=monkey&\
noblank=true&subject=sw.opencyc.org&predicate=NOT%20
    label
```

Another useful feature is to filter for strings whose content belongs to a particular natural language. The following only returns language-tagged strings that belong to the English language (language tag en):

```
http://lotus.lodlaundromat.org/retrieve?string=monkey&\
noblank=true&subject=sw.opencyc.org&predicate=NOT%20
    label&\
langtag=en
```

These sample queries in LOTUS show that the various filters can be easily combined to make fine-grained queries.

4.2   Find Statements

Statements within a single RDF document can be queried with LDF. Documents in LOD Laundromat are identified with unique MD5 hashes. The following requests all triples that appear in the RDF document with identifier 000000e0599d427f4d2245fc84f9d65b that have rdfs:label as their predicate term:

```
http://ldf.lodlaundromat.org/000000
    e0599d427f4d2245fc84f9d65b?\
predicate=http%3A%2F%2Fwww.w3.org%2F2000%2F01%2Frdf-
    schema%23label
```

LDF endpoints are self-descriptive: query responses contain Hydra[12]-formatted metadata that describe how follow-up queries can be performed. The first response contains 100 triples. Pagination is used in order to retrieve all statements. This requires issuing consecutive HTTP requests.

---

[12] http://www.hydra-cg.com/.

The same statements-based queries can also be performed over all RDF documents at once using the `frank statements` command. The command-line flags `--subject`, `--predicate` and `--object` allow the same Simple Graph Patterns [22] to be formulate. `frank statements` returns a continuous stream of result triples without requiring the user to page through results. The following query enumerates information about monkeys:

```
$ frank statements -s dbr:Monkey
```

The resulting stream of triples comes from multiple documents. This can be made explicit by including the -g flag (for graph). Using N-Quads notation, each triple is now followed by the name of the LOD Laundromat data document in which the triple appears:

```
$ frank statements -s dbr:Monkey -g
```

4.3 Find Documents

For Semantic Web evaluations we are often interested in collections of RDF statements that are published together with a certain purpose in mind. One granularity level where a meaningful collection is (often) present, is the level of the RDF document where the publisher has purposefully put a collection of RDF statements together in a file. LOD Laundromat includes the following two large-scale indexes for finding documents:

- A mapping from all unique IRIs (approximately two billion) to RDF documents in which they appear (http://index.lodlaundromat.org/r2d/).
- A mapping from all unique namespaces to RDF documents in which they appear (http://index.lodlaundromat.org/ns2d/).

In addition to these two indexes, LOD Laundromat also provides a SPARQL endpoint (http://lodlaundromat.org/sparql). The following query finds documents that are of specified size. Such queries allow algorithms to be run on 'bins' of documents of varying size, which is useful for evaluating algorithms to see whether their performance is linear, quadratic or otherwise:

```
SELECT ?doc {
    ?doc llo:triples ?n
    FILTER (?n >= 1000)
    FILTER (?n <= 10000)
}
```

`frank documents` allows individual documents to be retrieved. It interfaces with the two indexes and with the SPARQL backend. Each documents is identified with the following two URIs:

1. The URI from which the data document (serialization) can be downloaded (`--downloadUri`).
2. The URI that identifies the RDF document and about which metadata assertions are made (`--resourceUri`).

The following command prints all LOD Laundromat download URIs:

```
$ frank documents --downloadUri
http://download.lodlaundromat.org/fcf...b92
http://download.lodlaundromat.org/134...344
http://download.lodlaundromat.org/d4a...b85
...
```

`frank documents` comes with many flags for filtering RDF documents. All flags are enumerated with command `frank documents --help`. By using the flags `--minTriples` and `--maxTriples`, the above SPARQL query can be written as follows:

```
$ frank documents --minTriples 880 --maxTriples 900
```

We can also enumerate the triples that belong to documents of at least 880 and at most 900 triples, simply by combining two *Frank* commands with a Bash pipe:

```
$ frank documents --resourceUri --minTriples 100000 --
    maxTriples 1000000 |
  frank statements
dbp:1921Novels rdfs:label"1921 novels" .
dbp:1921Operas rdfs:label "1921 operas" .
...
```

The following SPARQL query finds documents based on their average in-degree:

```
SELECT ?doc ?x {
  ?doc llm:metrics/llm:inDegree/llm:mean ?x
  FILTER(?x >= 3)
}
```

The same query in *Frank*:

```
$ frank documents --minAvgInDegree 3
```

This allows algorithms to be evaluated on RDF documents that have specific structural properties. As was shown in the context of the LOD Lab, the performance of algorithms is sometimes related to structural properties of the data. These relationships are currently not very well explored. The LOD Laundromat makes it possible to not only perform Semantic Web evaluations on a much larger scale, but it also allows evaluation results to be more systematically related to (structural) properties of the data.

The following query finds all documents that contain the namespace often abbreviated as vcard:

```
http://index.lodlaundromat.org/ns2d/\
http%3A%2F%2Fwww.w3.org%2F2006%2Fvcard%2Fns%23
```

The same query in *Frank*:

```
$ frank documents --namespace http://www.w3.org/2006/vcard
  /ns#
```

4.4 Metadata

frank meta returns metadata descriptions of RDF documents using the N-Triples format. The following returns the metadata for all documents:

```
$ frank documents --resourceUri | frank meta
ll:85d...33c ll:triples "54"^^xsd:int .
ll:85d...33c llo:added "2014-10-10T00:23:56"^^xsd:
  dateTime .
```

Since detailed metadata requires the use of property paths, the SPARQL endpoint allows for more specific metadata queries than *Frank*. For instance, the data in Fig. 4 (Sect. 3.4) is generated with the following SPARQL query:

```
SELECT * {
   [] llm:outDegree/llm:mean ?mean
}
```

4.5 Combining Queries

The types of queries that we have shown can also be combined in order to cover a wider array of use cases. Since *Frank* uses the standard conventions for output handling, other processes can utilize the resultant triples by simply reading from standard input. The following prints pairs of people who know each other according to RDF documents that also contain VoID information and that contain at least one thousand statements:

```
$ frank documents --namespace void --minTriples 1000 |
  frank statements --predicate foaf:knows |
  awk '{ print $1, $3 }' |
  sed 's/.$//' |
  head -n 2
<http://florianjensen.com/micro/user/1> <http://identi.
   ca/user/568>
<http://fun.ohhh.no/user/1> <http://identi.ca/user/568>
```

As the above uses of *awk* and *sed* show, it is easy to combine use of the LOD Laundromat with your favorite scripts, tools and libraries. Firstly, the Web Services can be used from within a programming language that handles HTTP requests and that provides basic support for parsing N-Quads 1.1.

```
$ frank statements -p foaf:knows |
  grep last-fm |
  ./ntriplesToGml > last-fm.gml
```

5 Concluding Remarks

The LOD Laundromat provides added value for a wide range of use cases. We hereby offer a brief overview over some examples of potential application scenarios.

LOD Observatory. From the scraping process, metadata statistics can be extracted that provide a (partial) overview of the state of the LOD Cloud (e.g., how much data is disseminated and how well this is done).

Publishing support. Since all errors are stored as part of the metadata, the LOD Laundromat gives detailed feedback to dataset publishers as to how they can improve the quality of their data.

Evaluation of algorithms. By publishing very many datasets in exactly the same, standards-compliant way, and by making available the programmable interface *Frank*, LOD Laundromat supports the evaluation of Semantic Web algorithms on large-scale, heterogeneous and real-world data.

Data search. Data consumers can search for datasets based on a collection of desired properties.

Moreover the presented infrastructure can immediately be useful from a system engineering perspective:

Efficient data loading. Datasets can be loaded more efficiently from the LOD Wardrobe than from the original source, since all triples are sorted, duplicates are removed, and verbose serialization formats are avoided.

Load balancing. Semantic Web algorithms can use the SPARQL endpoint or the command-line tool *Frank* to download the data documents, according to metadata criteria that allow data to be evenly distributed over a given number of processing nodes.

From a Knowledge Representation & Reasoning perspective, the LOD Laundromat framework and the LOD Lab approach provide unique opportunities. In the long run, the vast Web of Data and can be made available for intelligent agents to reason and act upon. More directly, the data that is published, and the tools introduced in this chapter, allow an upscaling of the application and evaluation of reasoning methods and tools. With this work we seek to equip the informed computational logician with the necessary skills to dive into the fascinating and challenging endavour to apply their tools and methods to a large collection of heterogeneous data.

References

1. Hogan, A., Harth, A., Passant, A., Decker, S., Polleres, A.: Weaving the pedantic web. In: Linked Data on the Web Workshop (2010)
2. Hogan, A., Umbrich, J., Harth, A., Cyganiak, R., Polleres, A., Decker, S.: An empirical survey of linked data conformance. Web Semant.: Sci. Serv. Agents World Wide Web **14**, 14–44 (2012)
3. Beek, W., Rietveld, L., Bazoobandi, H.R., Wielemaker, J., Schlobach, S.: LOD laundromat: a uniform way of publishing other people's dirty data. In: Mika, P., et al. (eds.) ISWC 2014. LNCS, vol. 8796, pp. 213–228. Springer, Heidelberg (2014). doi:10.1007/978-3-319-11964-9_14
4. Rietveld, L., Verborgh, R., Beek, W., Vander Sande, M., Schlobach, S.: Linkeddata-as-a-service: the semantic web redeployed. In: Gandon, F., Sabou, M., Sack, H., d'Amato, C., Cudré-Mauroux, P., Zimmermann, A. (eds.) ESWC 2015. LNCS, vol. 9088, pp. 471–487. Springer, Heidelberg (2015). doi:10.1007/978-3-319-18818-8_29

5. Verborgh, R., et al.: Querying datasets on the web with high availability. In: Mika, P., et al. (eds.) ISWC 2014. LNCS, vol. 8796, pp. 180–196. Springer, Heidelberg (2014). doi:10.1007/978-3-319-11964-9_12

6. Ermilov, I., Martin, M., Lehmann, J., Auer, S.: Linked open data statistics: collection and exploitation. In: Klinov, P., Mouromtsev, D. (eds.) KESW 2013. CCIS, vol. 394, pp. 242–249. Springer, Heidelberg (2013). doi:10.1007/978-3-642-41360-5_19

7. Auer, S., Demter, J., Martin, M., Lehmann, J.: LODStats – an extensible framework for high-performance dataset analytics. In: Teije, A., Völker, J., Handschuh, S., Stuckenschmidt, H., d'Acquin, M., Nikolov, A., Aussenac-Gilles, N., Hernandez, N. (eds.) EKAW 2012. LNCS (LNAI), vol. 7603, pp. 353–362. Springer, Heidelberg (2012). doi:10.1007/978-3-642-33876-2_31

8. Buil-Aranda, C., Hogan, A., Umbrich, J., Vandenbussche, P.-Y.: SPARQL web-querying infrastructure: ready for action? In: Alani, H., et al. (eds.) ISWC 2013. LNCS, vol. 8219, pp. 277–293. Springer, Heidelberg (2013). doi:10.1007/978-3-642-41338-4_18

9. Cheng, G., Gong, S., Qu, Y.: An empirical study of vocabulary relatedness and its application to recommender systems. In: Aroyo, L., Welty, C., Alani, H., Taylor, J., Bernstein, A., Kagal, L., Noy, N., Blomqvist, E. (eds.) ISWC 2011. LNCS, vol. 7031, pp. 98–113. Springer, Heidelberg (2011). doi:10.1007/978-3-642-25073-6_7

10. Ge, W., Chen, J., Hu, W., Qu, Y.: Object link structure in the semantic web. In: Aroyo, L., Antoniou, G., Hyvönen, E., Teije, A., Stuckenschmidt, H., Cabral, L., Tudorache, T. (eds.) ESWC 2010. LNCS, vol. 6089, pp. 257–271. Springer, Heidelberg (2010). doi:10.1007/978-3-642-13489-0_18

11. Alexander, K., Cyganiak, R., Hausenbals, M., Zhao, J.: Describing linked datasets with the VoID vocabulary, March 2011. http://www.w3.org/TR/2011/NOTE-void-20110303/

12. Millard, I., Glaser, H., Salvadores, M., Shadbolt, N.: Consuming multiple Linked Data sources: challenges and experiences. In: First International Workshop on Consuming Linked Data (COLD), November 2010

13. Prud'hommeaux, E., Buil-Aranda, C.: SPARQL 1.1 Federated Query (2013). http://www.w3.org/TR/sparql11-federated-query/

14. Fernández, J.D., Martínez-Prieto, M.A., Gutiérrez, C., Polleres, A., Arias, M.: Binary RDF representation for publication and exchange (HDT). Web Seman.: Sci. Serv. Agents World Wide Web 19, 22–41 (2013)

15. Mäkelä, E.: Aether – generating and viewing extended void statistical descriptions of RDF datasets. In: Presutti, V., Blomqvist, E., Troncy, R., Sack, H., Papadakis, I., Tordai, A. (eds.) ESWC 2014. LNCS, vol. 8798, pp. 429–433. Springer, Heidelberg (2014). doi:10.1007/978-3-319-11955-7_61

16. Callahan, A., Cruz-Toledo, J., Ansell, P., Dumontier, M.: Bio2RDF release 2: improved coverage, interoperability and provenance of life science linked data. In: Cimiano, P., Corcho, O., Presutti, V., Hollink, L., Rudolph, S. (eds.) ESWC 2013. LNCS, vol. 7882, pp. 200–212. Springer, Heidelberg (2013). doi:10.1007/978-3-642-38288-8_14

17. Bazoobandi, H.R., Rooij, S., Urbani, J., Teije, A., Harmelen, F., Bal, H.: A compact in-memory dictionary for RDF data. In: Gandon, F., Sabou, M., Sack, H., d'Amato, C., Cudré-Mauroux, P., Zimmermann, A. (eds.) ESWC 2015. LNCS, vol. 9088, pp. 205–220. Springer, Heidelberg (2015). doi:10.1007/978-3-319-18818-8_13

18. Christophides, V., Efthymiou, V., Stefanidis, K.: Entity Resolution in the Web of Data. Morgan & Claypool Publishers, San Rafael (2015)

19. Rietveld, L., Beek, W., Schlobach, S.: LOD lab: experiments at LOD scale. In: Arenas, M., et al. (eds.) ISWC 2015. LNCS, vol. 9367, pp. 339–355. Springer, Heidelberg (2015). doi:10.1007/978-3-319-25010-6_23

20. Schmachtenberg, M., Bizer, C., Paulheim, H.: Adoption of the linked data best practices in different topical domains. In: Mika, P., et al. (eds.) ISWC 2014. LNCS, vol. 8796, pp. 245–260. Springer, Heidelberg (2014). doi:10.1007/978-3-319-11964-9_16

21. Isele, R., Umbrich, J., Bizer, C., Harth, A.: LDSpider: an open-source crawling framework for the Web of Linked Data. In: 9th International Semantic Web Conference. Citeseer (2010)

22. Harris, S., Seaborne, A.: SPARQL 1.1 query language, March 2013

Inconsistency-Tolerant Querying of Description Logic Knowledge Bases

Meghyn Bienvenu[1(✉)] and Camille Bourgaux[2]

[1] LIRMM - CNRS, Inria, and Université de Montpellier, Montpellier, France
meghyn@lirmm.fr
[2] LRI - CNRS and Université Paris-Sud, Orsay, France
bourgaux@lri.fr

Abstract. An important issue that arises when querying description logic (DL) knowledge bases is how to handle the case in which the knowledge base is inconsistent. Indeed, while it may be reasonable to assume that the TBox (ontology) has been properly debugged, the ABox (data) will typically be very large and subject to frequent modifications, both of which make errors likely. As standard DL semantics is useless in such circumstances (everything is entailed from a contradiction), several alternative inconsistency-tolerant semantics have been proposed with the aim of providing meaningful answers to queries in the presence of such data inconsistencies. In the first part of this chapter, we present and compare these inconsistency-tolerant semantics, which can be applied to any DL (or ontology language). The second half of the chapter summarizes what is known about the computational properties of these semantics and gives an overview of the main algorithmic techniques and existing systems, focusing on DLs of the DL-Lite family.

1 Introduction

Ontology-mediated query answering (OMQA) is a promising approach to data access that leverages the semantic knowledge provided by an ontology to improve query answering (see [1] for a recent survey and references). Much of the work on OMQA considers focuses on ontologies formulated using description logics (DLs) [2], a well-known class of decidable fragments of first-order logic that provide the logically underpinnings of the W3C-standardized OWL 2 ontology language [3]. Over the past decade, significant research efforts have been devoted to understanding how the complexity of OMQA varies depending on the choice of DL and query language, which has led to the identification of DLs with favourable computational properties. The DL-Lite family of DLs [4], which was specifically designed with OMQA in mind, has gained particular prominence, due to the fact that query answering can be reduced via first-order query rewriting to database query evaluation. In addition to theoretical work, there has been a lot of practically-oriented research aimed at developing and implementing efficient algorithms for OMQA, particularly for DL-Lite ontologies.

An important practical issue that arises in the context of DL-based OMQA is how to handle the case in which the dataset (or ABox, in DL parlance) is

© Springer International Publishing AG 2017
J.Z. Pan et al. (Eds.): Reasoning Web 2016, LNCS 9885, pp. 156–202, 2017.
DOI: 10.1007/978-3-319-49493-7_5

inconsistent with the ontology (TBox). Indeed, while it may be reasonable to assume that the TBox has been properly debugged, the ABox will typically be very large and/or subject to frequent modifications, which makes errors likely. Unfortunately, standard DL semantics is next to useless in such circumstances, as everything is entailed from a contradiction. It is therefore essential to devise robust methods for handling inconsistent data if OMQA is to be widely adopted in practice. Modifying the ABox to restore consistency may seem like the ideal solution, as it allows us to use existing query answering algorithms. However, this approach is not always feasible. First, it can be difficult and time-consuming to identify the erroneous parts of the data, and removing all potentially erroneous assertions will typically lead to unacceptable loss of information. Second, even if it can be determined which ABox assertions should be removed, the OMQA system may lack the authorization to make data modifications (e.g., in information integration applications involving external data sources).

As inconsistencies cannot always be eliminated, it is important to provide principled methods for obtaining meaningful answers to queries posed over inconsistent DL KBs. In this chapter, we will present a number of different ways of approaching this problem, based upon using different kinds of inconsistency-tolerant semantics for defining what tuples should be counted as query answers. Probably the most well-known and arguably the most natural such semantics is the AR semantics, introduced in [5]. The semantics is inspired by work on consistent query answering in relational databases [6–9], where the standard approach is to define a set of data repairs (which correspond to minimally changing the dataset to restore consistency), and to define the set of query results as those answers that can be obtained from each of the data repairs. In the DL setting, an analogous notion of ABox repair can be defined by considering the inclusion-maximal subsets of the ABox that are consistent w.r.t. the TBox. The AR semantics amounts to considering those answers that can be derived from each of the ABox repairs using the axioms in the TBox. As we shall see in this chapter, there are in fact many other inconsistency-tolerant semantics that can be defined in terms of ABox repairs, such as the IAR semantics [5], which defines query answers w.r.t. to the intersection of the ABox repairs, and the brave semantics [10] that only requires that an answer hold w.r.t. at least one repair.

The aim of this chapter is to provide an introduction to inconsistency-tolerant query answering in the DL setting. After some preliminaries, we will introduce in Sect. 3 a variety of different inconsistency-tolerant semantics and examine the relationships that hold between the semantics. In the following section, we will explore the computational properties of these semantics, providing a detailed complexity landscape for DLs of the DL-Lite family, as well as a short discussion of what happens when one considers other DLs. In Sect. 5, we will briefly describe the systems that have been implemented for inconsistency-tolerant query answering, before concluding in Sect. 6 with a discussion of current and future research directions.

2 Preliminaries

2.1 Description Logic Knowledge Bases

A DL *knowledge base (KB)* consists of an ABox and a TBox, which are constructed from a set N_C of *concept names* (unary predicates), a set of N_R of *role names* (binary predicates), and a set N_I of *individual names* (constants). The *ABox* (dataset) is a finite set of *concept assertions* of the form $A(a)$, with $A \in N_C$ and $a \in N_I$, and *role assertions* of the form $R(a, b)$, with $R \in N_R$, $a, b \in N_I$. The *TBox* (ontology) consists of a finite set of axioms whose form depends on the DL in question.

In this chapter, we will mainly focus on description logics from the DL-Lite family [4]. In the DL-Lite$_R$ dialect (which underlies the OWL 2 QL profile [11]), TBoxes are composed of *concept inclusions* $B \sqsubseteq C$ and *role inclusions* $Q \sqsubseteq S$, where B and C (resp. Q and S) are complex concepts (resp. roles) formed according to the following syntax:

$$B := A \mid \exists Q \qquad C := B \mid \neg B \qquad Q := R \mid R^- \qquad S := Q \mid \neg Q$$

where $A \in N_C$ and $R \in N_R$. Inclusions that have $\neg$ on the right-hand side are called *negative inclusions*; all other inclusions are called *positive inclusions*. In the DL-Lite$_{core}$ dialect, only concept inclusions are allowed in the TBox.

The next example introduces the DL-Lite$_{core}$ TBox that we will use throughout the chapter, which is adapted from the one in [10]:

Example 1. We consider the TBox $\mathcal{T}_{univ}$ consisting of the following axioms:

Prof $\sqsubseteq$ Faculty	Prof $\sqsubseteq$ $\exists$Teaches	Prof $\sqsubseteq$ $\neg$Lect	Faculty $\sqsubseteq$ $\neg$Course
Lect $\sqsubseteq$ Faculty	Lect $\sqsubseteq$ $\exists$Teaches	Prof $\sqsubseteq$ $\neg$Fellow	
Fellow $\sqsubseteq$ Faculty	$\exists$Teaches$^-$ $\sqsubseteq$ Course	Lect $\sqsubseteq$ $\neg$Fellow	

The inclusions in the first column state that professors, lecturers, and research fellows are three classes of faculty members. Due to the negative inclusions in the third column, we know that these are disjoint classes. In the second column, we state that professors and lecturers must teach something (that is, they must occur in the first argument of relation Teaches), and everything that is taught (that is, appears in the second argument of Teaches) is of type Course. The rightmost axiom states that Faculty and Course are disjoint.

We will also briefly mention the description logics $\mathcal{EL}_\perp$ and $\mathcal{ALC}$. In both DLs, the TBox consists of concept inclusions $C_1 \sqsubseteq C_2$, where C_1, C_2 are concepts built according to the syntax of the DL. In $\mathcal{EL}_\perp$, we use the grammar:

$$C := A \mid \top \mid \perp \mid C \sqcap C \mid \exists R.C$$

and in $\mathcal{ALC}$, concepts are constructed as follows:

$$C := A \mid \top \mid \perp \mid \neg C \mid C \sqcap C \mid C \sqcup C \mid \exists R.C \mid \forall R.C$$

where as before, $A \in \mathsf{N_C}$ and $R \in \mathsf{N_R}$. Thus, in $\mathcal{EL}_\perp$, we can build concepts using the top and bottom concepts $(\top, \perp)$, conjunction $(\sqcap)$, and qualified existential restrictions $(\exists R.C)$, and in $\mathcal{ALC}$, we may additionally use unrestricted negation $(\neg)$, disjunction $(\sqcup)$, and qualified universal restrictions $(\forall R.C)$.

The semantics of DL knowledge bases is defined using interpretations. An *interpretation* takes the form $\mathcal{I} = (\Delta^\mathcal{I}, \cdot^\mathcal{I})$, where $\Delta^\mathcal{I}$ is a non-empty set and $\cdot^\mathcal{I}$ maps each $a \in \mathsf{N_I}$ to $a^\mathcal{I} \in \Delta^\mathcal{I}$, each $A \in \mathsf{N_C}$ to $A^\mathcal{I} \subseteq \Delta^\mathcal{I}$, and each $R \in \mathsf{N_R}$ to $R^\mathcal{I} \subseteq \Delta^\mathcal{I} \times \Delta^\mathcal{I}$. The function $\cdot^\mathcal{I}$ is straightforwardly extended to general concepts and roles:

$$\top^\mathcal{I} = \Delta^\mathcal{I} \quad \perp^\mathcal{I} = \varnothing \qquad\qquad (\neg B)^\mathcal{I} = \Delta^\mathcal{I} \setminus B^\mathcal{I}$$
$$(C \sqcap D)^\mathcal{I} = C^\mathcal{I} \cap D^\mathcal{I} \qquad\qquad (C \sqcup D)^\mathcal{I} = C^\mathcal{I} \cup D^\mathcal{I}$$
$$(\exists Q)^\mathcal{I} = \{c \mid \exists d : (c,d) \in Q^\mathcal{I}\} \quad (\exists R.C)^\mathcal{I} = \{c \mid \exists d : (c,d) \in R^\mathcal{I}, d \in C^\mathcal{I}\}$$
$$(R^-)^\mathcal{I} = \{(c,d) \mid (d,c) \in R^\mathcal{I}\} \quad (\forall R.C)^\mathcal{I} = \{c \mid \forall d : (c,d) \in R^\mathcal{I} \Rightarrow d \in C^\mathcal{I}\}$$

An interpretation $\mathcal{I}$ satisfies an inclusion $G \sqsubseteq H$ if $G^\mathcal{I} \subseteq H^\mathcal{I}$; it satisfies $A(a)$ (resp. $R(a,b)$) if $a^\mathcal{I} \in A^\mathcal{I}$ (resp. $(a^\mathcal{I}, b^\mathcal{I}) \in R^\mathcal{I}$). An interpretation $\mathcal{I}$ is a *model* of $\mathcal{K} = \langle \mathcal{T}, \mathcal{A} \rangle$ if $\mathcal{I}$ satisfies all inclusions in $\mathcal{T}$ and assertions in $\mathcal{A}$. A KB $\mathcal{K}$ is *consistent* if it has a model, and we say that an ABox $\mathcal{A}$ is $\mathcal{T}$-*consistent* if the KB $\langle \mathcal{T}, \mathcal{A} \rangle$ is consistent. A subset $\mathcal{A}' \subseteq \mathcal{A}$ is called a *minimal $\mathcal{T}$-inconsistent subset* of $\mathcal{A}$ if (i) $\mathcal{A}'$ is $\mathcal{T}$-inconsistent, and (ii) every $\mathcal{A}'' \subsetneq \mathcal{A}'$ is $\mathcal{T}$-consistent.

Example 2. In our running example, we will consider the following ABox:

$$\mathcal{A}_\mathsf{univ} = \{\mathsf{Prof(sam), Lect(sam), Fellow(sam), Prof(kim), Lect(kim),}$$
$$\mathsf{Fellow(julie), Teaches(csc343, julie), Fellow(alex), Teaches(alex, csc486)}\}$$

Observe that $\mathcal{A}_\mathsf{univ}$ is $\mathcal{T}_\mathsf{univ}$-inconsistent. Indeed, we have the following minimal $\mathcal{T}_\mathsf{univ}$-inconsistent subsets:

- $\{\mathsf{Prof(sam), Lect(sam)}\}$ which contradicts $\mathsf{Prof} \sqsubseteq \neg\mathsf{Lect}$
- $\{\mathsf{Prof(sam), Fellow(sam)}\}$ which contradicts $\mathsf{Prof} \sqsubseteq \neg\mathsf{Fellow}$
- $\{\mathsf{Lect(sam), Fellow(sam)}\}$ which contradicts $\mathsf{Lect} \sqsubseteq \neg\mathsf{Fellow}$
- $\{\mathsf{Prof(kim), Lect(kim)}\}$ which contradicts $\mathsf{Prof} \sqsubseteq \neg\mathsf{Lect}$
- $\{\mathsf{Fellow(julie), Teaches(csc343, julie)}\}$ which contradicts $\mathsf{Faculty} \sqsubseteq \neg\mathsf{Course}$, due to the positive inclusions $\mathsf{Fellow} \sqsubseteq \mathsf{Faculty}$ and $\exists\mathsf{Teaches}^- \sqsubseteq \mathsf{Course}$

We say that an assertion or axiom α is *entailed* from a KB $\mathcal{K}$, written $\mathcal{K} \models \alpha$, if every model of $\mathcal{K}$ satisfies α. If $\mathcal{A}$ is $\mathcal{T}$-consistent, then the $\mathcal{T}$-*closure of $\mathcal{A}$*, denoted $\mathsf{cl}_\mathcal{T}(\mathcal{A})$, consists of all assertions α such that $\langle \mathcal{T}, \mathcal{A} \rangle \models \alpha$. Note that since $\langle \mathcal{T}, \mathcal{A} \rangle$ is finite and consistent, the closure $\mathsf{cl}_\mathcal{T}(\mathcal{A})$ is necessarily finite.

2.2 Querying DL KBs

We will generally assume that the user query is given as a *conjunctive query* (CQ), which takes the form $q(\mathbf{x}) = \exists\mathbf{y}\, \psi(\mathbf{x}, \mathbf{y})$, where $\mathbf{x}$ and $\mathbf{y}$ are tuples of variables, and ψ is a conjunction of atoms of the forms $A(t)$ or $R(t, t')$, where

t, t' are variables from $\mathbf{x} \cup \mathbf{y}$ or individual names. The variables $\mathbf{x}$ are called the *answer variables* of $q(\mathbf{x})$. A CQ without answer variables is called *Boolean*, a CQ without any existentially quantified variables is called a *ground CQ*, and a ground CQ consisting of a single atom is called an *instance query* (IQ).

We say that a Boolean CQ q is *entailed* from $\mathcal{K}$, written $\mathcal{K} \models q$, just in the case that q holds in all models of $\mathcal{K}$. For a non-Boolean CQ q with answer variables $\mathbf{x} = (x_1, \ldots, x_k)$, a tuple of individuals $\mathbf{a} = (a_1, \ldots, a_k)$ is a *certain answer* for q w.r.t. $\mathcal{K}$ just in the case that $\mathcal{K} \models q(\mathbf{a})$, where $q(\mathbf{a})$ is the Boolean query obtained by replacing each x_i by a_i. We will use the notation $\mathcal{K} \models q(\mathbf{a})$ to denote that $\mathbf{a}$ is a certain answer to q w.r.t. $\mathcal{K}$. Later in the chapter, we will introduce alternative semantics for defining query answers, and we will use 'classical semantics' to refer to the certain answer semantics we have just defined.

Example 3. Consider the following queries

$$q_1(x) = \mathsf{Faculty}(x) \qquad q_2(x) = \exists y\, \mathsf{Teaches}(x, y)$$
$$q_3(x) = \exists y\, \mathsf{Faculty}(x) \wedge \mathsf{Teaches}(x, y) \qquad q_4(x, y) = \mathsf{Faculty}(x) \wedge \mathsf{Teaches}(x, y)$$

If we compute the certain answers to these queries over our example KB $\mathcal{K}_{\mathsf{univ}} = \langle \mathcal{T}_{\mathsf{univ}}, \mathcal{A}_{\mathsf{univ}} \rangle$, then due to the inconsistency of $\mathcal{K}_{\mathsf{univ}}$, every individual appearing in $\mathcal{A}_{\mathsf{univ}}$ will be returned as a certain answer to q_1, q_2, and q_3, and every pair of individuals will be a certain answer for q_4.

To better illustrate the notion of certain answers, let us consider the following $\mathcal{T}_{\mathsf{univ}}$-consistent subset of $\mathcal{A}_{\mathsf{univ}}$:

$$\mathcal{A}_{\mathsf{univ}}^{\mathsf{cons}} \{\mathsf{Prof}(\mathsf{sam}), \mathsf{Lect}(\mathsf{kim}), \mathsf{Fellow}(\mathsf{julie}), \mathsf{Fellow}(\mathsf{alex}), \mathsf{Teaches}(\mathsf{alex}, \mathsf{csc486})\}$$

Evaluating the four queries over the KB consisting of $\mathcal{T}_{\mathsf{univ}}$ and the preceding ABox yields the following results:

- $q_1(x)$ has four certain answers: sam, kim, julie, and alex
- $q_2(x)$ has three certain answers: sam, kim, and alex
- $q_3(x)$ has three certain answers: sam, kim, and alex
- $q_4(x, y)$ has a single certain answer: (alex, csc486)

Indeed, for q_1, each of the individuals sam, kim, julie, and alex belongs to either Prof, Lect, or Fellow, which are declared to be subclasses of Faculty. For q_2, we obtain sam and kim using the axioms $\mathsf{Prof} \sqsubseteq \exists\mathsf{Teaches}$ and $\mathsf{Lect} \sqsubseteq \exists\mathsf{Teaches}$; note that julie is not an answer to q_2 as the TBox does not guarantee that every Fellow teaches something. The certain answers to q_3 can be obtained by intersecting the certain answers of q_1 and q_2. Finally, for q_4, there are no answers involving sam and kim, since although these individuals are known to teach some course, there is no information in the KB that allows us to identify the course(s) taught.

2.3   Query Rewriting

First-order (FO) query rewriting is an algorithmic technique that allows us to reduce ontology-mediated query answering to the evaluation of first-order

($\sim$ SQL) queries, which can be handled by relational database systems. The idea is as follows: the user query is first transformed (independently of the ABox) into an FO-query that incorporates the relevant information from the TBox, and in second step, the resulting query is evaluated over the ABox, viewed as a database.

Formally, given a CQ $q(\mathbf{x})$ and a TBox $\mathcal{T}$, we call an FO-query $q'(\mathbf{x})$ an *(FO)-rewriting* of $q(\mathbf{x})$ w.r.t. $\mathcal{T}$ if the following equivalence holds for every ABox $\mathcal{A}$ and every tuple of individuals $\mathbf{a}$ (of the same arity as $\mathbf{x}$):

$$\mathcal{T}, \mathcal{A} \models q(\mathbf{a}) \quad \Leftrightarrow \quad \mathcal{I}_\mathcal{A} \models q'(\mathbf{a})$$

where $\mathcal{I}_\mathcal{A}$ is the finite interpretation isomorphic to $\mathcal{A}$, i.e., the domain consists of the individuals occurring in $\mathcal{A}$ and every concept or role name P is interpreted by $\{\mathbf{a} \mid P(\mathbf{a}) \in \mathcal{A}\}$. Note that the symbol $\models$ is used differently on the two sides of the equivalence. On the left, we are checking whether $q(\mathbf{a})$ is entailed from the KB $\langle \mathcal{T}, \mathcal{A} \rangle$, which requires us to consider all models of the KB, whereas on the right, we only need to test whether the FO-sentence $q'(\mathbf{a})$ holds in a single interpretation, $\mathcal{I}_\mathcal{A}$.

We recall that DL-Lite$_\mathcal{R}$ (like most DL-Lite dialects) possesses the *FO-rewritability* property, meaning that for every CQ q and every DL-Lite$_\mathcal{R}$ TBox $\mathcal{T}$, we can effectively construct an FO-rewriting q' of q w.r.t. $\mathcal{T}$. Many of the rewriting algorithms developed for DL-Lite produce rewritings that belong to the more restricted class of *union of conjunctive queries* (UCQs), which are disjunctions of CQs $q_1(\mathbf{x}) \vee \ldots \vee q_n(\mathbf{x})$ having the same answer variables $\mathbf{x}$. In this case, we speak of UCQ-rewritings.

Example 4. The following queries

$$
\begin{aligned}
q_1'(x) \;&=\; \mathsf{Faculty}(x) \vee \mathsf{Prof}(x) \vee \mathsf{Lect}(x) \vee \mathsf{Fellow}(x) \\
q_2'(x) \;&=\; \exists y.\mathsf{Teaches}(x,y) \vee \mathsf{Prof}(x) \vee \mathsf{Lect}(x) \\
q_3'(x) \;&=\; (\exists y.\mathsf{Faculty}(x) \wedge \mathsf{Teaches}(x,y)) \vee \mathsf{Prof}(x) \vee \mathsf{Lect}(x) \vee \\
&\qquad (\exists y.\mathsf{Fellow}(x) \wedge \mathsf{Teaches}(x,y)) \\
q_4'(x,y) \;&=\; (\mathsf{Faculty}(x) \wedge \mathsf{Teaches}(x,y)) \vee (\mathsf{Prof}(x) \wedge \mathsf{Teaches}(x,y)) \\
&\qquad (\mathsf{Lect}(x) \wedge \mathsf{Teaches}(x,y)) \vee (\mathsf{Fellow}(x) \wedge \mathsf{Teaches}(x,y))
\end{aligned}
$$

are UCQ-rewritings, respectively, of the queries $q_1(x)$, $q_2(x)$, $q_3(x)$, and $q_4(x,y)$ w.r.t. the TBox $\mathcal{T}_{\mathsf{univ}}$. Observe that the disjuncts of q_i' correspond to all of the ways that a (pair of) individual(s) can be derived as a certain answer. The query $q_2'(x)$, for example, states that an individual is a certain answer to $q_2(x) = \exists y\,\mathsf{Teaches}(x,y)$ if the individual appears in the first argument of a $\mathsf{Teaches}$ assertion, or if it appears in a Prof assertion, or if it appears in a Lect assertion.

If we evaluate the rewriting q_i' over $\mathcal{I}_{\mathcal{A}_{\mathsf{univ}}^{\mathsf{cons}}}$, then we will obtain the certain answers of q_i over $\langle \mathcal{T}_{\mathsf{univ}}, \mathcal{A}_{\mathsf{univ}}^{\mathsf{cons}} \rangle$. For example, evaluating $q_2'(x)$ over $\mathcal{I}_{\mathcal{A}_{\mathsf{univ}}^{\mathsf{cons}}}$ returns the following three answers: sam (due to the disjunct $\mathsf{Prof}(x)$), kim (due to the disjunct $\mathsf{Lect}(x)$), and alex (due to the disjunct $\exists y.\mathsf{Teaches}(x,y)$).

2.4   Complexity of Query Answering

As usual, when we speak of the complexity of ontology-mediated query answering, we mean the computational complexity of the associated *decision problem*, which is to determine whether $\langle \mathcal{T}, \mathcal{A} \rangle \models q(\mathbf{a})$ (here $\mathbf{a}$ is a candidate answer, i.e. tuple of individuals of the same arity as q). There are two standard ways of measuring the complexity of query answering:

- *Combined complexity:* we measure the complexity in terms of the size of the entire input ($\mathcal{T}, \mathcal{A}$, and $q(\mathbf{a})$)
- *Data complexity:* we measure the complexity only in terms of the size of the ABox $\mathcal{A}$ (so $|\mathcal{T}|$ and $|q(\mathbf{a})|$ are treated as fixed constants)

Data complexity is often considered the more relevant measure as the ABox (data) is typically significantly larger than the size of the TBox and query. However, combined complexity is often more fine-grained, allowing us to distinguish between different problems with the same data complexity. By considering both measures, we get a more complete picture of the complexity landscape.

In our complexity analysis, we will refer to the following complexity classes:

- AC^0: problems that can be solved by a uniform family of circuits of constant depth and polynomial size, with unlimited fan-in AND gates and OR gates
- NL: problems solvable in non-deterministic logarithmic space
- coNL: problems whose complement is solvable in non-deterministic logarithmic space
- P: problems solvable in polynomial time
- NP: problems solvable in non-deterministic polynomial time
- coNP: problems whose complement is in NP
- Δ_2^p: problems solvable in polynomial time with access to an NP oracle
- $\Delta_2^p[O(\log n)]$: problems solvable in polynomial time with at most logarithmically many calls to an NP oracle
- Σ_2^p: problems solvable in non-deterministic polynomial time with access to an NP oracle
- Π_2^p: problems whose complement is in Σ_2^p
- EXP: problems solvable in deterministic single-exponential time

We recall that these complexity classes are related as follows:

$$AC^0 \subseteq NL \subseteq P \quad \begin{array}{c} \subsetneq \; NP \; \subsetneq \\ \subsetneq \; coNP \; \subsetneq \end{array} \; \Delta_2^p[O(\log n)] \subseteq \Delta_2^p \quad \begin{array}{c} \subsetneq \; \Sigma_2^p \; \subsetneq \\ \subsetneq \; \Pi_2^p \; \subsetneq \end{array} \; EXP$$

and that NL=coNL [12,13].

3   Inconsistency-Tolerant Semantics

Classical semantics does not allow us to obtain meaningful answers to queries posed over inconsistent DL knowledge bases. Indeed, if we have two inconsistent

KBs $\mathcal{K}_1$ and $\mathcal{K}_2$ sharing the same set of individuals, then for every n-ary query, we will obtain the same set of certain answers, namely the set of all n-tuples of individuals appearing in these KBs. This is clearly undesirable, as the next example illustrates:

Example 5. Consider the ABoxes $\mathcal{A}_1$ and $\mathcal{A}_2$ defined as follows:

$$\mathcal{A}_1 \;=\; \{\mathsf{Prof}(\mathsf{sam}), \mathsf{Lect}(\mathsf{sam}), \mathsf{Fellow}(\mathsf{alex})\}$$
$$\mathcal{A}_2 \;=\; \{\mathsf{Prof}(\mathsf{sam}), \mathsf{Fellow}(\mathsf{alex}), \mathsf{Lect}(\mathsf{alex})\}$$

Observe that the KBs $\langle \mathcal{T}_{\mathsf{univ}}, \mathcal{A}_1 \rangle$ and $\langle \mathcal{T}_{\mathsf{univ}}, \mathcal{A}_2 \rangle$ are both inconsistent and use the same set of individuals, so they will be treated identically under classical semantics. However, we naturally draw different conclusions from these two KBs. Indeed, from the first KB, it seems reasonable to infer that $\mathsf{Fellow}(\mathsf{alex})$ since the assertion $\mathsf{Fellow}(\mathsf{alex})$ is not involved in any contradictions. By contrast, there is conflicting information about the professional status of sam (is sam a professor or lecturer?), so we would probably choose not to return sam as an answer to a query asking for professors (similarly for lecturers). We might nevertheless conclude that sam is a faculty member ($\mathsf{Faculty}(\mathsf{sam})$) and that sam teaches ($\exists y\, \mathsf{Teaches}(\mathsf{sam}, y)$), since all of the information about the individual sam supports these statements. Applying a similar reasoning on the second KB, one would naturally return sam as an answer to the query $\mathsf{Prof}(x)$ (since the ABox asserts $\mathsf{Prof}(\mathsf{sam})$ and contains no information contradicting this assertion). However, but for the individual alex, we would probably conclude that $\mathsf{Faculty}(\mathsf{alex})$ but would not say more about the specific position alex occupies (nor whether alex teaches).

Several inconsistency-tolerant semantics have been proposed to overcome the inadequacy of classical semantics for querying inconsistent KBs. Each of these semantics provides a different way of defining the set of answers of a query over a (possibly inconsistent) KB. In what follows, we will use the notation $\mathcal{K} \models_S q(\mathbf{a})$ to denote that $\mathbf{a}$ is an answer to q over the KB $\mathcal{K}$ under semantics S. If no semantics S is indicated, then we mean the classical semantics.

In general, there is no single 'best' inconsistency-tolerant semantics, and indeed, it may be interesting to use multiple semantics together, e.g. to identify answers of different levels of confidence. In order to choose which (combinations of) semantics to use in a given application, it is important to understand the properties of the different semantics, as well as the relationships between them.

We propose two notions of consistency for an inconsistency-tolerant semantics. The first one requires that all answers returned by the semantics have an internally consistent justification. Formally, we say that a set $C \subseteq \mathcal{A}$ is a *(consistent) $\mathcal{T}$-support* of $q(\mathbf{a})$ if (i) C is $\mathcal{T}$-consistent, and (ii) $\langle \mathcal{T}, C \rangle \models q(\mathbf{a})$. The consistency condition can then be defined as follows:

Definition 1 (Consistent Support Property). *A semantics S is said to satisfy the* CONSISTENT SUPPORT *property if for every KB $\langle \mathcal{T}, \mathcal{A} \rangle$, query q, and tuple $\mathbf{a}$, if $\langle \mathcal{T}, \mathcal{A} \rangle \models_S q(\mathbf{a})$, then there exists a $\mathcal{T}$-support $C \subseteq \mathcal{A}$ of $q(\mathbf{a})$.*

The CONSISTENT SUPPORT property is important for explaining query results to users, as it ensures that we can always extract a consistent subset of the KB which yields a given query answer. Indeed, one could argue that CONSISTENT SUPPORT is a minimal requirement for an inconsistency-tolerant semantics to be meaningful.

Our second notion of consistency requires that the set of all query results obtainable from a KB are (simultaneously) consistent with the TBox:

Definition 2 (Consistent Results Property). *A semantics S is said to satisfy the* CONSISTENT RESULTS *property if for every KB $\langle \mathcal{T}, \mathcal{A} \rangle$, there exists a model $\mathcal{I}$ of $\mathcal{T}$ such that $\mathcal{I} \models q(\mathbf{a})$ for every $q(\mathbf{a})$ with $\langle \mathcal{T}, \mathcal{A} \rangle \models_S q(\mathbf{a})$.*

The CONSISTENT RESULTS property means that users can safely combine the query results together without any risk of obtaining a contradiction. If a semantics does not satisfy this property, then users need to be made aware of this (e.g. by informing users that the returned results should be considered as 'potential answers').

We point out that neither of the consistency properties implies the other. Indeed, we will see examples of semantics that satisfy CONSISTENT SUPPORT but not CONSISTENT RESULTS, and other semantics that verify CONSISTENT RESULTS but not CONSISTENT SUPPORT.

In addition to comparing semantics based upon the properties they satisfy, we can also compare them w.r.t. the set of answers they define.

Definition 3. *Given two semantics S and S', we say that:*

- *S' is an* under-approximation *(or: sound approximation) of S just in the case that*

$$\langle \mathcal{T}, \mathcal{A} \rangle \models_{S'} q(\mathbf{a}) \quad \Rightarrow \quad \langle \mathcal{T}, \mathcal{A} \rangle \models_S q(\mathbf{a})$$

for every KB $\langle \mathcal{T}, \mathcal{A} \rangle$, query q, and tuple $\mathbf{a}$.
- *S' is an* over-approximation *(or: complete approximation) of S just in the case that*

$$\langle \mathcal{T}, \mathcal{A} \rangle \models_S q(\mathbf{a}) \quad \Rightarrow \quad \langle \mathcal{T}, \mathcal{A} \rangle \models_{S'} q(\mathbf{a})$$

for every KB $\langle \mathcal{T}, \mathcal{A} \rangle$, query q, and tuple $\mathbf{a}$.

We observe that the two consistency properties are preserved by taking under-approximations:

Theorem 1. *Suppose that S' is an under-approximation of S, and let $P \in \{$CONSISTENT SUPPORT, CONSISTENT RESULTS$\}$. If S satisfies P, then S' also satisfies P.*

We shall see later in the chapter that under- and over-approximations are not only useful for gaining a better understanding of the relationships between different semantics, but also because they allow us to use the query results of a computationally well-behaved semantics in order to (partially) compute the answers to queries under a computationally more difficult semantics. Indeed, if

S' is an under-approximation of S, then we immediately have that every answer obtained using semantics S' is an answer under semantics S. Conversely, if S' is an over-approximation of S, then knowing that a tuple is not an answer under S' tells us that the tuple cannot be an answer under S.

3.1 Repairs

In order to extract reasonable answers from an inconsistent KB, it is useful to consider those portions of the data that are consistent with the TBox. This idea is captured by the notion of repair, defined as follows:

Definition 4. *A* repair *of an ABox A w.r.t. a TBox T is an inclusion-maximal subset of A that is T-consistent. We use $Rep(A, T)$ to denote the set of repairs of A w.r.t. T, which we abbreviate to $Rep(K)$ when $K = \langle T, A \rangle$.*

The repairs of an ABox correspond to the different ways of achieving consistency with the TBox while retaining as much of the original data as possible. By definition, every consistent KB has a single repair, consisting of the original ABox. When a KB is inconsistent, it is guaranteed to have at least one repair (take the empty set and keep adding assertions as long as consistency is preserved), but will more typically have more than one repair, reflecting the incomplete knowledge of which assertions are to blame for the inconsistency. We illustrate the notion of repair in the following example:

Example 6. As seen in an earlier example, the KB $K_{univ} = \langle T_{univ}, A_{univ} \rangle$ is inconsistent. There are twelve repairs of A_{univ} w.r.t. T_{univ}:

$$
\begin{aligned}
R_1 &= \{\mathsf{Prof(sam), Prof(kim), Fellow(julie)}\} \cup A_{Int} \\
R_2 &= \{\mathsf{Lect(sam), Lect(kim), Fellow(julie)}\} \cup A_{Int} \\
R_3 &= \{\mathsf{Fellow(sam), Prof(kim), Fellow(julie)}\} \cup A_{Int} \\
R_4 &= \{\mathsf{Prof(sam), Lect(kim), Fellow(julie)}\} \cup A_{Int} \\
R_5 &= \{\mathsf{Lect(sam), Prof(kim), Fellow(julie)}\} \cup A_{Int} \\
R_6 &= \{\mathsf{Fellow(sam), Lect(kim), Fellow(julie)}\} \cup A_{Int} \\
R_7 &= \{\mathsf{Prof(sam), Prof(kim), Teaches(csc343, julie)}\} \cup A_{Int} \\
R_8 &= \{\mathsf{Lect(sam), Lect(kim), Teaches(csc343, julie)}\} \cup A_{Int} \\
R_9 &= \{\mathsf{Fellow(sam), Prof(kim), Teaches(csc343, julie)}\} \cup A_{Int} \\
R_{10} &= \{\mathsf{Prof(sam), Lect(kim), Teaches(csc343, julie)}\} \cup A_{Int} \\
R_{11} &= \{\mathsf{Lect(sam), Prof(kim), Teaches(csc343, julie)}\} \cup A_{Int} \\
R_{12} &= \{\mathsf{Fellow(sam), Lect(kim), Teaches(csc343, julie)}\} \cup A_{Int}
\end{aligned}
$$

where the ABox A_{Int} that is common to all the repairs is as follows:

$$
A_{Int} = \{\mathsf{Fellow(alex), Teaches(alex, csc486)}\}
$$

Indeed, it is easily verified that every $\mathcal{R}_i$ is $\mathcal{T}_{\mathsf{univ}}$-consistent and is maximal in the sense that adding any additional assertion from $\mathcal{A}_{\mathsf{univ}}$ leads to a contradiction. Moreover, every $\mathcal{T}_{\mathsf{univ}}$-consistent subset of $\mathcal{A}_{\mathsf{univ}}$ is contained in one of the $\mathcal{R}_i$.

In the following sections, we will introduce several different inconsistency-tolerant semantics based upon the notion of repair.

3.2   AR Semantics

The most well-known, and arguably the most natural, inconsistency-tolerant semantics is the AR semantics, which was first introduced in [5], inspired by earlier work on consistent query answering in relational databases [6,8,9].

Note that in the following definition, and throughout this section, we assume that $\mathcal{K} = \langle \mathcal{T}, \mathcal{A} \rangle$ is a KB, q is a conjunctive query, and $\mathbf{a}$ is a tuple of constants from $\mathcal{A}$ of the same arity as q.

Definition 5 (AR semantics). *A tuple* $\mathbf{a}$ *is a certain answer of q over* $\mathcal{K} = \langle \mathcal{T}, \mathcal{A} \rangle$ *under the* AR (ABox Repair) *semantics, written* $\mathcal{K} \models_{\mathsf{AR}} q(\mathbf{a})$, *just in the case that* $\langle \mathcal{T}, \mathcal{B} \rangle \models q(\mathbf{a})$ *for every repair* $\mathcal{B} \in Rep(\mathcal{K})$.

The intuition behind the AR semantics is as follows. Under the assumption that the ABox is mostly correct, it seems reasonable to suppose that one of the repairs reflects the correct part of the ABox. In the absence of further information, we cannot identify which repair is the "correct" one, and so we only consider a tuple to be a query answer if it can be obtained from every repair.

We remark that AR semantics can be viewed a natural generalization of classical semantics. Indeed, classical semantics requires that a query answer hold in *all models* of a KB, since it is unknown which interpretation correctly describes the actual situation. With AR semantics, we additionally have uncertainty on the repair, so we consider *all models of all repairs* of the KB.

We illustrate the AR semantics on our running example:

Example 7. We determine the answers for our example queries q_1, q_2, q_3 over the KB $\mathcal{K}_{\mathsf{univ}}$ using the AR semantics. For query $q_1 = \mathsf{Faculty}(x)$, we observe that for each of the repairs $\mathcal{R}_i \in Rep(\mathcal{K}_{\mathsf{univ}})$ and each of the individuals $\mathsf{ind} \in \{\mathsf{sam}, \mathsf{kim}, \mathsf{alex}\}$, we have $\langle \mathcal{T}_{\mathsf{univ}}, \mathcal{R}_i \rangle \models \mathsf{Faculty}(\mathsf{ind})$:

- each $\mathcal{R}_i$ contains one of $\mathsf{Prof}(\mathsf{sam})$, $\mathsf{Lect}(\mathsf{sam})$, and $\mathsf{Fellow}(\mathsf{sam})$, and each of these assertions allows us to infer $\mathsf{Faculty}(\mathsf{sam})$;
- each $\mathcal{R}_i$ contains either $\mathsf{Prof}(\mathsf{kim})$ or $\mathsf{Lect}(\mathsf{sam})$, and both assertions imply $\mathsf{Faculty}(\mathsf{kim})$ in the presence of $\mathcal{T}_{\mathsf{univ}}$;
- each $\mathcal{R}_i$ contains $\mathsf{Fellow}(\mathsf{alex})$, which allows us to derive $\mathsf{Faculty}(\mathsf{alex})$.

It follows that the individuals sam, kim, and alex are all answers to q_1 under AR semantics:

$$\mathcal{K}_{\mathsf{univ}} \models_{\mathsf{AR}} q_1(\mathsf{sam}) \quad \mathcal{K}_{\mathsf{univ}} \models_{\mathsf{AR}} q_1(\mathsf{kim}) \quad \mathcal{K}_{\mathsf{univ}} \models_{\mathsf{AR}} q_1(\mathsf{alex})$$

Note that these are the only answers to q_1 under AR semantics:

$$\mathcal{K}_{\text{univ}} \not\models_{\text{AR}} q_1(\text{julie}) \quad \mathcal{K}_{\text{univ}} \not\models_{\text{AR}} q_1(\text{csc486}) \quad \mathcal{K}_{\text{univ}} \not\models_{\text{AR}} q_1(\text{csc343})$$

To see why julie does not count as an answer, observe that there exist repairs (like $\mathcal{R}_7$) from which Faculty(julie) cannot be derived. These repairs describe possible states of affairs in which julie may not be a faculty member.

Next consider the query $q_2 = \exists y\, \text{Teaches}(x, y)$. It is not hard to see that for every repair $\mathcal{R}_i$, we have $\langle \mathcal{T}_{\text{univ}}, \mathcal{R}_i \rangle \models q_2(\text{kim})$ and $\langle \mathcal{T}_{\text{univ}}, \mathcal{R}_i \rangle \models q_2(\text{alex})$. Indeed:

- each $\mathcal{R}_i$ contains either Prof(kim) or Lect(sam), both of which allow us to infer $\exists y\, \text{Teaches}(\text{kim}, y)$;
- each $\mathcal{R}_i$ contains Teaches(alex, csc486), which yields $\exists y\, \text{Teaches}(\text{alex}, y)$.

We thus obtain

$$\langle \mathcal{T}_{\text{univ}}, \mathcal{A}_{\text{univ}} \rangle \models_{\text{AR}} q_2(\text{kim}) \quad \langle \mathcal{T}_{\text{univ}}, \mathcal{A}_{\text{univ}} \rangle \models_{\text{AR}} q_2(\text{alex})$$

However, it can be verified that sam is *not* a certain answer to q_2 over the KB $\langle \mathcal{T}_{\text{univ}}, \mathcal{R}_3 \rangle$, and hence not an answer to q_2 under AR semantics. We can similarly show that julie, csc486, and csc343 are not answers to q_2 under AR semantics.

We can show that the AR semantics verifies both consistency properties.

Theorem 2. *AR semantics satisfies the properties* CONSISTENT SUPPORT *and* CONSISTENT RESULTS.

Proof. For CONSISTENT SUPPORT, it suffices to observe that every KB possesses at least one repair, and so if a query result can be obtained from all repairs, then it can be obtained from at least one repair. For CONSISTENT RESULTS, pick some repair $\mathcal{R}$ of $\mathcal{K}$, and let $\mathcal{I}$ be any model of the consistent KB $\langle \mathcal{T}, \mathcal{R} \rangle$. By definition, if $\mathcal{K} \models_{\text{AR}} q(\mathbf{a})$, then $\langle \mathcal{T}, \mathcal{R} \rangle \models q(\mathbf{a})$, and hence $\mathcal{I} \models q(\mathbf{a})$. □

While the AR semantics is very natural, there are at least two reasons to consider other semantics. First, it may be interesting to use an alternative semantics instead of (or conjointly with) the AR semantics in order to restrict the set of answers further, or inversely, to be more liberal and obtain more possible answers. Second, we shall see in the next section that query answering under AR semantics is very often intractable, thus motivating the interest of considering alternative semantics with better computational properties.

3.3 IAR Semantics

The AR semantics corresponds to querying each of the repairs separately, then intersecting the sets of answers. If we instead start by intersecting the repairs, then querying the resulting ABox, we obtain the IAR semantics [5]:

Definition 6 (IAR semantics). *A tuple* $\mathbf{a}$ *is a certain answer of q over $\mathcal{K}$ under the* IAR *(Intersection of ABox Repairs) semantics, written* $\mathcal{K} \models_{\text{IAR}} q(\mathbf{a})$, *just in the case that* $\langle \mathcal{T}, \mathcal{D} \rangle \models q(\mathbf{a})$ *where* $\mathcal{D} = \bigcap_{\mathcal{B} \in Rep(\mathcal{K})} \mathcal{B}$.

Observe that if an assertion does not appear in every repair, it is because the assertion belongs to a minimal inconsistent subset of the ABox. Thus, the IAR semantics computes the query answers that can be obtained from the 'surest' assertions in the ABox, i.e., those that are not involved in any contradiction. This is in contrast to the AR semantics, which allows inferences based upon potentially incorrect facts so long as all the alternatives support the conclusion.

The following theorem, which directly follows from the definitions, formalizes the relationship between AR and IAR semantics:

Theorem 3. *IAR semantics is an under-approximation of AR semantics.*

The difference between the IAR and AR semantics is illustrated by the next example:

Example 8. To determine the answers to q_1, q_2, and q_3 on $\mathcal{K}_{univ}$ under IAR semantics, it suffices to compute the certain answers to these queries over the intersection

$$\mathcal{A}_{Int} \quad = \quad \{\mathsf{Fellow}(\mathsf{alex}), \mathsf{Teaches}(\mathsf{alex}, \mathsf{csc486})\}$$

of the repairs of $\mathcal{K}_{univ}$. For all three queries, we obtain only alex as an answer:

$$\mathcal{K}_{univ} \models_{IAR} q_1(\mathsf{alex}) \quad \mathcal{K}_{univ} \models_{IAR} q_2(\mathsf{alex}) \quad \mathcal{K}_{univ} \models_{IAR} q_3(\mathsf{alex})$$

Observe that sam and kim, which were answers to q_1 under AR semantics, are not considered answers under the stricter IAR semantics, since obtaining these answers requires reasoning by cases (e.g. sam is either a professor, lecturer, or research fellow; kim is either a professor or lecturer). By contrast, to derive $q_1(\mathsf{alex})$, we only need to consider the assertion Fellow(alex), which belongs to every repair and thus can be viewed as 'uncontroversial'. Similarly, for query q_2, kim is an answer under AR semantics, but we cannot find a subset of facts common to all repairs from which we can derive these query answers. For this reason, kim is not returned as an answer under the stricter IAR semantics.

By combining Theorems 1, 2, and 3, we can show that IAR semantics satisfies both consistency properties:

Theorem 4. *IAR semantics satisfies the properties* CONSISTENT SUPPORT *and* CONSISTENT RESULTS.

3.4   Brave Semantics

Instead of considering those tuples that are answers w.r.t. every repair, one may consider those which are answers in at least one repair. This idea is captured by the brave semantics [10]:

Definition 7 (Brave semantics). *A tuple* **a** *is a certain answer of q over* $\mathcal{K} = \langle \mathcal{T}, \mathcal{A} \rangle$ *under the* brave *semantics, written* $\mathcal{K} \models_{brave} q(\mathbf{a})$, *just in the case that* $\langle \mathcal{T}, \mathcal{B} \rangle \models q(\mathbf{a})$ *for some repair* $\mathcal{B} \in Rep(\mathcal{K})$.

The following theorem, which relates the brave and AR semantics, is immediate from the definition:

Theorem 5. *Brave semantics is an over-approximation of AR semantics.*

The following example shows that the brave and AR semantics can differ:

Example 9. We continue our running example, this time using the brave semantics. For query q_1, moving from AR to brave semantics yields one additional answer, namely julie:

$$\mathcal{K}_{\mathsf{univ}} \models_{\mathsf{brave}} q_1(\mathsf{sam}) \qquad \mathcal{K}_{\mathsf{univ}} \models_{\mathsf{brave}} q_1(\mathsf{kim})$$

$$\mathcal{K}_{\mathsf{univ}} \models_{\mathsf{brave}} q_1(\mathsf{alex}) \qquad \mathcal{K}_{\mathsf{univ}} \models_{\mathsf{brave}} q_1(\mathsf{julie})$$

Indeed, while julie is not a certain answer to q_1 in every repair (as required by AR semantics), it is a certain answer to q_1 w.r.t. repairs $\mathcal{R}_1$ - $\mathcal{R}_6$, and thus holds under brave semantics. Each of these repairs describes a possible world in which $q_1(\mathsf{julie})$ holds. Observe that the individuals csc486 and csc343 do not count as answers to q_1 under brave semantics, as they are not certain answers to q_1 in any of the repairs.

For query q_2, we obtain the following four answers under brave semantics:

$$\mathcal{K}_{\mathsf{univ}} \models_{\mathsf{brave}} q_2(\mathsf{sam}) \qquad \mathcal{K}_{\mathsf{univ}} \models_{\mathsf{brave}} q_2(\mathsf{kim})$$

$$\mathcal{K}_{\mathsf{univ}} \models_{\mathsf{brave}} q_2(\mathsf{alex}) \qquad \mathcal{K}_{\mathsf{univ}} \models_{\mathsf{brave}} q_2(\mathsf{csc343})$$

The individual csc343, which was not an answer under AR semantics, is returned as a brave answer because of the assertion Teaches(csc343, julie) which is present in repairs $\mathcal{R}_7$ - $\mathcal{R}_{12}$. The individuals julie and csc486 are not brave answers, as none of the repairs contains evidence that these individuals teach something.

We next show that the brave semantics satisfies one of the consistency properties, but not the other.

Theorem 6. *Brave semantics satisfies* CONSISTENT SUPPORT *but does not satisfy* CONSISTENT RESULTS.

Proof. The property CONSISTENT SUPPORT holds by definition: every query answer holds in some repairs, and repairs are consistent subsets of the original ABox. To see why CONSISTENT RESULTS does not hold, we observe that for our example KB $\mathcal{K}_{\mathsf{univ}}$, we have both $\mathcal{K}_{\mathsf{univ}} \models_{\mathsf{brave}} \mathsf{Prof}(\mathsf{sam})$ and $\mathcal{K}_{\mathsf{univ}} \models_{\mathsf{brave}} \mathsf{Lect}(\mathsf{sam})$. Due to the presence of the negative inclusion $\mathsf{Prof} \sqsubseteq \neg\mathsf{Lect}$ in $\mathcal{T}_{\mathsf{univ}}$, no model of $\mathcal{K}_{\mathsf{univ}}$ can satisfy both $\mathsf{Prof}(\mathsf{sam})$ and $\mathsf{Lect}(\mathsf{sam})$. □

In fact, it is not hard to see that brave semantics is the weakest semantics satisfying CONSISTENT SUPPORT:

Theorem 7. *For every semantics S satisfying* CONSISTENT SUPPORT, *brave semantics is an over-approximation of S.*

In other words, the answers returned by the brave semantics are guaranteed to contain all of the answers that can be obtained with any semantics that satisfies the CONSISTENT SUPPORT property. Thus, the brave answers can very naturally be considered as the set of possible or maybe answers.

3.5 k-Support and k-Defeater Semantics

We have seen in the preceding subsections that the IAR semantics and brave semantics provide natural under- and over-approximations, respectively, of the AR semantics. The families of k-support and k-defeater semantics [10] were introduced to generalize these semantics and obtain more fine-grained approximations.

We begin by considering the family of k-support semantics, which approximate the AR semantics from below:

Definition 8 (k-support semantics). *A tuple* **a** *is a certain answer of q over* $\mathcal{K} = \langle \mathcal{T}, \mathcal{A} \rangle$ *under the k-support semantics, written* $\langle \mathcal{T}, \mathcal{A} \rangle \models_{k\text{-supp}} q(\mathbf{a})$, *if there exist (not necessarily distinct) subsets* $S_1, \ldots, S_k$ *of* $\mathcal{A}$ *that satisfy the following conditions:*

- *each S_i is a $\mathcal{T}$-support for $q(\mathbf{a})$ in $\mathcal{A}$*
- *for every $R \in Rep(\mathcal{K})$, there is some S_i with $S_i \subseteq R$*

The intuition is as follows: if $\langle \mathcal{T}, \mathcal{A} \rangle \models_{\text{AR}} q(\mathbf{a})$, then there must exist a set $\{S_1, \ldots, S_n\}$ of $\mathcal{T}$-supports for $q(\mathbf{a})$ such that every repair contains at least one of the sets S_i. The k-support semantics can thus be seen as a restriction of the AR semantics in which a maximum of k different supports can be used.

The following example illustrates how the k-support semantics changes as we vary the value of k:

Example 10. When $k = 1$, we obtain the same query answers as for IAR semantics (cf. Theorem 8):

$$\mathcal{K}_{\text{univ}} \models_{1\text{-supp}} q_1(\text{alex}) \quad \mathcal{K}_{\text{univ}} \models_{1\text{-supp}} q_2(\text{alex})$$

To see why, observe that $\{\text{Fellow}(\text{alex}), \text{Teaches}(\text{alex}, \text{csc486})\}$ is a support for all both query answers, and it is contained in every repair.

Evaluating q_1 under 2-support semantics yields one additional answer:

$$\mathcal{K}_{\text{univ}} \models_{2\text{-supp}} q_1(\text{kim}) \quad \mathcal{K}_{\text{univ}} \models_{2\text{-supp}} q_1(\text{alex})$$

To show that kim is an answer, we can take the supports $S_1 = \{\text{Prof}(\text{kim})\}$ and $S_2 = \{\text{Lect}(\text{kim})\}$ for $q_1(\text{kim})$ and observe that every repair $\mathcal{R}_i \in Rep(\mathcal{K}_{\text{univ}})$ contains either S_1 or S_2.

When $k = 3$, we obtain the same answers for q_1 as under AR semantics:

$$\mathcal{K}_{\text{univ}} \models_{3\text{-supp}} q_1(\text{sam}) \quad \mathcal{K}_{\text{univ}} \models_{3\text{-supp}} q_1(\text{kim}) \quad \mathcal{K}_{\text{univ}} \models_{3\text{-supp}} q_1(\text{alex})$$

It can be verified that this same set of answers is obtained for every $k \geq 3$.

We next consider the query q_2. When $k = 2$, we additionally obtain kim as an answer:

$$\mathcal{K}_{\text{univ}} \models_{2\text{-supp}} q_2(\text{kim}) \quad \mathcal{K}_{\text{univ}} \models_{2\text{-supp}} q_2(\text{alex})$$

Indeed, we can take the same pair of supports S_1 and S_2 as used for q_1. We obtain exactly the same result for every higher value of k.

The key properties of the k-support semantics are summarized in the following theorem. The first point states that the k-support semantics coincides with the IAR semantics when $k = 1$. The second item shows that for every value of k, the k-support semantics is an under-approximation of the AR semantics. Moreover, there exists a value of k for which the AR semantics and k-support semantics coincide. By the following property, there is a monotone convergence of the k-support semantics to the AR semantics: when we move to a higher value of k, we keep all of the answers obtained for lower k values and possibly add some further answers, until we reach the AR semantics. Note that the value of k needed to converge to the AR semantics will depend on the particular KB and query under consideration. The final item states that, like the IAR semantics, the k-support semantics satisfy both consistency conditions.

Theorem 8. *For every KB $\mathcal{K}$:*

1. *$\mathcal{K} \models_{\mathsf{IAR}} q(\mathbf{a})$ if and only if $\mathcal{K} \models_{\text{1-supp}} q(\mathbf{a})$;*
2. *there exists some $k \geq 1$ such that $\mathcal{K} \models_{\mathsf{AR}} q(\mathbf{a})$ if and only if $\mathcal{K} \models_{k\text{-supp}} q(\mathbf{a})$;*
3. *for every $k \geq 1$, if $\mathcal{K} \models_{k\text{-supp}} q(\mathbf{a})$, then $\mathcal{K} \models_{k+1\text{-supp}} q(\mathbf{a})$;*
4. *for every $k \geq 1$, the k-support semantics satisfies properties CONSISTENT SUPPORT and CONSISTENT RESULTS.*

Proof. The first item follows directly from the definitions of the semantics. For the second point, it suffices to take $k = 2^{|\mathcal{A}|}$. Indeed, there cannot be more than $2^{|\mathcal{A}|}$ different repairs of $\mathcal{A}$ w.r.t. $\mathcal{T}$, and thus, at most $2^{|\mathcal{A}|}$ different supports are needed to cover all repairs. For the third property, we note that the definition allows us to use a support more than once. Thus, if $\mathcal{K} \models_{k\text{-supp}} q(\mathbf{a})$ with supports $S_1, \ldots, S_k$, then we can use the sequence of supports $S_1, \ldots, S_k, S_{k+1}$ with $S_{k+1} = S_k$ to show that $\mathcal{K} \models_{k+1\text{-supp}} q(\mathbf{a})$. The final item follows from Theorems 1 and 2, together with the second item of the theorem. $\square$

We next introduce the family of k-defeater semantics, which approximate the AR semantics from above:

Definition 9 (k-defeater semantics). *A tuple $\mathbf{a}$ is a certain answer of q over $\mathcal{K} = \langle \mathcal{T}, \mathcal{A} \rangle$ under the k-defeater semantics, written $\mathcal{K} \models_{k\text{-def}} q(\mathbf{a})$, if there does not exist a $\mathcal{T}$-consistent subset S of $\mathcal{A}$ with $|S| \leq k$ such that $\langle \mathcal{T}, S \cup C \rangle \models \bot$ for every inclusion-minimal $\mathcal{T}$-support $C \subseteq \mathcal{A}$ of $q(\mathbf{a})$.*

We remark that if an ABox does not contain any $\mathcal{T}$-support for $q(\mathbf{a})$, then $\mathbf{a}$ is not a certain answer under 0-defeater semantics since one can simply take $S = \varnothing$ as the defeating set.

Example 11. For all three queries, the set of answers under 0-defeater semantics is the same as for brave semantics. One can show that for every $k \geq 1$, the answers to these queries under k-defeater semantics is the same as for AR semantics.

To illustrate, let us consider the query q_2. When $k = 0$, we have the following four answers: sam, kim, alex, and csc343. When $k = 1$, we 'lose' the answers sam and csc343:

$$\mathcal{K}_{\mathsf{univ}} \models_{\text{1-def}} q_2(\mathsf{sam}) \qquad \mathcal{K}_{\mathsf{univ}} \models_{\text{1-def}} q_2(\mathsf{csc343})$$

Indeed, $q_2(\mathsf{sam})$ has two minimal supports $\{\mathsf{Prof}(\mathsf{sam})\}$ and $\{\mathsf{Lect}(\mathsf{sam})\}$, both of which are contradicted by the 1-element set $\{\mathsf{Fellow}(\mathsf{sam})\}$. As for csc343, it is easy to see that $\{\mathsf{Teaches}(\mathsf{csc343}, \mathsf{julie})\}$ is the only minimal support for $q_2(\mathsf{csc343})$, and it can be contradicted by the single assertion $\mathsf{Fellow}(\mathsf{julie})$. The two other individuals (kim and alex) are still answers under 1-defeater semantics:

$$\mathcal{K}_{\mathsf{univ}} \models_{\mathsf{1\text{-}def}} q_2(\mathsf{kim}) \quad \mathcal{K}_{\mathsf{univ}} \models_{\mathsf{1\text{-}def}} q_2(\mathsf{alex})$$

This is because there is no single assertion that simultaneously contradicts the two supports ($\{\mathsf{Prof}(\mathsf{kim})\}$ and $\{\mathsf{Lect}(\mathsf{kim})\}$) of $q_2(\mathsf{kim})$, and nor any assertion in conflict with the unique support of $q_2(\mathsf{alex})$ (which is $\{\mathsf{Fellow}(\mathsf{alex}), \mathsf{Teaches}(\mathsf{alex}, \mathsf{csc486})\}$).

The properties of the k-defeater semantics are resumed in the following theorem. The first two statements show that the k-defeater semantics equals the brave semantics when $k = 0$ and converges to the AR semantics in the limit. The third statement shows that the convergence to the AR semantics is anti-monotone in k: as we increase the value of k, the set of answers can only decrease, until we reach the set of AR answers. Note that this means in particular that the k-defeater semantics is an under-approximation of the brave semantics, for every $k \geq 0$. Like the brave semantics, the k-defeater semantics satisfy the first consistency property, but not the second.

Theorem 9

1. $\mathcal{K} \models_{\mathsf{brave}} q(\mathbf{a})$ if and only if $\mathcal{K} \models_{\mathsf{0\text{-}def}} q(\mathbf{a})$;
2. $\mathcal{K} \models_{\mathsf{AR}} q(\mathbf{a})$ if and only if $\mathcal{K} \models_{k\text{-}\mathsf{def}} q(\mathbf{a})$ for every k;
3. for every $k \geq 0$, if $\mathcal{K} \models_{k+1\text{-}\mathsf{def}} q(\mathbf{a})$, then $\mathcal{K} \models_{k\text{-}\mathsf{def}} q(\mathbf{a})$;
4. for every $k \geq 0$, the k-defeater semantics satisfies CONSISTENT SUPPORT but not CONSISTENT RESULTS.

Proof. The first two statements follow easily from the definitions. We prove the third statement by contraposition: if $\mathcal{K} \not\models_{k\text{-}\mathsf{def}} q(\mathbf{a})$, then there exists a set $S \subseteq \mathcal{A}$ with $|S| \leq k$ that contradicts every support for $q(\mathbf{a})$, and this same set S can be used to witness that $\mathcal{K} \not\models_{k+1\text{-}\mathsf{def}} q(\mathbf{a})$. Finally, regarding the consistency properties, to show that the k-defeater semantics satisfies CONSISTENT SUPPORT it suffices to recall that the brave semantics does not satisfy this property (Theorem 6) and to apply Theorem 1. To see why CONSISTENT RESULTS is not satisfied by the k-defeater semantics, take the TBox $\mathcal{T} = \{\exists R_j \sqsubseteq A_j, \exists R_j^- \sqsubseteq \neg B_j \mid j \in \{1, 2\}\}$ and the ABox $\mathcal{A} = \{R_j(a, b_i), B_j(b_i) \mid j \in \{1, 2\}, 1 \leq i \leq k + 1\}$. It can be verified that for both $j \in \{1, 2\}$, we have $\langle \mathcal{T}, \mathcal{A} \rangle \models_{k\text{-}\mathsf{def}} A_j(a)$, as any defeating set for $A_j(a)$ must contain $B_j(b_i)$ for every $1 \leq i \leq k + 1$. However, there can be no model $\mathcal{I}$ of $\mathcal{T}$ that satisfies both $A_1(a)$ and $A_2(a)$, because $\mathcal{T}$ contains the axiom $A_1 \sqsubseteq \neg A_2$. □

3.6 ICR Semantics

Another way of obtaining a finer under-approximation of AR semantics than the IAR semantics is to close the repairs before intersecting them. This idea is formalized in the ICR semantics, proposed in [14]:

Definition 10 (ICR semantics). *A tuple* **a** *is a certain answer of q over* $\mathcal{K} = \langle \mathcal{T}, \mathcal{A} \rangle$ *under the* ICR *(Intersection of Closed Repairs) semantics, written* $\mathcal{K} \models_{\mathsf{ICR}} q(\mathbf{a})$, *just in the case that* $\langle \mathcal{T}, \mathcal{D} \rangle \models q(\mathbf{a})$ *where* $\mathcal{D} = \bigcap_{\mathcal{B} \in \mathit{Rep}(\mathcal{K})} \mathsf{cl}_{\mathcal{T}}(\mathcal{B})$.

Example 12. To compute the answers to q_1, q_2, and q_3 under ICR semantics, we first compute the closure of the repairs of $\mathcal{K}_{\mathsf{univ}}$:

$$\mathsf{cl}_{\mathcal{T}_{\mathsf{univ}}}(\mathcal{R}_1) = \{\mathsf{Prof}(\mathsf{sam}), \mathsf{Prof}(\mathsf{kim}), \mathsf{Fellow}(\mathsf{julie})\} \cup \mathcal{A}'_{\mathsf{Int}} \cup \{\mathsf{Faculty}(\mathsf{julie})\}$$

$$\mathsf{cl}_{\mathcal{T}_{\mathsf{univ}}}(\mathcal{R}_2) = \{\mathsf{Lect}(\mathsf{sam}), \mathsf{Lect}(\mathsf{kim}), \mathsf{Fellow}(\mathsf{julie})\} \cup \mathcal{A}'_{\mathsf{Int}} \cup \{\mathsf{Faculty}(\mathsf{julie})\}$$

$$\mathsf{cl}_{\mathcal{T}_{\mathsf{univ}}}(\mathcal{R}_3) = \{\mathsf{Fellow}(\mathsf{sam}), \mathsf{Prof}(\mathsf{kim}), \mathsf{Fellow}(\mathsf{julie})\} \cup \mathcal{A}'_{\mathsf{Int}} \cup \{\mathsf{Faculty}(\mathsf{julie})\}$$

$$\mathsf{cl}_{\mathcal{T}_{\mathsf{univ}}}(\mathcal{R}_4) = \{\mathsf{Prof}(\mathsf{sam}), \mathsf{Lect}(\mathsf{kim}), \mathsf{Fellow}(\mathsf{julie})\} \cup \mathcal{A}'_{\mathsf{Int}} \cup \{\mathsf{Faculty}(\mathsf{julie})\}$$

$$\mathsf{cl}_{\mathcal{T}_{\mathsf{univ}}}(\mathcal{R}_5) = \{\mathsf{Lect}(\mathsf{sam}), \mathsf{Prof}(\mathsf{kim}), \mathsf{Fellow}(\mathsf{julie})\} \cup \mathcal{A}'_{\mathsf{Int}} \cup \{\mathsf{Faculty}(\mathsf{julie})\}$$

$$\mathsf{cl}_{\mathcal{T}_{\mathsf{univ}}}(\mathcal{R}_6) = \{\mathsf{Fellow}(\mathsf{sam}), \mathsf{Lect}(\mathsf{kim}), \mathsf{Fellow}(\mathsf{julie})\} \cup \mathcal{A}'_{\mathsf{Int}} \cup \{\mathsf{Faculty}(\mathsf{julie})\}$$

$$\mathsf{cl}_{\mathcal{T}_{\mathsf{univ}}}(\mathcal{R}_7) = \{\mathsf{Prof}(\mathsf{sam}), \mathsf{Prof}(\mathsf{kim}), \mathsf{Teaches}(\mathsf{csc343}, \mathsf{julie})\} \cup \mathcal{A}'_{\mathsf{Int}} \cup \{\mathsf{Course}(\mathsf{julie})\}$$

$$\mathsf{cl}_{\mathcal{T}_{\mathsf{univ}}}(\mathcal{R}_8) = \{\mathsf{Lect}(\mathsf{sam}), \mathsf{Lect}(\mathsf{kim}), \mathsf{Teaches}(\mathsf{csc343}, \mathsf{julie})\} \cup \mathcal{A}'_{\mathsf{Int}} \cup \{\mathsf{Course}(\mathsf{julie})\}$$

$$\mathsf{cl}_{\mathcal{T}_{\mathsf{univ}}}(\mathcal{R}_9) = \{\mathsf{Fellow}(\mathsf{sam}), \mathsf{Prof}(\mathsf{kim}), \mathsf{Teaches}(\mathsf{csc343}, \mathsf{julie})\} \cup \mathcal{A}'_{\mathsf{Int}}$$
$$\cup \{\mathsf{Course}(\mathsf{julie})\}$$

$$\mathsf{cl}_{\mathcal{T}_{\mathsf{univ}}}(\mathcal{R}_{10}) = \{\mathsf{Prof}(\mathsf{sam}), \mathsf{Lect}(\mathsf{kim}), \mathsf{Teaches}(\mathsf{csc343}, \mathsf{julie})\} \cup \mathcal{A}'_{\mathsf{Int}} \cup \{\mathsf{Course}(\mathsf{julie})\}$$

$$\mathsf{cl}_{\mathcal{T}_{\mathsf{univ}}}(\mathcal{R}_{11}) = \{\mathsf{Lect}(\mathsf{sam}), \mathsf{Prof}(\mathsf{kim}), \mathsf{Teaches}(\mathsf{csc343}, \mathsf{julie})\} \cup \mathcal{A}'_{\mathsf{Int}} \cup \{\mathsf{Course}(\mathsf{julie})\}$$

$$\mathsf{cl}_{\mathcal{T}_{\mathsf{univ}}}(\mathcal{R}_{12}) = \{\mathsf{Fellow}(\mathsf{sam}), \mathsf{Lect}(\mathsf{kim}), \mathsf{Teaches}(\mathsf{csc343}, \mathsf{julie})\} \cup \mathcal{A}'_{\mathsf{Int}}$$
$$\cup \{\mathsf{Course}(\mathsf{julie})\}$$

where $\mathcal{A}'_{\mathsf{Int}}$ is the following ABox:

$$\mathcal{A}'_{\mathsf{Int}} = \mathcal{A}_{\mathsf{Int}} \cup \{\mathsf{Faculty}(\mathsf{sam}), \mathsf{Faculty}(\mathsf{kim}), \mathsf{Faculty}(\mathsf{alex}), \mathsf{Course}(\mathsf{csc486})\}$$

and $\mathcal{A}_{\mathsf{Int}}$ is defined as before as the intersection of repairs. It can be verified that $\mathcal{A}'_{\mathsf{Int}}$ is the ABox resulting from intersecting the twelve closed repairs.

It can be verified that q_1 has three certain answers over the KB $\langle \mathcal{T}_{\mathsf{univ}}, \mathcal{A}'_{\mathsf{Int}} \rangle$: sam, kim, alex. We thus obtain exactly the same set of answers as was obtained under AR semantics:

$$\mathcal{K}_{\mathsf{univ}} \models_{\mathsf{ICR}} q_1(\mathsf{sam}) \quad \mathcal{K}_{\mathsf{univ}} \models_{\mathsf{ICR}} q_1(\mathsf{kim}) \quad \mathcal{K}_{\mathsf{univ}} \models_{\mathsf{ICR}} q_1(\mathsf{alex})$$

Observe that ICR semantics returns sam and kim as answers, whereas neither individual is an answer under IAR semantics.

For q_2, there is a single certain answer over $\langle \mathcal{T}_{\mathsf{univ}}, \mathcal{A}'_{\mathsf{Int}} \rangle$, namely the individual alex. It follows that we have:

$$\mathcal{K}_{\mathsf{univ}} \models_{\mathsf{ICR}} q_2(\mathsf{alex})$$

and that all other individuals are not answers to q_2 under ICR semantics. Indeed, while every repair allows us to infer $\exists y\, \mathsf{Teaches}(\mathsf{kim}, y)$, the closures of the repairs (hence their intersection) do not contain any assertions of the form $\mathsf{Teaches}(\mathsf{kim}, \_)$.

As mentioned earlier, the ICR semantics is positioned between the AR and IAR semantics. Formally, we have the following relationships:

Theorem 10

1. *ICR semantics is an under-approximation of AR semantics;*
2. *ICR semantics is an over-approximation of IAR semantics;*
3. *ICR semantics is equal to AR semantics for the class of ground CQs (i.e. CQs without existential quantifiers).*

Using Theorem 1, we can show that the ICR semantics satisfies both consistency properties.

Theorem 11. *ICR semantics satisfies the properties* CONSISTENT SUPPORT *and* CONSISTENT RESULTS.

3.7   CAR and ICAR Semantics

Like the ICR semantics, the CAR and ICAR semantics [5] also involve a closure operation, but this time the closure is performed directly on the input ABox (rather than on the repairs).

In order to be able to distinguish between different inconsistent KBs, the authors of [5] introduce a special closure operator $(\mathsf{cl}_\mathcal{T}^*)$ for inconsistent KBs, defined as follows:

$$\mathsf{cl}_\mathcal{T}^*(\mathcal{A}) \;=\; \{\beta \mid \exists S \subseteq \mathcal{A} \text{ such that } S \text{ is } \mathcal{T}\text{-consistent and } \langle \mathcal{T}, S \rangle \models \beta\}$$

We observe that $\mathsf{cl}_\mathcal{T}^*(\mathcal{A})$ corresponds to the set of ABox assertions that are entailed under brave semantics.

An alternative notion of repair is introduced in order to incorporate assertions from $\mathsf{cl}_\mathcal{T}^*(\mathcal{A})$.

Definition 11 (Closed ABox repair). *A subset $\mathcal{R} \subseteq \mathsf{cl}_\mathcal{T}^*(\mathcal{A})$ is a closed ABox repair of $\mathcal{A}$ w.r.t. $\mathcal{T}$ if (i) it is $\mathcal{T}$-consistent, and (ii) there is no $\mathcal{T}$-consistent $\mathcal{R}' \subseteq \mathsf{cl}_\mathcal{T}^*(\mathcal{A})$ such that $\mathcal{R} \cap \mathcal{A} \subsetneq \mathcal{R}' \cap \mathcal{A}$ or $\mathcal{R} \cap \mathcal{A} = \mathcal{R}' \cap \mathcal{A}$ and $\mathcal{R} \subsetneq \mathcal{R}'$. If $\mathcal{K} = \langle \mathcal{T}, \mathcal{A} \rangle$, the set of closed ABox repairs of $\mathcal{A}$ w.r.t. $\mathcal{T}$ is denoted $ClosedRep(\mathcal{K})$.*

It should be noted that every closed ABox repair is a repair of the KB $\langle \mathcal{T}, \mathsf{cl}_\mathcal{T}^*(\mathcal{A}) \rangle$. However, the converse does not hold in general (i.e., some repairs of $\langle \mathcal{T}, \mathsf{cl}_\mathcal{T}^*(\mathcal{A}) \rangle$ are not closed ABox repairs), as the following example demonstrates:

Example 13. We consider another example KB about the university domain:

$$
\begin{aligned}
\mathcal{T} \;&=\; \{\mathsf{Prof} \sqsubseteq \mathsf{Employee}, \mathsf{UnderGrad} \sqsubseteq \mathsf{Student}, \\
&\qquad \mathsf{Student} \sqsubseteq \neg\mathsf{Prof}, \mathsf{UnderGrad} \sqsubseteq \neg\mathsf{Employee}\} \\
\mathcal{A} \;&=\; \{\mathsf{Prof}(\mathsf{sam}), \mathsf{UnderGrad}(\mathsf{sam})\}
\end{aligned}
$$

The TBox states that professors are employees, that undergraduate students are students, and that students and professors and undergraduate students and employees are disjoint classes. Applying the closure operator cl_T^* to $\mathcal{A}$ gives:

$$\mathsf{cl}_T^*(\mathcal{A}) \quad = \quad \mathcal{A} \cup \{\mathsf{Employee(sam)}, \mathsf{Student(sam)}\}$$

and it can be easily verified that $\langle T, \mathsf{cl}_T^*(\mathcal{A})\rangle$ has the following three repairs:

$$
\begin{aligned}
\mathcal{R}_1 &= \{\mathsf{Prof(sam)}, \mathsf{Employee(sam)}\} \\
\mathcal{R}_2 &= \{\mathsf{UnderGrad(sam)}, \mathsf{Student(sam)}\} \\
\mathcal{R}_3 &= \{\mathsf{Employee(sam)}, \mathsf{Student(sam)}\}
\end{aligned}
$$

The repair $\mathcal{R}_3$ is not a closed ABox repair because it has an empty intersection with $\mathcal{A}$, unlike $\mathcal{R}_1$ and $\mathcal{R}_2$.

However, closed ABox repairs can be equivalently defined as the subsets of $\mathsf{cl}_T^*(\mathcal{A})$ that can obtained as follows: (i) take some $\mathcal{R} \in Rep(\langle T, \mathcal{A}\rangle)$, and (ii) add to $\mathcal{R}$ an inclusion-maximal subset C of $\mathsf{cl}_T^*(\mathcal{A}) \setminus \mathcal{A}$ such that $\mathcal{R} \cup C$ is T-consistent. Thus, closed ABox repairs can be seen as maximally 'completing' the standard ABox repairs with additional facts from $\mathsf{cl}_T^*(\mathcal{A}) \setminus \mathcal{A}$.

We can now take any of the semantics from earlier and substitute closed ABox repairs for the standard notion of repair. In particular, if we use closed ABox repairs in conjunction with the AR and IAR semantics, then we obtain the CAR and ICAR semantics introduced in [5]:

Definition 12 (CAR semantics). *A tuple* $\mathbf{a}$ *is a certain answer of* q *over* $\mathcal{K} = \langle T, \mathcal{A}\rangle$ *under the CAR (Closed ABox Repair) semantics, written* $\mathcal{K} \models_{\mathsf{CAR}} q(\mathbf{a})$, *just in the case that* $\langle T, \mathcal{R}\rangle \models q(\mathbf{a})$ *for every* $\mathcal{R} \in ClosedRep(\mathcal{K})$.

Definition 13 (ICAR semantics). *A tuple* $\mathbf{a}$ *is a certain answer of* q *over* $\mathcal{K} = \langle T, \mathcal{A}\rangle$ *under the ICAR (Intersection of Closed ABox Repairs) semantics, written* $\mathcal{K} \models_{\mathsf{ICAR}} q(\mathbf{a})$, *just in the case that* $\langle T, \mathcal{D}\rangle \models q(\mathbf{a})$ *where* $\mathcal{D}$ *is the intersection of the closed ABox repairs of* $\mathcal{A}$ *w.r.t.* T.

We return to our running example.

Example 14. Applying the special closure operator to our example KB yields:

$$
\begin{aligned}
\mathsf{cl}_{T_{\mathsf{univ}}}^*(\mathcal{A}_{\mathsf{univ}}) \quad = \quad &\mathcal{A}_{\mathsf{univ}} \cup \{\mathsf{Faculty(sam)}, \mathsf{Faculty(kim)}, \mathsf{Faculty(alex)}, \\
&\mathsf{Course(csc486)}, \mathsf{Faculty(julie)}, \mathsf{Course(julie)}\}
\end{aligned}
$$

We next compute the closed ABox repairs of our example KB $\langle T_{\mathsf{univ}}, \mathcal{A}_{\mathsf{univ}}\rangle$. As mentioned earlier, these can be obtained by taking each of the repairs $\mathcal{R}_1 - \mathcal{R}_{12}$ (see Example 6) and adding as many facts from $\mathsf{cl}_{T_{\mathsf{univ}}}^*(\mathcal{A}_{\mathsf{univ}}) \setminus \mathcal{A}$ as possible while preserving consistency. For our example KB, it turns out that this gives us precisely the same result as closing each of the original repairs of the KB:

$$ClosedRep(\langle T_{\mathsf{univ}}, \mathcal{A}_{\mathsf{univ}}\rangle) = \{\mathsf{cl}_{T_{\mathsf{univ}}}(\mathcal{R}_i) \mid \mathcal{R}_i \in Rep(\langle T_{\mathsf{univ}}, \mathcal{A}_{\mathsf{univ}}\rangle)\}$$

It follows that on this KB, the CAR semantics gives the same answers as the AR semantics, and the ICAR semantics coincides with the ICR semantics.

In general, however, the CAR and ICAR semantics can yield different results. To demonstrate this, let us slightly modify our example KB by adding $\exists \mathsf{Teaches} \sqsubseteq \mathsf{Faculty}$ to the TBox. Call the modified TBox $\mathcal{T}'_{\mathsf{univ}}$ and modified KB $\mathcal{K}'_{\mathsf{univ}}$. The special closure operator gives almost the same result as before:

$$\mathsf{cl}^*_{\mathcal{T}'_{\mathsf{univ}}}(\mathcal{A}_{\mathsf{univ}}) = \mathcal{A}_{\mathsf{univ}} \cup \{\mathsf{Faculty}(\mathsf{sam}), \mathsf{Faculty}(\mathsf{kim}), \mathsf{Faculty}(\mathsf{alex}), \mathsf{Course}(\mathsf{csc486}),$$
$$\mathsf{Faculty}(\mathsf{julie}), \mathsf{Course}(\mathsf{julie}), \mathsf{Faculty}(\mathsf{csc343})\}$$

except that we now also have the fact $\mathsf{Faculty}(\mathsf{csc343})$, which is derived from $\mathsf{Teaches}(\mathsf{csc343}, \mathsf{julie})$ using the new inclusion. Since $\mathsf{Faculty}(\mathsf{csc343})$ is not involved in any contradictions, it can be added to each repair while preserving consistency, and so it belongs to every closed ABox repair. It follows that csc343 is an answer to q_1 under ICAR and CAR semantics. Observe that csc343 is not an answer under AR semantics since the assertion $\mathsf{Teaches}(\mathsf{csc343}, \mathsf{julie})$ which is required to infer $\mathsf{Faculty}(\mathsf{csc343})$ is not present in all (standard) repairs.

The following theorem, which easily follows from the definitions, resumes the key relationships holding with respect to the CAR and ICAR semantics:

Theorem 12

1. *The CAR semantics is an over-approximation of the AR semantics.*
2. *The ICAR semantics is an over-approximation of the ICR semantics.*
3. *The ICAR semantics is an under-approximation of the CAR semantics.*
4. *The CAR semantics and ICAR semantics coincide for ground CQs.*

The next theorem shows that the CAR and ICAR semantics satisfy the second consistency property, but not the first.

Theorem 13. *CAR and ICAR semantics satisfy* CONSISTENT RESULTS *but do not satisfy* CONSISTENT SUPPORT.

Proof. A counter-example for CONSISTENT SUPPORT is given in the next example. To show that CONSISTENT RESULTS holds for CAR, we can use essentially the same proof as for Theorem 2: pick a closed ABox repair $\mathcal{R}$ of $\mathcal{K}$, and let $\mathcal{I}$ be any model of the consistent KB $\langle \mathcal{T}, \mathcal{R} \rangle$. Then by the definition of the CAR semantics, if $\mathcal{K} \models_{\mathsf{CAR}} q(\mathbf{a})$, then $\langle \mathcal{T}, \mathcal{R} \rangle \models q(\mathbf{a})$, and hence $\mathcal{I} \models q(\mathbf{a})$. For the ICAR semantics, we can use Theorem 1 and the fact that ICAR is a sound approximation of CAR.   □

Example 15. To see why the CAR and ICAR semantics do not satisfy CONSISTENT SUPPORT, we consider another example KB set in the university domain:

$$\mathcal{T} = \{\mathsf{Prof} \sqsubseteq \mathsf{Employee}, \mathsf{Student} \sqsubseteq \mathsf{GetDiscount}, \mathsf{Student} \sqsubseteq \neg\mathsf{Prof}\}$$
$$\mathcal{A} = \{\mathsf{Prof}(\mathsf{sam}), \mathsf{Student}(\mathsf{sam})\}$$

The TBox states that professors are employees, that students get discounted access to some university services, and that students and professors are disjoint classes. The ABox states that sam is both a student and a professor. This KB is inconsistent and has two repairs:

$$\mathcal{B}_1 = \{\mathsf{Prof}(\mathsf{sam})\} \qquad \mathcal{B}_2 = \{\mathsf{Student}(\mathsf{sam})\}$$

We compute the result of applying the special closure operator on this KB:

$$\mathsf{cl}_{\mathcal{T}}^*(\mathcal{A}) \quad = \quad \mathcal{A} \cup \{\mathsf{Employee}(\mathsf{sam}), \mathsf{GetDiscount}(\mathsf{sam})\}$$

As both new assertions in the closure can be added to the repairs without introducing any contradiction, we obtain two closed ABox repairs, namely:

$$\begin{aligned}
\mathcal{R}_1 \quad &= \quad \{\mathsf{Prof}(\mathsf{sam}), \mathsf{Employee}(\mathsf{sam}), \mathsf{GetDiscount}(\mathsf{sam})\} \\
\mathcal{R}_2 \quad &= \quad \{\mathsf{Student}(\mathsf{sam}), \mathsf{Employee}(\mathsf{sam}), \mathsf{GetDiscount}(\mathsf{sam})\}
\end{aligned}$$

Now consider the query $q(x) = \mathsf{Employee}(x) \wedge \mathsf{GetDiscount}(x)$, which asks for employees that get the discount. If we use either of the CAR and ICAR semantics, then we return sam as an answer:

$$\langle \mathcal{T}, \mathcal{A} \rangle \models_{\mathsf{CAR}} q(\mathsf{sam}) \qquad \langle \mathcal{T}, \mathcal{A} \rangle \models_{\mathsf{ICAR}} q(\mathsf{sam})$$

since $\mathsf{Employee}(\mathsf{sam})$ and $\mathsf{GetDiscount}(\mathsf{sam})$ appear in both closed ABox repairs. Observe however that there is no consistent $\mathcal{T}$-support for $q(\mathsf{sam})$ in $\mathcal{A}$. Indeed, we needed $\mathsf{Prof}(\mathsf{sam})$ to infer that sam is an employee and $\mathsf{Student}(\mathsf{sam})$ to infer that sam gets the discount, but according to the information in the TBox, these two assertions cannot both hold.

3.8 k-Lazy Semantics

Another parameterized family of inconsistency-tolerant semantics, the k-lazy semantics, was proposed in [15]. The semantics are based upon an alternative notion of a repair, called a k-lazy repair, which is obtained from the ABox by removing for each *cluster* of contradictory assertions (defined next), either a minimal subset of at most k assertions that restores the consistency of the cluster, or the whole cluster if there is no such subset. Because of the possible removal of some entire clusters, lazy repairs need not be standard repairs.

Definition 14 (Clusters, k-lazy repairs). *Given a KB $\mathcal{K} = \langle \mathcal{T}, \mathcal{A} \rangle$, we let $\equiv_{\perp}^{\mathcal{K}}$ be the smallest equivalence relation over the assertions in $\mathcal{A}$ satisfying the following condition: if assertions α, β appear together in a minimal $\mathcal{T}$-inconsistent subset, then $\alpha \equiv_{\perp}^{\mathcal{K}} \beta$. The clusters of $\mathcal{K}$ are the equivalence classes of $\equiv_{\perp}^{\mathcal{K}}$.*

A k-lazy repair of $\mathcal{K}$ is obtained by removing from $\mathcal{A}$ for each cluster $\mathcal{C}_i$ of $\mathcal{K}$: either (i) an inclusion-minimal subset $\mathcal{C}_i' \subseteq \mathcal{C}_i$ such that $|\mathcal{C}_i'| \leq k$ and $\mathcal{C}_i \setminus \mathcal{C}_i'$ is $\mathcal{T}$-consistent, or (ii) the whole cluster $\mathcal{C}_i$ if there does not exist a set $\mathcal{C}_i' \subseteq \mathcal{C}_i$ satisfying these conditions.

The k-lazy semantics is then defined like the AR semantics, except that we use k-lazy repairs instead of the usual notion of repair.

Definition 15 (k-lazy semantics). *A tuple* **a** *is an answer for a query q over a KB $\mathcal{K} = \langle \mathcal{T}, \mathcal{A} \rangle$ under k-lazy semantics, written $\langle \mathcal{T}, \mathcal{A} \rangle \models_{k\text{-lazy}} q(\mathbf{a})$, if and only if $\langle \mathcal{T}, \mathcal{R}_k \rangle \models q(\mathbf{a})$ for every k-lazy repair $\mathcal{R}_k$ of $\mathcal{K}$.*

We illustrate the k-lazy semantics with an example:

Example 16. Returning to our running example, we first compute the clusters of $\mathcal{K}_{\text{univ}}$ (see Example 2 for the minimal $\mathcal{T}_{\text{univ}}$-inconsistent subsets of $\mathcal{A}_{\text{univ}}$):

- $\{\text{Prof}(\text{sam}), \text{Lect}(\text{sam}), \text{Fellow}(\text{sam})\}$
- $\{\text{Prof}(\text{kim}), \text{Lect}(\text{kim})\}$
- $\{\text{Fellow}(\text{julie}), \text{Teaches}(\text{csc343}, \text{julie})\}$

There is only one 0-lazy repair, obtained by removing all of the clusters. Observe that removing all clusters is the same as taking the intersection of the standard repairs, so we obtain the same results as with the IAR semantics. In the case of query q_1, we get a single answer: $\langle \mathcal{T}_{\text{univ}}, \mathcal{A}_{\text{univ}} \rangle \models_{0\text{-lazy}} q_1(\text{alex})$.

Restoring the consistency of the cluster $\{\text{Prof}(\text{kim}), \text{Lect}(\text{kim})\}$ is done by removing one of its assertions. Thus, every k-lazy repair of $\langle \mathcal{T}_{\text{univ}}, \mathcal{A}_{\text{univ}} \rangle$ for $k \geq 1$ contains either $\text{Prof}(\text{kim})$ or $\text{Lect}(\text{kim})$, and thus, $\langle \mathcal{T}_{\text{univ}}, \mathcal{A}_{\text{univ}} \rangle \models_{k\text{-lazy}} q_1(\text{kim})$.

To restore the consistency of the cluster $\{\text{Prof}(\text{sam}), \text{Lect}(\text{sam}), \text{Fellow}(\text{sam})\}$, it is sufficient and necessary to remove two assertions. The 2-lazy repairs all contain some cause for $q_1(\text{sam})$. It follows that $\langle \mathcal{T}_{\text{univ}}, \mathcal{A}_{\text{univ}} \rangle \models_{k\text{-lazy}} q_1(\text{sam})$ for $k \geq 2$, but $\langle \mathcal{T}_{\text{univ}}, \mathcal{A}_{\text{univ}} \rangle \not\models_{1\text{-lazy}} q_1(\text{sam})$ since the 1-lazy repairs are obtained by removing the whole cluster $\{\text{Prof}(\text{sam}), \text{Lect}(\text{sam}), \text{Fellow}(\text{sam})\}$.

The following theorem relates the k-lazy semantics to other inconsistency-tolerant semantics. In particular, it shows that 0-lazy semantics coincides with the IAR semantics and that k-lazy semantics is an under-approximation of brave semantics, for every value of k:

Theorem 14

1. $\mathcal{K} \models_{\text{IAR}} q(\mathbf{a})$ *if and only if* $\mathcal{K} \models_{0\text{-lazy}} q(\mathbf{a})$;
2. *for every $k \geq 0$, if $\mathcal{K} \models_{k\text{-lazy}} q(\mathbf{a})$, then $\mathcal{K} \models_{\text{brave}} q(\mathbf{a})$;*
3. *for every KB $\mathcal{K}$, there exists some $k \geq 0$ such that for every $k' \geq k$:*
   $\mathcal{K} \models_{\text{AR}} q(\mathbf{a})$ *if and only if* $\mathcal{K} \models_{k'\text{-lazy}} q(\mathbf{a})$.

The last item of the theorem shows that the k-lazy semantics converge to the AR semantics. However, these semantics are not sound approximations of the AR semantics, and unlike the k-support semantics, the convergence is not monotone in k. The following example illustrates these points:

Example 17. To show that $\mathcal{K} \models_{k\text{-lazy}} q$ does not implies that $\mathcal{K} \models_{k+1\text{-lazy}} q$, we consider the following KB:

$$\mathcal{T} = \{\exists\text{Teaches} \sqsubseteq \text{Faculty}, \text{Student} \sqsubseteq \neg\text{Faculty}\}$$
$$\mathcal{A} = \{\text{Teaches}(\text{sam}, \text{csc343}), \text{Teaches}(\text{sam}, \text{csc236}), \text{Student}(\text{sam})\}$$

All assertions belong to the same cluster. There is only one 1-lazy repair

$$\{\mathsf{Teaches}(\mathsf{sam}, \mathsf{csc343}), \mathsf{Teaches}(\mathsf{sam}, \mathsf{csc236})\}$$

so $\mathcal{K} \models_{\text{1-lazy}} q_1(\mathsf{sam})$. However, for $k \geq 2$, there are two k-lazy repairs:

$$\{\mathsf{Teaches}(\mathsf{sam}, \mathsf{csc343}), \mathsf{Teaches}(\mathsf{sam}, \mathsf{csc236})\} \qquad \{\mathsf{Student}(\mathsf{sam})\}$$

We thus have $\mathcal{K} \not\models_{k\text{-lazy}} q_1(\mathsf{sam})$ for every $k \geq 2$. Note that under AR semantics, sam is not an answer to q_1, and so the answers obtained with 1-lazy semantics are not included in the answers obtained with AR semantics.

The k-lazy semantics do however verify both consistency properties.

Theorem 15. *The k-lazy semantics satisfy* CONSISTENT SUPPORT *and* CONSISTENT RESULTS.

Proof. For CONSISTENT SUPPORT, it suffices to notice that if $q(\mathbf{a})$ holds under k-lazy semantics, then there is a k-lazy repair $\mathcal{R}$ for which $\langle \mathcal{T}, \mathcal{R} \rangle \models q(\mathbf{a})$. Since k-lazy repairs are $\mathcal{T}$-consistent by definition, it follows that every query answer has a consistent support.

To show CONSISTENT RESULTS, we can proceed in the same manner as for the AR and CAR semantics: pick any a k-lazy repair $\mathcal{R}$ of $\mathcal{K}$, and let $\mathcal{I}$ be any model of the consistent KB $\langle \mathcal{T}, \mathcal{R} \rangle$. It follows from the definition of k-lazy semantics that if $\mathcal{K} \models_{k\text{-lazy}} q(\mathbf{a})$, then $\langle \mathcal{T}, \mathcal{R} \rangle \models q(\mathbf{a})$, and hence $\mathcal{I} \models q(\mathbf{a})$. □

3.9 Preferred Repair Semantics

The notion of repair in Definition 4 integrates a very simple preference relation, namely set inclusion. When additional information on the reliability of ABox assertions is available, it is natural to use this information to identify preferred repairs, and to use these repairs as the basis for inconsistency-tolerant query answering. This idea has been explored in [16], where a preorder is used to identify the most preferred repairs:

Definition 16. *Let* $\mathcal{K} = \langle \mathcal{T}, \mathcal{A} \rangle$ *be a KB, and let* $\preceq$ *be a preorder over subsets of* $\mathcal{A}$. *A subset* $\mathcal{A}' \subseteq \mathcal{A}$ *is a* $\preceq$-*repair of* $\mathcal{K}$ *if it satisfies the following two conditions:*

- $\mathcal{A}'$ *is* $\mathcal{T}$-*consistent*
- *there is no* $\mathcal{T}$-*consistent* $\mathcal{A}'' \subseteq \mathcal{A}$ *such that* $\mathcal{A}'' \preceq \mathcal{A}'$ *and* $\mathcal{A}' \not\preceq \mathcal{A}''$

The set of $\preceq$-*repairs of* $\mathcal{K}$ *is denoted* $Rep_{\preceq}(\mathcal{K})$.

With this new notion of $\preceq$-repair, we can revisit the different inconsistency-tolerant semantics. Following [16,17], we will consider variants of the AR, IAR, and brave semantics based upon preferred repairs.

Definition 17. *A tuple* **a** *is an answer for a query* q *over* $\mathcal{K} = \langle \mathcal{T}, \mathcal{A} \rangle$

- *under the* $\preceq$-*AR semantics if* $\langle \mathcal{T}, \mathcal{B} \rangle \models q(\mathbf{a})$ *for every* $\mathcal{B} \in Rep_{\preceq}(\mathcal{K})$;
- *under the* $\preceq$-*IAR semantics if* $\langle \mathcal{T}, \mathcal{B}_{\cap} \rangle \models q(\mathbf{a})$ *where* $\mathcal{B}_{\cap} = \bigcap_{\mathcal{B} \in Rep_{\preceq}(\mathcal{K})} \mathcal{B}$;
- *under the* $\preceq$-*brave semantics if* $\langle \mathcal{T}, \mathcal{B} \rangle \models q(\mathbf{a})$ *for some* $\mathcal{B} \in Rep_{\preceq}(\mathcal{K})$.

The preceding definitions can be instantiated with different preference relations $\preceq$. We consider the four types of preference relations studied in [16], which correspond to well-known methods of defining preferences over subsets, cf. [26].

Cardinality ($\leq$). A first possibility is to compare subsets using set cardinality:

$$\mathcal{A}_1 \leq \mathcal{A}_2 \quad \text{iff} \quad |\mathcal{A}_1| \leq |\mathcal{A}_2|$$

The resulting notion of $\leq$-repair is appropriate when all assertions are believed to have the same (small) likelihood of being erroneous, in which case repairs with the largest number of assertions are most likely to be correct.

Example 18. Reconsider the TBox $\mathcal{T}'_{\text{univ}}$ (which adds $\exists \mathsf{Teaches} \sqsubseteq \mathsf{Faculty}$ to $\mathcal{T}_{\text{univ}}$), and consider the ABox:

$$\mathcal{A} = \{\mathsf{Fellow}(\mathsf{julie}), \mathsf{Prof}(\mathsf{julie}), \mathsf{Teaches}(\mathsf{julie}, \mathsf{csc486}), \mathsf{Teaches}(\mathsf{julie}, \mathsf{csc236}),$$
$$\mathsf{Teaches}(\mathsf{csc343}, \mathsf{julie}), \mathsf{Course}(\mathsf{julie})\}$$

The repairs of $\langle \mathcal{T}'_{\text{univ}}, \mathcal{A} \rangle$ are:

$$\mathcal{R}_1 = \{\mathsf{Fellow}(\mathsf{julie}), \mathsf{Teaches}(\mathsf{julie}, \mathsf{csc486}), \mathsf{Teaches}(\mathsf{julie}, \mathsf{csc236})\}$$
$$\mathcal{R}_2 = \{\mathsf{Prof}(\mathsf{julie}), \mathsf{Teaches}(\mathsf{julie}, \mathsf{csc486}), \mathsf{Teaches}(\mathsf{julie}, \mathsf{csc236})\}$$
$$\mathcal{R}_3 = \{\mathsf{Teaches}(\mathsf{csc343}, \mathsf{julie}), \mathsf{Course}(\mathsf{julie})\}$$

Only $\mathcal{R}_1$ and $\mathcal{R}_2$ are $\leq$-repairs. It follows that $\langle \mathcal{T}'_{\text{univ}}, \mathcal{A} \rangle \models_{\leq\text{-IAR}} q_2(\mathsf{julie})$, while $\langle \mathcal{T}'_{\text{univ}}, \mathcal{A} \rangle \not\models_{\subseteq\text{-AR}} q_2(\mathsf{julie})$.

Priority Levels ($\subseteq_P, \leq_P$). The next preference relations assume that the ABox assertions can be partitioned into priority levels $\mathcal{P}_1, \ldots, \mathcal{P}_n$ based on their perceived reliability, with assertions in $\mathcal{P}_1$ considered most reliable, and those in $\mathcal{P}_n$ least reliable. Such a prioritization can be used to separate a part of the dataset that has already been validated from more recent additions. Alternatively, one might assign assertions to priority levels based upon the concept or role names they use (when some predicates are known to be more reliable), or the data sources from which they originate (in information integration applications). Given such a prioritization $P = \langle \mathcal{P}_1, \ldots, \mathcal{P}_n \rangle$ of $\mathcal{A}$, we can refine the $\subseteq$ and $\leq$ preorders by considering each priority level in turn:

- *Prioritized set inclusion:* $\mathcal{A}_1 \subseteq_P \mathcal{A}_2$ iff $\mathcal{A}_1 \cap \mathcal{P}_i = \mathcal{A}_2 \cap \mathcal{P}_i$ for every $1 \leq i \leq n$, or there is some $1 \leq i \leq n$ such that $\mathcal{A}_1 \cap \mathcal{P}_i \subsetneq \mathcal{A}_2 \cap \mathcal{P}_i$ and for all $1 \leq j < i$, $\mathcal{A}_1 \cap \mathcal{P}_j = \mathcal{A}_2 \cap \mathcal{P}_j$.
- *Prioritized cardinality:* $\mathcal{A}_1 \leq_P \mathcal{A}_2$ iff $|\mathcal{A}_1 \cap \mathcal{P}_i| = |\mathcal{A}_2 \cap \mathcal{P}_i|$ for every $1 \leq i \leq n$, or there is some $1 \leq i \leq n$ such that $|\mathcal{A}_1 \cap \mathcal{P}_i| < |\mathcal{A}_2 \cap \mathcal{P}_i|$ and for all $1 \leq j < i$, $|\mathcal{A}_1 \cap \mathcal{P}_j| = |\mathcal{A}_2 \cap \mathcal{P}_j|$.

Notice that a single assertion on level $\mathcal{P}_i$ is preferred to any number of assertions from $\mathcal{P}_{i+1}$, so these preorders are best suited for cases in which there is a significant difference in the perceived reliability of adjacent priority levels.

Example 19. We use the same KB has in Example 18 with the following prioritization:

$$\begin{aligned} \mathcal{P}_1 &= \{\mathsf{Prof}(\mathsf{julie}), \mathsf{Teaches}(\mathsf{julie}, \mathsf{csc486}), \mathsf{Course}(\mathsf{julie})\} \\ \mathcal{P}_2 &= \{\mathsf{Fellow}(\mathsf{julie})\} \\ \mathcal{P}_3 &= \{\mathsf{Teaches}(\mathsf{csc343}, \mathsf{julie}), \mathsf{Teaches}(\mathsf{julie}, \mathsf{csc236})\} \end{aligned}$$

The $\subseteq_P$-repairs are $\mathcal{R}_2$ and $\mathcal{R}_3$ ($\mathcal{R}_1$ is not a $\subseteq_P$-repair because $\mathcal{R}_1 \cap \mathcal{P}_1 \subsetneq \mathcal{R}_2 \cap \mathcal{P}_1$). Note that $\langle \mathcal{T}'_{\mathsf{univ}}, \mathcal{A} \rangle \not\models_{\subseteq_P\text{-brave}} \mathsf{Fellow}(\mathsf{julie})$.

There is only one $\leq_P$-repair, namely $\mathcal{R}_2$. It follows that julie is an answer to q_1, q_2 and q_3 under $\leq_P$-IAR semantics.

Weights ($\leq_w$). The reliability of different assertions can also be modelled quantitatively by using a function $w : \mathcal{A} \to \mathbb{N}$ to assign weights to the ABox assertions. Such a weight function w induces a preorder $\leq_w$ over subsets of $\mathcal{A}$ in the expected way:

$$\mathcal{A}_1 \leq_w \mathcal{A}_2 \quad \text{iff} \quad \sum_{\alpha \in \mathcal{A}_1} w(\alpha) \leq \sum_{\alpha \in \mathcal{A}_2} w(\alpha)$$

If the ABox is populated using information extraction techniques, the weights may be derived from the confidence levels output by the extraction tool. Weight-based preorders can also be used in place of the $\leq_P$ preorder to allow for compensation between the priority levels.

Example 20. If we assign a weight of 1 to the assertions of $\mathcal{P}_3$, 2 to the assertions of $\mathcal{P}_2$ and 3 to the assertions of $\mathcal{P}_1$ in the ABox of Example 19, the weights of the repairs are 6 for $\mathcal{R}_1$, 7 for $\mathcal{R}_2$, and 4 for $\mathcal{R}_3$. Only $\mathcal{R}_2$ is a $\leq_w$-repair.

3.10   Summary and Discussion

Figure 1 summarizes the relationships holding between the different semantics considered in this section. An arrow from S to S' indicates that S is a sound approximation of S' (or equivalently, that S' is an over-approximation of S). All relationships that hold can be derived from those indicated in the figure, so if there is no directed path from S to S', this means that S is *not* an under-approximation of S'. We also indicate by means of the three shaded regions which consistency properties hold for the different semantics.

A recent work [27] has proposed a general framework for specifying different inconsistency-tolerant semantics in terms of pairs composed of a *modifier*, which creates a set of ABoxes from the original ABox, and an *inference strategy*, which states how query results are computed from these ABoxes. The authors consider (combinations of) three basic modifiers: *positive closure* (C), which adds to each ABox all facts that can be derived by applying the positive inclusions of

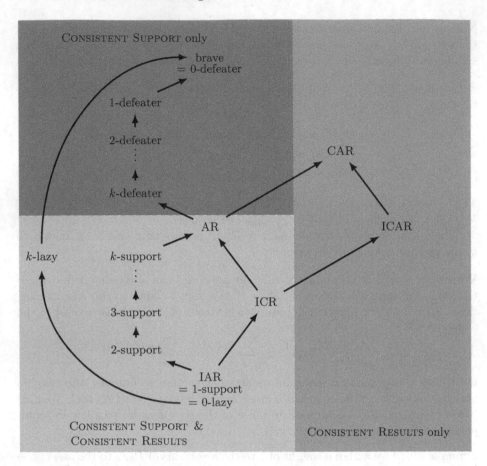

Fig. 1. Relationships between inconsistency-tolerant semantics.

the TBox, *splitting into repairs* (*R*) of each ABox, and *selecting the cardinality-maximal ABoxes* (*M*). Four inference strategies are considered: *universal* (∀), which requires a query answer to hold w.r.t. every ABox in the set, *existential* (∃), which requires an answer to hold w.r.t. some ABox, *safe* (∩), which considers the query answers that can be derived in the intersection of the ABoxes, *majority-based* (*maj*), which considers answers holding w.r.t. the majority of the ABoxes We can thus express the AR semantics as the pair (R, ∀): we first split into the different repairs, and then apply the universal inference strategy. The IAR, brave, and ICR semantics can be expressed respectively by the pairs (R, ∩), (R, ∃), and (CR, ∩). For ICAR, CAR and *k*-lazy semantics, it is necessary to introduce further modifiers (e.g. adding the special closure operator).

4   Complexity of Inconsistency-Tolerant Query Answering

In this section, we examine the computational properties of the inconsistency-tolerant semantics presented in the preceding subsection.

Our focus will be on description logics of the DL-Lite family [4], as they are a popular choice for OMQA and their computational properties have been more exhaustively studied than for other ontology languages. However, we will conclude the section with a brief overview of what is known about inconsistency-tolerant query answering with DLs outside the DL-Lite family.

As for the query language, we consider both CQs, which are the standard query language for OMQA, and the more restricted IQs. We note that all upper bounds obtained for IQs hold also for ground CQs.

4.1   Complexity Landscape for DL-Lite

In this subsection, we explore the complexity of inconsistency-tolerant query answering when the TBox is expressed in DL-Lite. The results we present (displayed in Fig. 2) hold for common DL-Lite dialects like DL-Lite$_{core}$, DL-Lite$_{\mathcal{R}}$, and DL-Lite$_{\mathcal{A}}$, and more generally, for every DL-Lite dialect that possesses the following properties:

- every minimal support for $q(\mathbf{a})$ contains at most $|q|$ assertions;
- every minimal $\mathcal{T}$-inconsistent subset has cardinality at most two;
- CQ answering, IQ answering, and KB consistency can be performed by first-order query rewriting [4], and are thus in AC^0 in data complexity;
- CQ answering is NP-complete for combined complexity [4];
- IQ answering is NL-complete in combined complexity [28].

In what follows, we use the generic term 'DL-Lite' to mean any dialect with the preceding properties. We point out that the Horn versions of DL-Lite [28], which allow for conjunction ($\sqcap$), do not satisfy these conditions; interested readers can consult [10] for complexity results concerning those dialects.

We begin with the following lemma that gives the complexity of identifying assertions that occur in every repair and determining whether a subset of the ABox is a repair or equal to the intersection of all repairs.

Lemma 1. *In DL-Lite, it can be decided in NL in combined complexity whether an assertion $\alpha \in \mathcal{A}$ belongs to every repair of $\mathcal{A}$ w.r.t. $\mathcal{T}$, and in P whether a subset $\mathcal{A}' \subseteq \mathcal{A}$ is (i) a repair of $\mathcal{A}$ w.r.t. $\mathcal{T}$, or (ii) is equal to the intersection of all repairs of $\mathcal{A}$ w.r.t. $\mathcal{T}$.*

Proof. For the first statement, we give an NL algorithm for deciding whether α does *not* belong to every repair of $\mathcal{A}$ w.r.t. $\mathcal{T}$, and then use the fact that NL=coNL to obtain an NL algorithm for the complement. To show that α is not present in every repair, it suffices to guess an assertion $\beta \in \mathcal{A}$ and show that $\{\alpha, \beta\}$ is $\mathcal{T}$-inconsistent. To see why, note that an assertion is omitted from a repair only if it introduction leads to a contradiction, and for DL-Lite KBs, it

Semantics	Data complexity		Combined complexity	
	CQs	IQs	CQs	IQs
classical	in AC^0	in AC^0	NP	NL
AR	coNP	coNP	Π_2^p	coNP
IAR	in AC^0	in AC^0	NP	NL
brave	in AC^0	in AC^0	NP	NL
k-support $(k \geq 1)$	in AC^0	in AC^0	NP	NL
k-defeater $(k \geq 0)$	in AC^0	in AC^0	NP	NL
ICR	coNP	coNP	$\Delta_2^p[O(\log n)]$	coNP
CAR	coNP	in AC^0	Π_2^p	NL
ICAR	in AC^0	in AC^0	NP	NL
k-lazy $(k \geq 1)$	coNP	in P	Π_2^p	in P

Fig. 2. Complexity of inconsistency-tolerant query answering in DL-Lite. All results are completeness results unless otherwise indicated.

is well known that every minimal $\mathcal{T}$-inconsistent subset of an ABox contains at most two assertions. As consistency checking can be tested in NL, we obtain the required NL procedure.

To decide whether $\mathcal{A}' \subseteq \mathcal{A}$ is a repair of $\mathcal{A}$ w.r.t. $\mathcal{T}$, it suffices to verify that $\mathcal{A}'$ is $\mathcal{T}$-consistent and that for every $\alpha \in \mathcal{A}\backslash\mathcal{A}'$, the ABox $\mathcal{A}' \cup \{\alpha\}$ is $\mathcal{T}$-inconsistent. As consistency checking can be performed in NL (hence P) in combined complexity, we obtain a P procedure for repair checking. To decide whether $\mathcal{A}'$ is equal to the intersection of all repairs, we use the NL procedure described in the previous paragraph to verify that every assertion in $\mathcal{A}'$ belongs to every repair and that every assertion in $\mathcal{A} \setminus \mathcal{A}'$ does not belong to every repair. □

We now pass in review the different semantics from the preceding section, starting with the AR semantics, which we show to be intractable even for instance queries:

Theorem 16 [5]. *Under AR semantics, CQ answering over DL-Lite KBs is coNP-complete in data complexity. The same holds for IQs.*

Proof. For the upper bound, we consider the following non-deterministic procedure for showing that $\mathcal{K} \not\models_{\mathsf{AR}} q(\mathbf{a})$: guess a subset $\mathcal{R} \subseteq \mathcal{A}$ and check that it is a repair of $\mathcal{K}$ and such that $\langle \mathcal{T}, \mathcal{A} \rangle \not\models q(\mathbf{a})$. As repair checking and CQ answering (under classical semantics) are both in P for data complexity for DL-Lite KBs, the described procedure runs in non-deterministic polynomial time in $|\mathcal{A}|$.

The lower bound can be shown by a simple reduction from propositional unsatisfiability (UNSAT) [14]. Let $\varphi = c_1 \wedge \ldots \wedge c_m$ be a propositional formula over variables $v_1, \ldots, v_k$, where each c_i is a propositional clause. We use the following DL-Lite knowledge base, where the ABox encodes φ and the TBox does not depend on φ:

$$\mathcal{T} = \{ \exists P^- \sqsubseteq \neg \exists N^-, \ \exists P \sqsubseteq \neg \exists U^-, \exists N \sqsubseteq \neg \exists U^-, \exists U \sqsubseteq A \}$$
$$\mathcal{A} = \{ U(a, c_i) \mid 1 \le i \le m \} \cup \{ P(c_i, v_j) \mid v_j \in c_i \} \cup \{ N(c_i, v_j) \mid \neg v_j \in c_i \}$$

It is not hard to verify that φ is unsatisfiable if and only if $\langle \mathcal{T}, \mathcal{A} \rangle \models_{\mathsf{AR}} A(a)$. Indeed, because of the axiom $\exists P^- \sqsubseteq \neg \exists N^-$, each repair corresponds to a valuation of the variables in which v_j assigned true if it has an incoming P-edge in the repair. If a clause c_i is not satisfied by the valuation encoded by the repair, then the individual c_i will have no outgoing P- or N-edges, and hence the repair will contain the assertion $U(a, c_i)$. Thus, if φ is unsatisfiable, every repair will contain an assertion $U(a, c_i)$, and so $A(a)$ will be entailed under AR semantics. Conversely, if $\langle \mathcal{T}, \mathcal{A} \rangle \models_{\mathsf{AR}} A(a)$, then the assertions $U(a, c_i)$ that are present in the repairs witness that the corresponding valuations each contain at least one unsatisfied clause. □

We observe that in the preceding coNP-hardness proof, we needed to use DL-Lite axioms involving existential concepts. However, if we consider CQs, then we need only a single disjointness statement to obtain intractability.

Theorem 17 [14]. *For any ontology language that allows for class disjointness ($A \sqsubseteq \neg B$), CQ answering under AR semantics is coNP-hard in data complexity.*

Proof. This time we use a restricted version of UNSAT, called 2+2UNSAT, proven coNP-hard in [29], in which each clause has 2 positive and 2 negative literals, and any of the four positions in a clause can be filled instead by one of the truth constants `true` and `false`. Consider an instance $\varphi = c_1 \wedge \ldots \wedge c_m$ of 2+2-UNSAT over $v_1, \ldots, v_k$, `true`, and `false`. We let $\mathcal{T} = \{ T \sqsubseteq \neg F \}$, and encode φ in the ABox as follows:

$$\{ P_1(c_i, u), P_2(c_i, x), N_1(c_i, y), N_2(c_i, z) \mid c_i = u \vee x \vee \neg y \vee \neg z, 1 \le i \le m \}$$
$$\cup \ \{ T(v_j), F(v_j) \mid 1 \le j \le k \} \cup \{ T(\mathtt{true}), F(\mathtt{false}) \}$$

It can be shown that φ is unsatisfiable just in the case that the following Boolean query is entailed from $\langle \mathcal{T}, \mathcal{A} \rangle$ under AR semantics:

$$\exists x, y_1, \ldots, y_4 \ P_1(x, y_1) \wedge F(y_1) \wedge P_2(x, y_2) \wedge F(y_2)$$
$$\wedge N_1(x, y_3) \wedge T(y_3) \wedge N_2(x, y_4) \wedge T(y_4)$$

Intuitively, the axiom $T \sqsubseteq \neg F$ selects a truth value for each variable, so the repairs of $\mathcal{A}$ correspond exactly to the set of valuations. Importantly, there is only one way to avoid satisfying a 2+2-clause: the first two variables must be assigned false and the last two variables must be assigned true. Thus, q holds in a repair iff there is an unsatisfied clause in the corresponding valuation. □

For CQ answering, switching from data to combined complexity takes us one level higher in the polynomial hierarchy, while for IQs, we have the same complexity for both measures:

Theorem 18 [10]. *Under AR semantics, CQ answering over DL-Lite KBs is Π_2^p-complete in combined complexity; for IQ answering, the combined complexity drops to coNP-complete.*

Proof. For the membership results, we can use the same non-deterministic procedure as in Theorem 16 for testing whether a tuple **a** is *not* an answer: guess a subset $\mathcal{R}$ of $\mathcal{A}$ and check that it is a repair and satisfies $\langle \mathcal{T}, \mathcal{R} \rangle \not\models q(\mathbf{a})$. Since the latter checks can be performed in coNP (resp. in P if q is an IQ) in combined complexity, we obtain a Σ_2^p (resp. NP) procedure for deciding $\mathcal{K} \not\models_{\mathsf{AR}} q(\mathbf{a})$ when q is a CQ (resp. IQ). It follows that the original problem is in Π_2^p for CQs, and in coNP for IQs.

The lower bound for instance queries follows from Theorem 16. For CQs, Π_2^p-hardness is shown in [10] by the following reduction from validity of $\mathrm{QBF}_{2,\forall}$ formulas. Let $\varphi = \forall x_1, ..., x_n \exists y_1, ..., y_m \bigwedge_{j=1}^{k} c_j$ where $\bigwedge_{j=1}^{k} c_j$ is a 3CNF formula over the variables $x_1, ..., x_n, y_1, ..., y_m$, where every c_j is a clause of the form $\ell_j^1 \vee \ell_j^2 \vee \ell_j^3$. The variable of literal ℓ_j^p is denoted by $v(\ell_j^p)$. For example, for the clause $\neg x_1 \vee y_2 \vee \neg y_1$, we have $v(\neg x_1) = x_1$, $v(y_2) = y_2$, and $v(\neg y_1) = y_1$. To decide whether φ is valid, we construct in polynomial time the following KB and Boolean query:

$$\mathcal{T} = \{\exists GX_i \sqsubseteq GX_i^- \mid 1 \leq i \leq n\}$$

$$\mathcal{A} = \bigcup_{j=1}^{k} \{L_j^1(c_j^V, V(v(\ell_j^1))), L_j^2(c_j^V, V(v(\ell_j^2))), L_j^3(c_j^V, V(v(\ell_j^3))) \mid$$
$$V \text{ is a valuation of } v(\ell_j^1), v(\ell_j^2), v(\ell_j^3) \text{ satisfying } C_j\}$$
$$\cup \{GX_i(0,1), GX_i(1,0) \mid 1 \leq i \leq n\}$$

$$q = \bigwedge_{j=1}^{k} \bigwedge_{h=1}^{3} L_j^h(w_j, v(\ell_j^h)) \wedge \bigwedge_{i=1}^{n} GX_i(x_i, z_i)$$

where all variables in q are existentially quantified. It is shown in [10] that φ is valid iff $\langle \mathcal{T}, \mathcal{A} \rangle \models_{\mathsf{AR}} q$. Intuitively, the assertion $GX_i(0,1)$ (resp. $GX_i(1,0)$) means that $x_i = 0$ (resp. $x_i = 1$), and the repairs of $\langle \mathcal{T}, \mathcal{A} \rangle$ are in one-to-one correspondence with the valuations of $x_1, ..., x_n$ since every repair must contain exactly one of assertions $GX_i(0,1)$ and $GX_i(1,0)$. The query q looks for a valuation of $x_1, ..., x_n, y_1, ..., y_m$ that satisfies $\bigwedge_{j=1}^{k} c_j$. More precisely, the variables $w_1, ..., w_k$ correspond to partial valuations that satisfy $c_1, ..., c_k$ respectively and will be mapped to some individuals $c_1^{V_1}, ..., c_k^{V_k}$, and each $v(\ell_j^h)$ is a variable x_i or y_i that will be mapped to either 0 or 1. If q is entailed under AR semantics, then for every valuation μ of $x_1, ..., x_n$, we can define a satisfying valuation μ' for $x_1, ..., x_n, y_1, ..., y_m$ by examining the way that q is mapped into the repair corresponding to μ. Conversely, if the QBF is valid, then every valuation μ for $x_1, ..., x_n$ can be extended to a valuation μ' of $x_1, ..., x_n, y_1, ..., y_m$ that satisfies φ, so in every repair of $\mathcal{K}$, there will be a way of mapping the query. □

The intractability of querying under AR semantics motivated the introduction of other semantics with more favourable computational properties. In particular, it led to the proposal of the IAR semantics, for which the complexity of query answering matches that of classical semantics:

Theorem 19 [30,32]. *Under IAR semantics, CQ answering over DL-Lite KBs is in AC^0 in data complexity and NP-complete in combined complexity, and IQ answering is NL-complete in combined complexity.*

Proof. The AC^0 membership result is proven by means of a query rewriting procedure. As shown in [30,32], it is possible, for every DL-Lite TBox $\mathcal{T}$ and CQ q, to construct an FO-query q' such that for every ABox $\mathcal{A}$ and candidate answer $\mathbf{a}$, we have the following:

$$\langle \mathcal{T}, \mathcal{A} \rangle \models_{\mathsf{IAR}} q(\mathbf{a}) \qquad \Longleftrightarrow \qquad \mathcal{I}_\mathcal{A} \models q'(\mathbf{a})$$

As the rewriting is independent of the ABox, the preceding equivalence shows that the data complexity of CQ answering under IAR semantics is the same as for FO-query evaluation, namely, in AC^0. It is beyond the scope of this chapter to detail the rewriting algorithm, but we give the intuition in Example 21.

The NP upper bound in combined complexity can be obtained as follows: guess a subset of $\mathcal{A}' \subseteq \mathcal{A}$ and a polynomial-size proof that $\langle \mathcal{T}, \mathcal{A}' \rangle \models q(\mathbf{a})$ and check that every assertion α in $\mathcal{A}'$ belongs to the intersection of repairs. By Lemma 1, we know that the latter check can be performed in polynomial time. If q is a instance query, then $\langle \mathcal{T}, \mathcal{A} \rangle \models_{\mathsf{IAR}} q(\mathbf{a})$ iff there is an assertion $\alpha \in \mathcal{A}$ such that (i) $\langle \mathcal{T}, \{\alpha\} \rangle \models q(\mathbf{a})$, and (ii) α appears in every repair in $Rep\langle \mathcal{T}, \mathcal{A} \rangle$. It thus suffices to iterate (in non-logarithmic space) over the (binary encodings of) ABox assertions and check (in NL) whether the latter conditions are verified. □

The following example illustrates the use of query rewriting to compute the answers under IAR semantics:

Example 21. The general idea is to add to the classical rewriting expressions that ensure that the assertions used to derive the query are not contradicted by other assertions in the ABox. For instance, we have seen in Example 4 that

$$q_2'(x) = \mathsf{Prof}(x) \vee \mathsf{Lect}(x) \vee \exists y.\mathsf{Teaches}(x, y)$$

is a rewriting of $q_2(x) = \exists y\, \mathsf{Teaches}(x, y)$ w.r.t. $\mathcal{T}_{\mathsf{univ}}$. To transform q_2' into a rewriting of q_2 for the IAR semantics, we add to every atom in the rewriting a formula that ensures that there is no assertion in the ABox that will conflict with the (instantiation of) the atom. This yields the following query:

$$
\begin{aligned}
q_2''(x) \;=\; & \mathsf{Prof}(x) \wedge (\neg \mathsf{Lect}(x) \wedge \neg \mathsf{Fellow}(x) \wedge \neg \mathsf{Course}(x) \wedge \neg \exists z.\, \mathsf{Teaches}(z, x)) \\
& \vee\, \mathsf{Lect}(x) \wedge (\neg \mathsf{Prof}(x) \wedge \neg \mathsf{Fellow}(x) \wedge \neg \mathsf{Course}(x) \wedge \neg \exists z.\, \mathsf{Teaches}(z, x)) \\
& \vee\, \exists y.\, (\mathsf{Teaches}(x, y) \wedge (\neg \mathsf{Prof}(y) \wedge \neg \mathsf{Lect}(y) \wedge \neg \mathsf{Fellow}(y)))
\end{aligned}
$$

Observe, for example, that in the first line, we have added negations of the four types of atoms that can contradict the first disjunct $\mathsf{Prof}(x)$. This disjunct in the first line states that an individual a is an answer to $q_2(x)$ under the IAR semantics if the assertion $\mathsf{Prof}(a)$ is present in the ABox *and* the ABox does not contain $\mathsf{Lect}(a)$, $\mathsf{Fellow}(a)$, $\mathsf{Course}(a)$, or any assertion of the form $\mathsf{Teaches}(\_, a)$.

Like the IAR semantics, we can show that querying DL-Lite KBs under brave semantics can be carried out by means of query rewriting and has the same data and combined complexity as classical semantics.

Theorem 20 [10]. *Under brave semantics, CQ answering over DL-Lite KBs is in AC^0 in data complexity and NP-complete in combined complexity, and IQ answering is NL-complete in combined complexity.*

We omit the proof (which is a special case of Theorems 21 and 22) and instead give an example which illustrates how query rewriting works for brave semantics:

Example 22. As with the IAR semantics, we start with a standard rewriting of the query for classical semantics, represented as a union of CQs. We then add to each disjunct the necessary constraints to force that the set of ABox assertions witnessing the satisfaction of the disjunct be consistent with the TBox. For example, consider again the modified TBox $\mathcal{T}'_{univ}$ (which adds the inclusion $\exists$Teaches $\sqsubseteq$ Faculty to $\mathcal{T}_{univ}$) and the query $q_2 = \exists y$ Teaches(x, y). It can be verified that the query $q'_2(x) = \text{Prof}(x) \vee \text{Lect}(x) \vee \exists y.\text{Teaches}(x, y)$ is a rewriting of q_2 w.r.t. $\mathcal{T}'_{univ}$. We observe that if we find an assertion Prof(a) or Lect(a) in the ABox, then a is indeed an answer to q_2 under brave semantics, as the single-element set $\{\text{Prof}(a)\}$ and $\{\text{Lect}(a)\}$ are both $\mathcal{T}'_{univ}$-consistent. Suppose however that we match the third disjunct using the assertion Teaches(a, a). This does *not* prove that a is a brave answer to q_2, as $\{\text{Teaches}(a, a)\}$ is $\mathcal{T}'_{univ}$-inconsistent (as this would imply that a belongs to both Faculty and Course). To prevent such false matches, we must add an extra condition to the final disjunct to prevent x and y from being mapped to the same individual. One can show that the resulting query

$$q'''_2(x) \quad = \quad \text{Prof}(x) \vee \text{Lect}(x) \vee \exists y. (\text{Teaches}(x, y) \wedge x \neq y)$$

is a rewriting of q_2 w.r.t. $\mathcal{T}'_{univ}$ under the brave semantics.

The k-support and k-defeater semantics were introduced to obtain finer approximations of the AR semantics, while keeping the same desirable computational properties as the IAR and brave semantics. Indeed, as the following theorems show, for every value of k, query answering under the k-support and k-defeater semantics has the same complexity as for classical semantics and can be performed using query rewriting.

Theorem 21 [10]. *In DL-Lite, CQ answering under the k-support semantics is in AC^0 w.r.t. data complexity, for every $k \geq 1$. The same holds for the k-defeater semantics, for every $k \geq 0$.*

Proof (idea). We consider first the k-support semantics. In [10], the authors show how to construct, for every $k \geq 1$, every DL-Lite[1] $\mathcal{T}$, and for every CQ q, an FO-query q' with the following properties:

---

[1] In fact, the rewritability results in [10] are proven for all ontology languages for which CQ answering and unsatisfiability testing can be performed via $UCQ_{\neq}$-rewriting. As DL-Lite satisfies these conditions, the results apply to DL-Lite KBs.

$$\langle \mathcal{T}, \mathcal{A} \rangle \models_{k\text{-supp}} q(\mathbf{a}) \qquad \Longleftrightarrow \qquad \mathcal{I}_{\mathcal{A}} \models q'(\mathbf{a})$$

Membership in AC^0 follows immediately. The rewriting q' takes the form of a big disjunction $\psi_1 \vee \ldots \vee \psi_n$, where the disjuncts ψ_i correspond to the different possible choices of k supports for q of cardinality at most $|q|$, and each ψ_i asserts that the chosen supports are present in $\mathcal{A}$ and that there is no $\mathcal{T}$-consistent subset of $\mathcal{A}$ of cardinality at most k which conflicts with each of the supports. Since the rewriting is ABox-independent, the disjuncts do not mention the individuals involved in the supports, but just the types of atoms involved, and the (in)equalities holding between arguments of different atoms.

An analogous result is shown in [10] for the k-defeater semantics. In this case, the rewriting takes the form $\neg(\kappa_1 \vee \ldots \vee \kappa_n)$, where every κ_i asserts the existence of a $\mathcal{T}$-consistent set of facts of cardinality at most k which conflicts with every minimal $\mathcal{T}$-support for q (the negation ensures that no such set is present in the ABox). Here we again utilize the fact that for DL-Lite KBs, the size of minimal $\mathcal{T}$-supports for q is bounded by $|q|$, and hence there are only finitely many types of supports to consider. $\qquad \square$

Theorem 22 [10]. *In DL-Lite, CQ (resp. IQ) answering under the k-support semantics is in NP-complete (resp. NL-complete) w.r.t. combined complexity, for every $k \geq 1$. The same holds for the k-defeater semantics, for every $k \geq 0$.*

Proof. The lower bounds are inherited from the analogous results for classical semantics, so we only present proofs for the upper bounds, closely following the treatment in [10].

We begin by defining an NP procedure for deciding whether $\langle \mathcal{T}, \mathcal{A} \rangle \models_{k\text{-supp}} q(\mathbf{a})$ when q is a CQ. In the first step, we guess subsets $S_1, \ldots, S_k$ of $\mathcal{A}$, as well as polysize certificates that $\langle \mathcal{T}, S_i \rangle \models q(\mathbf{a})$, for each S_i. We next verify (in P) that each S_i is $\mathcal{T}$-consistent and that each certificate is valid. Finally, we check that every repair contains some S_i by considering all subsets $U \subseteq \mathcal{A}$ with $|U| \leq k$ and testing whether U is (i) $\mathcal{T}$-consistent and (ii) $\langle \mathcal{T}, U \cup S_i \rangle \models \bot$ for every $1 \leq i \leq k$. If no such set is found, the procedure outputs yes, and otherwise it outputs no. Since we take k to be a fixed constant, the described procedure runs in non-deterministic polynomial time. Regarding correctness, recall that since $\mathcal{T}$ is a DL-Lite TBox, every minimal $\mathcal{T}$-inconsistent subset of $\mathcal{A}$ contains at most 2 assertions. Hence, if there is some repair $\mathcal{R}$ that does not contain any S_i, then we can find a subset $U \subseteq \mathcal{R}$ of cardinality at most k which contradicts every S_i.

Next we define an NL procedure for IQ answering under k-support semantics. The procedure begins by guessing k assertions $\alpha_1, \ldots, \alpha_k$ from $\mathcal{A}$ (intuitively: these are the k supports). Observe that by using a binary encoding, we need only logarithmic space to store this guess. We verify next that $\{\alpha_i\}$ is $\mathcal{T}$-consistent and $\langle \mathcal{T}, \{\alpha_i\} \rangle \models q(\mathbf{a})$ for every $1 \leq i \leq k$, outputting no if one of these conditions fails for some α_i. Both checks can be done in NL, since satisfiability and instance checking in DL-Lite are NL-complete w.r.t. combined complexity. Finally we need to make sure that every repair $R \in Rep(\mathcal{K})$ contains one of the α_i. As every minimal $\mathcal{T}$-inconsistent subset contains at most two assertions, it follows that there exists some repair which *does not* contain any α_i iff there is a subset

$U \subseteq \mathcal{A}$ of cardinality at most k which is $\mathcal{T}$-consistent and such that $\langle \mathcal{T}, U \cup \{\alpha_i\} \rangle \models \bot$ for every α_i. We can therefore decide in NL whether such a repair exists by guessing such a subset U and checking whether it satisfies the required conditions. Using the fact that NL=coNL, we obtain a NL procedure for testing whether there is no such subset. If the check succeeds, we return yes, else no. It is easy to see that $\langle \mathcal{T}, \mathcal{A} \rangle \models_{k\text{-supp}} q(\mathbf{a})$ iff some execution of the described procedure returns yes.

We next consider CQ answering under k-defeater semantics. Let $S_1, \ldots, S_m$ be an enumeration of the $\mathcal{T}$-consistent subsets of $\mathcal{A}$ having cardinality at most k; observe that m is polynomial in $|\mathcal{A}|$ since k is fixed. We know that $\langle \mathcal{T}, \mathcal{A} \rangle \models_{k\text{-def}} q(\mathbf{a})$ iff for every S_i, there is some $\mathcal{T}$-support C of $q(\mathbf{a})$ such that $S_i \cup C$ is $\mathcal{T}$-consistent. The first step in the procedure is thus to guess a sequence $C_1, \ldots, C_m$ of subsets of $\mathcal{A}$, one for each S_i, together with polysize certificates that $\langle \mathcal{T}, C_i \rangle \models q(\mathbf{a})$ for each C_i. In a second step, the procedure checks in polytime that for every $1 \leq i \leq m$, the certificate is valid and $S_i \cup C_i$ is $\mathcal{T}$-consistent. If all checks succeed, the procedure outputs yes, and otherwise it outputs no. It is easily verified that $\langle \mathcal{T}, \mathcal{A} \rangle \models_{k\text{-def}} q(\mathbf{a})$ iff there is an execution of this procedure that returns yes.

To complete the proof, we describe an NL procedure for checking whether $\langle \mathcal{T}, \mathcal{A} \rangle \not\models_{k\text{-def}} q(\mathbf{a})$ when q is an IQ (we can then use NL=coNL to obtain an NL algorithm for the complement). The first step is to guess a (binary encoding of a) subset $S \subseteq \mathcal{A}$ of cardinality at most k. We then test in NL whether S is $\mathcal{T}$-consistent, and return no if not. Next we need to verify that $S \cup \{\alpha\}$ is $\mathcal{T}$-inconsistent for every minimal $\mathcal{T}$-support $\{\alpha\}$ of q, returning yes if this is the case, and no otherwise. This can be done by defining a NL procedure for the complementary problem, which works by guessing $\alpha \in \mathcal{A}$ and verifying that $\{\alpha\}$ is a $\mathcal{T}$-support of q and that $S \cup \{\alpha\}$ is $\mathcal{T}$-consistent.   □

We next consider the ICR semantics. While query answering remains coNP-complete in data complexity, we observe that the combined complexity for CQ answering is lower than for AR semantics ($\Delta_2^p[O(\log n)]$ rather than Π_2^p). Intuitively, this difference can be explained by the fact that ICR semantics is amenable to preprocessing. Indeed, if we compute and store in a query-independent offline phase the intersection of the closure of the repairs, then at query time we can use classical querying algorithms on the resulting ABox to obtain the answers under the ICR semantics.

Theorem 23. *Under the ICR semantics, CQ answering over DL-Lite KBs is coNP-complete in data complexity, and $\Delta_2^p[O(\log n)]$-complete in combined complexity; IQ answering is coNP-complete for both data and combined complexity.*

Proof. As the ICR and AR semantics coincide for IQs (Theorem 10), we can reuse the proof of Theorem 16 to show that IQ answering is coNP-complete for both data and combined complexity. To decide if a CQ is entailed under ICR semantics, we first use a coNP-oracle to decide, for every assertion that can be constructed using the vocabulary and individuals present in the KB, whether the assertion is entailed under AR semantics. Then we use a NP-oracle to decide if the query holds w.r.t. the ABox consisting of all assertions that are entailed under AR semantics. This yields a Δ_2^p upper bound. We further observe that the oracle calls can be structured as a tree, and thus we can apply results in [40] to obtain membership in $\Delta_2^p[O(log\ n)]$.

A matching $\Delta_2^p[O(log\ n)]$ lower bound can be proven by a rather involved reduction from the Parity(3SAT) problem (cf. [41,42]), which builds upon a similar reduction from [10]. A detailed proof is given in Chap. 2 of [17].   □

It turns out that if we instead adopt the ICAR semantics, then we can once again employ query rewriting and match the complexity of classical semantics.

Theorem 24 [5,30]. *Under ICAR semantics, CQ answering over DL-Lite KBs is in* AC^0 *in data complexity, and NP-complete in combined complexity; IQ answering is NL-complete in combined complexity.*

Proof. The lower bounds are again inherited from existing results for classical semantics. To show membership in AC^0 for data complexity, the authors of [30] show how to define, for every TBox $\mathcal{T}$ and every CQ q, an FO-query q' with the following properties:

$$\langle \mathcal{T}, \mathcal{A} \rangle \models_{\mathsf{ICAR}} q(\mathbf{a}) \qquad \Longleftrightarrow \qquad \mathcal{I}_\mathcal{A} \models q'(\mathbf{a})$$

The idea is similar to the rewriting for IAR semantics, except that in addition to adding constraints to each atom in the standard rewriting, we must also rewrite each atom in order to take into account the different assertions that could cause the atom to appear in $\mathsf{cl}_\mathcal{T}^*(\mathcal{A})$. We refer the reader to [30] for details.

Analogously to Lemma 1, one can show that deciding whether an assertion appears in every closed ABox repair can be done in NL in combined complexity. It follows that IQ answering under ICAR semantics is in NL: simply check that the assertion in question belongs to every closed ABox repair. For CQs, we can first compute the set of assertions appearing in every closed ABox repair (this can be done in P), and then perform standard CQ answering on the resulting ABox (whose combined complexity is in NP).   □

For the CAR semantics, the results are mixed: for IQs, the complexity is the same as for classical semantics, while for CQs, we have the same (high) complexity as with AR semantics.

Theorem 25 [5,30]. *Under CAR semantics, CQ answering over DL-Lite KBs is coNP-complete in data complexity and Π_2^p-complete in combined complexity; IQ answering is in* AC^0 *in data complexity and NL-complete in combined complexity.*

Proof. As the CAR semantics coincides with the ICAR semantics on ground CQs (Theorem 12), the results for IQs follow immediately from the analogous results for the ICAR semantics (Theorem 24).

For the coNP membership result, we observe that to show $\langle \mathcal{T}, \mathcal{A} \rangle \not\models_{\mathsf{CAR}} q(\mathbf{a})$, it suffices to guess a set $\mathcal{R}$ of ABox assertions built using the vocabulary and individuals from $\mathcal{K}$, and then check (in P in data complexity) that $\mathcal{R}$ is a closed ABox repair and $\langle \mathcal{T}, \mathcal{R} \rangle \not\models q(\mathbf{a})$. For the coNP lower bound, we can reuse the reduction from the proof of Theorem 17. Indeed, since the TBox from that reduction contains no positive inclusions, the CAR and AR semantics give the same answers. The combined complexity of CQ answering is not considered in [5,30]), but membership in Π_2^p can be shown using the same approach as for data complexity: guess a set of assertions, and verify that the guessed set is a closed ABox repair that does not yield the query answer. The latter verifications involve a polynomial number of NL consistency checks, as well as a coNP check that the considered tuple is not an answer under classical semantics. For the lower bound, we can reuse the reduction used to show Π_2^p-hardness of CQ answering under AR semantics (Theorem 18), as the TBox only contains negative inclusions. □

A tractability result for IQ answering[2] under k-lazy semantics is presented in [15] for ontologies expressed in linear Datalog$^{+/-}$, which generalizes DL-Lite.

Theorem 26. *For every $k \geq 0$, IQ answering over DL-Lite KBs under k-lazy semantics is in P for data and combined complexity.*

Proof. Following the approach from [15], we first compute the clusters $\mathcal{C}_1, \ldots, \mathcal{C}_m$ of the KB $\mathcal{K}$, and then for each cluster $\mathcal{C}_i$, we compute all inclusion-minimal subsets $\mathcal{C}_i' \subseteq \mathcal{C}_i$ of size at most k for which $\mathcal{C}_i \setminus \mathcal{C}_i'$ is $\mathcal{T}$-consistent. Let $\mathcal{C}_i^1, \ldots, \mathcal{C}_i^{\ell_i}$ be the subsets computed for cluster $\mathcal{C}_i$, and if no such subset exists, set $\mathcal{C}_i^1 = \mathcal{C}_i$. We then observe that because we are in DL-Lite (and so every IQ answer can inferred from a single fact), there exists a k-lazy repair $\mathcal{R}$ such that $\langle \mathcal{T}, \mathcal{R} \rangle \not\models q(\mathbf{a})$ if and only if:

- $\langle \mathcal{T}, \mathcal{A} \setminus \bigcup_i \mathcal{C}_i \rangle \not\models q(\mathbf{a})$, and
- for every cluster $\mathcal{C}_i$, there exists some $\mathcal{C}_i^j$ such that $\langle \mathcal{T}, \mathcal{C}_i \setminus \mathcal{C}_i^j \rangle \not\models q(\mathbf{a})$.

We can thus examine each cluster $\mathcal{C}_i$ in turn and see whether such a $\mathcal{C}_i^j$ exists for every cluster $\mathcal{C}_i$. To complete the argument, we note that the described procedure runs in polynomial time in combined complexity, when k is a fixed constant. □

However, for conjunctive queries, there is no improvement in complexity compared to AR semantics:

Theorem 27. *For every $k \geq 1$, CQ answering over DL-Lite KBs under k-lazy semantics is coNP-complete in data complexity and Π_2^p-complete in combined complexity.*

Proof. The upper bounds employ the usual strategy: to show $\langle \mathcal{T}, \mathcal{A} \rangle \not\models_{k\text{-lazy}} q(\mathbf{a})$, we guess a subset $\mathcal{R} \subseteq \mathcal{A}$ and verify that $\mathcal{R}$ is a k-lazy repair and $\langle \mathcal{T}, \mathcal{R} \rangle \not\models q(\mathbf{a})$. We note that for DL-Lite KBs, the latter checks can be performed in P in data complexity and in NP in combined complexity.

---

[2] The formulation of the results in [15] suggests that CQ answering is also in P for data complexity, but as Theorem 27 shows, this is not the case.

As noted in [10], the reduction from [14] used to show coNP-hardness of CQ answering under AR semantics (presented in the proof of Theorem 17) can be used without any modification to prove coNP-hardness of CQ answering under k-lazy semantics, for every $k \geq 1$. Indeed, the clusters of the KB used in that reduction take the form $\{T(v_j), F(v_j)\}$, and since a single assertion needs to be removed from each cluster to restore consistency, the k-lazy repairs are the same as the usual repairs when $k \geq 1$.

For the Π_2^p-hardness result, we can similarly reuse the proof of the corresponding result for AR semantics. Indeed, the KB used in the reduction from [10] (see proof of Theorem 18) has clusters of the form $\{GX_i(0,1), GX_i(1,0)\}$, and thus for this KB, the k-lazy and AR semantics coincide starting from $k = 1$.   □

We close the subsection with the following theorem which shows how the data complexity of the AR, brave, and IAR semantics changes depending on the notion of preferred repair. Interestingly, the complexity depends mainly on the choice of preference relation, rather than on the base semantics. In all cases, adding preferences leads to intractability, but in the case of prioritized set inclusion ($\subseteq_P$), the coNP data complexity is no worse than for (plain) AR semantics.

Theorem 28 [16,17]. *For $S \in \{AR, IAR, brave\}$, the data complexity of CQ answering over DL-Lite KBs under S semantics with preferred repairs is:*

- $\Delta_2^p[O(log\, n)]$-*complete if we use cardinality ($\leq$)*
- coNP-*complete (resp. NP-complete) if we use prioritized set inclusion ($\subseteq_P$) and $S \in \{AR, IAR\}$ (resp. $S = brave$*
- Δ_2^p-*complete if we use prioritized cardinality ($\leq_P$)*
- Δ_2^p-*complete if we use weights ($\leq_w$)*

For the last two cases ($\leq_P$ and $\leq_w$), the complexity drops to $\Delta_2^p[O(log\, n)]$-complete under the assumption that there is a bound on the number of priority classes (resp. maximal weight).

4.2   Beyond DL-Lite

In this subsection, we provide a brief glimpse at the complexity landscape for DLs outside the DL-Lite family. We will consider two representative DLs: $\mathcal{EL}_\perp$, which extends the lightweight $\mathcal{EL}$ description logic with the ability to express disjointness using $\perp$, and $\mathcal{ALC}$, the prototypical expressive description logic. We recall that concepts in $\mathcal{EL}_\perp$ are built from concept names and role names using the constructors $\top$, $\perp$, $\sqcap$, and $\exists R.C$. In $\mathcal{ALC}$, we have all of the preceding constructors as well as $\sqcup$, $\neg$, and $\forall R.C$.

In Fig. 3, we display the complexity of querying $\mathcal{EL}_\perp$ KBs under the AR, IAR, and brave semantics (for the sake of comparison, we also recall the complexity results for classical semantics [18–22]). The key thing to observe is that for $\mathcal{EL}_\perp$, query answering under the IAR and brave semantics is intractable in data complexity, even for IQs. This contrasts sharply with the tractability of these semantics in the DL-Lite setting. If we consider combined complexity, then the IAR semantics is bit better behaved than AR semantics, which can be explained by the fact that we can compute in advance the intersection of repairs, and then exploit standard querying algorithms. Interestingly, query answering under the brave semantics is NP-complete, irrespectively of whether we consider CQs or IQs, or whether we adopt data complexity or combined complexity.

The results for the AR and IAR semantics were proven in [23][3]. For the brave semantics, the upper bounds can be proven in the 'usual' way: guess a subset $\mathcal{A}' \subseteq \mathcal{A}$ together with a proof that $\langle \mathcal{T}, \mathcal{A}' \rangle \models q(\mathbf{a})$, and verify that $\mathcal{A}'$ is $\mathcal{T}$-consistent. We provide a proof of the NP-hardness of IQ answering under brave semantics (which we can adapt to show coNP-hardness for the IAR semantics):

Semantics	Data complexity		Combined complexity	
	CQs	IQs	CQs	IQs
classical	P	P	NP	P
AR	coNP	coNP	Π_2^p	coNP
IAR	coNP	coNP	$\Delta_2^p[O(log\,n)]$	coNP
brave	NP	NP	NP	NP

Fig. 3. Complexity of inconsistency-tolerant query answering in $\mathcal{EL}_\perp$. All results are completeness results.

Theorem 29. *In $\mathcal{EL}_\perp$, IQ answering under the IAR (resp. brave) semantics is coNP-hard (resp. NP-hard) in data complexity.*

Proof. The NP lower bound for IQ answering under brave semantics is by reduction from the satisfiability problem for propositional formulas in negation normal form (NNF). Let φ be an NNF formula over the variables $v_1, \ldots, v_m$. We define an $\mathcal{EL}_\perp$ KB $(\mathcal{T}, \mathcal{A})$ as follows:

$$\mathcal{T} = \{T \sqcap F \sqsubseteq \perp, \quad A_\neg \sqcap \exists r_1.F \sqsubseteq T, \quad A_\wedge \sqcap \exists r_1.T \sqcap \exists r_2.T \sqsubseteq T,$$
$$A_\vee \sqcap \exists r_1.T \sqsubseteq T, \quad A_\vee \sqcap \exists r_2.T \sqsubseteq T\}$$

$$\mathcal{A} = \{A_\wedge(a_\psi), r_1(a_\psi, a_{\chi_1}), r_2(a_\psi, a_{\chi_2}) \mid \psi = \chi_1 \wedge \chi_2 \text{ is a subformula of } \varphi\} \cup$$
$$\{A_\vee(a_\psi), r_1(a_\psi, a_{\chi_1}), r_2(a_\psi, a_{\chi_2}) \mid \psi = \chi_1 \vee \chi_2 \text{ is a subformula of } \varphi\} \cup$$
$$\{A_\neg(a_\psi), r_1(a_\psi, a_\chi) \mid \psi = \neg\chi \text{ is a subformula of } \varphi\} \cup$$
$$\{T(a_{v_i}), F(a_{v_i}) \mid 1 \le i \le m\}$$

The structure of the formula is encoded in the ABox, with every subformula ξ of φ (including the variables) represented by a corresponding individual a_ξ. The

---

[3] The results in [23] are formulated for UCQs rather than CQs, but the same results are obtained for CQs.

TBox axiom $T \sqsubseteq \neg F$ forces a choice of truth value for every variable, and the remaining TBox axioms serve to compute the truth value of a formula based upon the truth values of its subformula(s). Using these ideas, one can show that $\langle \mathcal{T}, \mathcal{A} \rangle \models_{\mathsf{brave}} T(a_\varphi)$ if and only if φ is satisfiable.

Semantics	Data complexity		Combined complexity	
	CQs	IQs	CQs	IQs
classical	coNP	coNP	EXP	EXP
AR	Π_2^p	Π_2^p	EXP	EXP
IAR	Π_2^p	Π_2^p	EXP	EXP
brave	Σ_2^p	Σ_2^p	EXP	EXP

Fig. 4. Complexity of inconsistency-tolerant query answering in $\mathcal{ALC}$. All results are completeness results.

The coNP lower bound for IAR semantics can be proven by a minor modification of the above reduction. Let $\mathcal{T}' = \mathcal{T} \cup \{T \sqsubseteq \neg B\}$ and $\mathcal{A}' = \mathcal{A} \cup \{B(a_\varphi)\}$. We claim that $\langle \mathcal{T}, \mathcal{A} \rangle \models_{\mathsf{IAR}} B(a_\varphi)$ if and only if φ is unsatisfiable. Indeed, if φ is satisfiable, then by the preceding paragraph, there must exist a repair of the KB from which $T(a_\varphi)$ is entailed, and hence we can find a repair in which $B(a_\varphi)$ does not hold. Conversely, if φ is satisfiable, then we can use the satisfying valuation to construct such a repair. □

In Fig. 4, we give the complexity of querying $\mathcal{ALC}$ KBs under the same three semantics. If we consider data complexity, then we again find that there is no apparent benefit to using the IAR and brave semantics rather than the AR semantics. If we consider combined complexity, then we obtain exactly the same complexity (EXP) as for classical semantics, intuitively because it is possible to enumerate in exponential-time all subsets of the ABox.

For the results concerning AR and IAR semantics, consult [23][4]. For the brave semantics, the upper bounds and combined complexity lower bound are straightforward. We give a proof of the Σ_2^p-hardness in data complexity, which holds already for the extension of $\mathcal{EL}_\perp$ with disjunction.

Theorem 30. *In $\mathcal{ELU}_\perp$ (hence, in $\mathcal{ALC}$), IQ answering under the brave semantics is Σ_2^p-hard in data complexity.*

Proof We give a reduction from the validity problem for $\exists\forall$-QBF, a well-known Σ_2^p-complete problem. Consider a 2QBF instance

$$\Phi = \exists x_1, \ldots, x_n \forall y_1, \ldots, y_m \ \tau_1 \vee \ldots \vee \tau_p$$

---

[4] The results in [23] are formulated for UCQs rather than CQs, but the same results are obtained for CQs.

where each $\tau_i = \ell_i^1 \wedge \ell_i^2 \wedge \ell_i^3$ is a propositional 3-term whose variables are drawn from $X = \{x_1, \ldots, x_n\}$ and $Y = \{y_1, \ldots, y_m\}$. We construct the following KB:

$$
\begin{aligned}
\mathcal{T} &= \{T \sqsubseteq \neg F, \quad V \sqsubseteq T \sqcup F, \quad \exists R.(S_1 \sqcap S_2 \sqcap S_3) \sqsubseteq A\} \cup \\
&\quad \{\exists P_k.T \sqsubseteq S_k \mid 1 \leq k \leq 3\} \cup \{\exists N_k.F \sqsubseteq S_k \mid 1 \leq k \leq 3\}\} \\
\mathcal{A}_\Phi &= \{P_k(t_i, z) \mid \ell_i^k = z\} \cup \{N_k(t_i, z) \mid \ell_i^k = \neg z\} \cup \\
&\quad \{T(x_j), F(x_j) \mid 1 \leq j \leq n\} \cup \{V(y_i) \mid 1 \leq j \leq m\} \cup \\
&\quad \{R(a, t_i) \mid 1 \leq i \leq p\}
\end{aligned}
$$

It can be shown that Φ is valid iff $\langle \mathcal{T}, \mathcal{A}_\Phi \rangle \models_{\mathsf{brave}} A(a)$. We give the main ideas of the argument. Similarly to earlier proofs, the terms and variables are used as ABox individuals, and the roles P_k and N_k are used respectively to link a term t_i to the variables that occur (positively, P, or negatively, N) in the kth literal of the term. For every variable x_j, the inclusion $T \sqsubseteq \neg F$ will force us to choose which of the assertions $T(x_j)$ and $F(x_j)$ to keep. It follows that the repairs will correspond to the different valuations of the x-variables. For each variable y_j, the assertion $V(x_j)$ and inclusions $V \sqsubseteq T \sqcup F$ and $T \sqsubseteq \neg F$ will ensure that in every model of a given repair, we have either $T(y_j)$ or $F(y_j)$. It then remains to check that there is some repair for which every model corresponds to a valuation that satisfies one of the terms. To this end, we use the inclusions $\exists P_k.T \sqsubseteq S_k$ and $\exists N_k.F \sqsubseteq S_k$ to mark with S_k those terms whose kth literal is satisfied by the valuation. The individual a is linked via R to all of the terms, and the inclusion $\exists R.(S_1 \sqcap S_2 \sqcap S_3) \sqsubseteq A$ adds A to a if there is a term marked with S_1, S_2, and S_3, meaning that the three literals in the term are satisfied in the considered valuation.   □

Finally, we should point out that the complexity of inconsistency-tolerant query answering has also been investigated for ontologies formulated using existential rules (aka Datalog$^{+/-}$), see e.g., [24,25].

5   Systems for Inconsistency-Tolerant Query Answering

To the best of our knowledge, there are currently three systems for querying DL KBs under inconsistency-tolerant semantics:

- the QuID system [31,32] that performs CQ answering under IAR semantics in an extension of DL-Lite$_\mathcal{A}$ with denial and identification constraints,
- the CQAPri system [16] that implements CQ answering under the $(\sqsubseteq_P\text{-})$IAR, $(\sqsubseteq_P\text{-})$AR, and brave semantics for $DL\text{-}Lite^\mathcal{R}$ KBs
- the system of Du et al. [33] for ground CQ answering under $\leq_w$-AR semantics for the highly expressive DL $\mathcal{SHIQ}$.

We mention in passing that there are also a few systems for querying inconsistent relational databases under AR semantics, including ConQuer [34,35] and Hippo [36,37], which implement polynomial-time procedures for tractable subcases, and the ConsEx [38] and EQUIP [39] systems, which reduce AR query answering to answer set programming and binary integer programming, respectively.

In what follows, we provide some further details on QuID and CQAPri, as these systems target DLs of the DL-Lite family, our focus in this chapter.

5.1 The QuID System

The QuID system implements three different approaches for querying DL-Lite KBs using the IAR semantics: first-order query rewriting, ABox annotation, and ABox cleaning.

Query Rewriting. A first-order query rewriting procedure for the IAR semantics was proposed in [30] for the logic DL-Lite$_\mathcal{A}$, which extends DL-Lite$_\mathcal{R}$ with functionality assertions and attributes. In a later work [32], the rewriting procedure was extended to DL-Lite$_{\mathcal{A},id,den}$, an expressive member of the DL-Lite family offering identification and denial constraints. The basic idea underlying these rewriting procedures was presented in Example 21.

ABox Annotation. A second approach to IAR query answering [31] annotates the ABox in an offline phase, using an extra argument to record whether or not an assertion is involved in a minimal inconsistent set. At query time, the query is rewritten using an existing rewriting algorithm for classical semantics, and the rewritten query is modified by adding an extra argument to every atom with the 'not in any conflict' value. The modified query can then be evaluated over the annotated ABox.

ABox Cleaning. The third method [31] computes in an offline phase the assertions appearing in some minimal inconsistent subset and removes these assertions from the ABox. The user query is then handled using an existing query rewriting algorithm for classical semantics.

· The three methods have been compared in [31], and the experiments showed that the ABox cleaning and annotation approaches often significantly outperform the method based on query rewriting. This is because the IAR-rewritings, which require rather complex conditions to block 'unsafe' assertions, are more difficult to evaluate than rewritings for classical semantics. However, as noted in [31], the annotation and cleaning approaches both involve making modifications to the ABox, which may not be feasible in applications where data is read-only.

5.2 The CQAPri System

The CQAPri system, which accepts DL-Lite$_\mathcal{R}$ KBs, uses the IAR, brave, and AR semantics in combination to identify three types of query answer:

- **Possible:** $\mathcal{K} \models_{\mathsf{brave}} q(\mathbf{a})$ and $\mathcal{K} \not\models_{\mathsf{AR}} q(\mathbf{a})$
- **Likely:** $\mathcal{K} \models_{\mathsf{AR}} q(\mathbf{a})$ and $\mathcal{K} \not\models_{\mathsf{IAR}} q(\mathbf{a})$
- **(Almost) Sure:** $\mathcal{K} \models_{\mathsf{IAR}} q(\mathbf{a})$

We briefly describe the steps needed to obtain such a classification.

In an offline phase, CQAPri computes and stores the set conflicts($\mathcal{K}$) of minimal $\mathcal{T}$-inconsistent subsets (called *conflicts* in [16]) of the KB $\mathcal{K} = \langle \mathcal{T}, \mathcal{A} \rangle$. As $\mathcal{K}$ is a DL-Lite$_{\mathcal{R}}$ KB, every set in conflicts($\mathcal{K}$) contains at most two assertions.

When a query q arrives, CQAPri uses an off-the-shelf rewriting system to compute a UCQ-rewriting $q_1 \vee \ldots \vee q_n$ of q under classical semantics. The rewriting is evaluated over the ABox to obtain the *candidate answers* of q. For each candidate answer $\mathbf{a}$, we store the *images* of $q(\mathbf{a})$ in $\mathcal{A}$, i.e., the sets of assertions corresponding to the homomorphic image of some $q_i(\mathbf{a})$. Candidate answers define a superset of the answers holding under the brave semantics.

Among the candidate answers, CQAPri identifies those holding under IAR semantics by checking whether there is some image of the candidate answer whose assertions are not involved in any conflict. It also identifies the candidate answers which are not brave-answers by discarding images which are $\mathcal{T}$-inconsistent: an answer that has only such images does not hold under brave semantics.

For the tuples that have been shown to be answers under brave semantics but not under IAR semantics, it remains to decide whether the tuple is an AR-answer. For every such tuple $\mathbf{a}$, a propositional formula $\varphi_{q(\mathbf{a})?}$ is constructed (in polynomial time in $|\mathcal{A}|$) and passed on to a SAT solver for evaluation. As shown in Theorem 31, the formula $\varphi_{q(\mathbf{a})?}$ is satisfiable iff $\mathcal{K} \not\models_{\mathsf{AR}} q(\mathbf{a})$.

The formula $\varphi_{q(\mathbf{a})?}$ is defined as the conjunction of the formulas $\varphi_{\neg q(\mathbf{a})}$ and φ_{cons} displayed in Fig. 5. The encoding makes use of the sets causes($q(\mathbf{a}), \mathcal{K}$) and confl($\mathcal{C}, \mathcal{K}$) defined as follows:

$$\mathsf{causes}(q(\mathbf{a}), \mathcal{K}) = \{ \mathcal{C} \subseteq \mathcal{A} \mid \mathcal{C} \text{ is minimal } \mathcal{T}\text{-consistent support for } q(\mathbf{a}) \}$$
$$\mathsf{confl}(\mathcal{C}, \mathcal{K}) = \{ \beta \mid \exists \alpha \in \mathcal{C}, \{\alpha, \beta\} \in \mathsf{conflicts}(\mathcal{K}) \}$$

and the notation vars($\varphi_{\neg q}$), which denotes the set of variables appearing in $\varphi_{\neg q}$.

Both formulas use variables of the form x_α, where α is an assertion from $\mathcal{A}$, so every valuation of the variables corresponds to a subset of $\mathcal{A}$. The formula $\varphi_{\neg q(\mathbf{a})}$ states that this subset must include, for every support $\mathcal{C}$ for $q(\mathbf{a})$, some assertion that contradicts $\mathcal{C}$. The second formula φ_{cons} formula ensures that the set of selected assertions is consistent with the TBox. Taken together, these formulas guarantee the existence of a consistent subset of the ABox that can be extended to a repair of the KB from which $q(\mathbf{a})$ cannot be derived.

$$\varphi_{\neg q} = \bigwedge_{\mathcal{C} \in \mathsf{causes}(q, \mathcal{K})} \bigvee_{\beta \in \mathsf{confl}(\mathcal{C}, \mathcal{K})} x_\beta$$

$$\varphi_{cons} = \bigwedge_{x_\alpha, x_\beta \in \mathsf{vars}(\varphi_{\neg q}), \beta \in \mathsf{confl}(\{\alpha\}, \mathcal{K})} \neg x_\alpha \vee \neg x_\beta$$

Fig. 5. SAT encoding for AR query answering.

Theorem 31 [16]. *Let $\varphi_{\neg q(a)}$ and φ_{cons} be defined as in Fig. 5. Then $\mathcal{K} \not\models_{AR} q(\mathbf{a})$ if and only if $\varphi_{\neg q(\mathbf{a})} \wedge \varphi_{cons}$ is satisfiable.*

The SAT encoding in Fig. 5 can be adapted to handle the $\subseteq_P$-IAR and $\subseteq_P$-AR semantics by adding a third set of clauses that ensure that the set of selected assertions is maximal w.r.t. $\subseteq_P$. See [16] for details.

The experiments conducted in [16] show that the IAR semantics generally constitutes a very good approximation of the AR semantics and that query answering scales well on realistic cases, when a few percents of the ABox assertions are involved in a conflict. Indeed, in such cases, a large portion of the brave -answers hold under the IAR semantics, so CQAPri does not need to call a SAT solver to decide whether they are entailed under AR semantics.

6   Summary and Outlook

In this chapter, we have surveyed a number of different inconsistency-tolerant semantics that have been proposed to obtain meaningful query results from inconsistent DL knowledge bases. In order to provide a basis for selecting the appropriate semantic(s) for a given application, we compared the different semantics in terms of the consistency properties they satisfy, the set of query answers they return (using the notion of under- and over-approximations), and the computational complexity of the associated query answering task, focusing mainly on DL-Lite KBs. We have also described how the IAR, brave, and AR semantics can be implemented in a query answering system. Preliminary experiments reported in [16] suggest that for DL-Lite knowledge bases, it is possible to efficiently compute the answers holding under the appealing but intractable AR semantics by combining tractable approximations in the form of the IAR and brave semantics with the power of modern SAT solvers.

With the emergence of the inconsistency-tolerant query answering systems for DL-Lite comes a set of new challenges related to improving both the efficiency and the usability of such systems. Regarding efficiency, both of the systems that have been implemented for querying inconsistent DL-Lite KBs employ UCQ-rewriting algorithms. It would be interesting to see whether other approaches to DL-Lite query answering, like the combined approach [43] or rewritings into positive existential or non-recursive Datalog queries, can be suitably adapted for use in inconsistency-tolerant querying systems. As for usability, one problem that has begun to be explored is how to explain query results to users, that is, how to justify why a given tuple appears as an answer under the considered inconsistency tolerant semantics and why some other tuples do not appear in the results. Some first results on explaining query (non)answers under the AR, IAR, and brave semantics have been presented in [44]. Interaction with the user is also important if we want to improve the quality of the ABox. A recent work [45] proposes to allow users to provide feedback on which query results are missing or erroneous and then interact with the user in order to identify a set of ABox modifications (additions and deletions of assertions) that fix the identified flaws.

Another important challenge for future work is to devise practical methods for inconsistency-tolerant OMQA that work for DLs outside the DL-Lite family. Indeed, as shown in Sect. 4, the complexity landscape changes rather dramatically when we leave the DL-Lite world. In particular, we have seen that even for $\mathcal{EL}_\perp$, a relatively simple DL that has good computational properties, IQ answering under the IAR and brave semantics is intractable in data complexity. Thus, a first research direction is to invent new inconsistency-tolerant semantics that permit tractable query answering in DLs beyond DL-Lite. A second possibility is to design algorithms for existing inconsistency-tolerant semantics that work well in practice, even if they are not guaranteed to run in polynomial time.

Acknowledgments. This work has been supported by ANR project PAGODA (ANR-12-JS02-007-01).

References

1. Bienvenu, M., Ortiz, M.: Ontology-mediated query answering with data-tractable description logics. In: Faber, W., Paschke, A. (eds.) Reasoning Web 2015. LNCS, vol. 9203, pp. 218–307. Springer, Heidelberg (2015). doi:10.1007/978-3-319-21768-0_9
2. Baader, F., Calvanese, D., McGuinness, D., Nardi, D., Patel-Schneider, P.F. (eds.): The Description Logic Handbook: Theory, Implementation, and Applications. Cambridge University Press, Cambridge (2003)
3. OWL Working Group. OWL 2 Web Ontology Language: Document Overview. W3C Recommendation (2009). http://www.w3.org/TR/owl2-overview/
4. Calvanese, D., De Giacomo, G., Lembo, D., Lenzerini, M., Rosati, R.: Tractable reasoning and efficient query answering in description logics: the DL-Lite family. J. Autom. Reason. (JAR) **39**(3), 385–429 (2007)
5. Lembo, D., Lenzerini, M., Rosati, R., Ruzzi, M., Savo, D.F.: Inconsistency-tolerant semantics for description logics. In: Hitzler, P., Lukasiewicz, T. (eds.) RR 2010. LNCS, vol. 6333, pp. 103–117. Springer, Heidelberg (2010). doi:10.1007/978-3-642-15918-3_9
6. Arenas, M., Bertossi, L.E., Chomicki, J.: Consistent query answers in inconsistent databases. In: Proceedings of the 18th Symposium on Principles of Database Systems (PODS), pp. 68–79 (1999)
7. Arenas, M., Bertossi, L., Kifer, M.: Applications of annotated predicate calculus to querying inconsistent databases. In: Lloyd, J., et al. (eds.) CL 2000. LNCS (LNAI), vol. 1861, pp. 926–941. Springer, Heidelberg (2000). doi:10.1007/3-540-44957-4_62
8. Chomicki, J.: Consistent query answering: five easy pieces. In Proceedings of the 10th International Conference on Database Theory (ICDT), pp. 1–17 (2007)
9. Bertossi, L.E.: Database Repairing and Consistent Query Answering: Synthesis Lectures on Data Management. Morgan & Claypool Publishers, San Rafael (2011)
10. Bienvenu, M., Rosati, R.: Tractable approximations of consistent query answering for robust ontology-based data access. In: Proceedings of the 23rd International Joint Conference on Artificial Intelligence (IJCAI) (2013)
11. Motik, B., Cuenca Grau, B., Horrocks, I., Wu, Z., Fokoue, A., Lutz, C.: OWL 2 Web Ontology Language Profiles. W3C Recommendation (2012). http://www.w3.org/TR/owl2-profiles/

12. Immerman, N.: Nondeterministic space is closed under complementation. SIAM J. Comput. **17**(5), 935–938 (1988)
13. Szelepcsényi, R.: The method of forcing for nondeterministic automata. Bull. EATCS **33**, 96–99 (1987)
14. Bienvenu, M.: On the complexity of consistent query answering in the presence of simple ontologies. In: Proceedings of the 26th AAAI Conference on Artificial Intelligence (2012)
15. Lukasiewicz, T., Martinez, M.V., Simari, G.I.: Inconsistency handling in datalog +/− ontologies. In: Proceedings of the 20th European Conference on Artificial Intelligence (ECAI) (2012)
16. Bienvenu, M., Bourgaux, C., Goasdoué, F.: Querying inconsistent description logic knowledge bases under preferred repair semantics. In: Proceedings of the 28th AAAI Conference on Artificial Intelligence (AAAI) (2014)
17. Bourgaux,C.: Inconsistency handling in ontology-mediated query answering. Ph.D. thesis, University of Paris-Sud (2016)
18. Baader, F., Brandt, S., Lutz, C.: Pushing the $\mathcal{EL}$ envelope. In: Proceedings of the 19th International Joint Conference on Artificial Intelligence (IJCAI), pp. 364–369 (2005)
19. Calvanese, D., Giacomo, G.D., Lembo, D., Lenzerini, M., Rosati, R.: Data complexity of query answering in description logics. In: Proceedings of the 10th International Conference on the Principles of Knowledge Representation and Reasoning (KR), pp. 260–270 (2006)
20. Rosati, R.: On conjunctive query answering in $\mathcal{EL}$. In: Proceedings of the 20th International Workshop on Description Logics (DL) (2007)
21. Krötzsch, M., Rudolph, S.: Conjunctive queries for $\mathcal{EL}$ with composition of roles. In: Proceedings of the 20th International Workshop on Description Logics (DL) (2007)
22. Krisnadhi, A., Lutz, C.: Data complexity in the $\mathcal{EL}$ family of DLs. In: Proceedings of the 20th International Workshop on Description Logics (DL) (2007)
23. Rosati, R.: On the complexity of dealing with inconsistency in description logic ontologies. In: Proceedings of the 22nd International Joint Conference on Artificial Intelligence (IJCAI) (2011)
24. Lukasiewicz, T., Martinez, M.V., Simari, G.I.: Complexity of inconsistency-tolerant query answering in datalog+/−. In: Meersman, R., Panetto, H., Dillon, T., Eder, J., Bellahsene, Z., Ritter, N., Leenheer, P., Dou, D. (eds.) OTM 2013. LNCS, vol. 8185, pp. 488–500. Springer, Heidelberg (2013). doi:10.1007/978-3-642-41030-7_35
25. Lukasiewicz, T., Martinez, M.V., Pieris, A., Simari, G.I.: From classical to consistent query answering under existential rules. In: Proceedings of the 29th AAAI Conference on Artificial Intelligence, pp. 1546–1552 (2015)
26. Eiter, T., Gottlob, G.: The complexity of logic-based abduction. J. ACM **42**(1), 3–42 (1995)
27. Baget, J., Benferhat, S., Bouraoui, Z., Croitoru, M., Mugnier, M., Papini, O, Rocher, S., Tabia, K.: A general modifier-based framework for inconsistency-tolerant query answering. In: Proceedings of the 15th International Conference on the Principles of Knowledge Representation and Reasoning (KR), pp. 513–516 (2016)
28. Artale, A., Calvanese, D., Kontchakov, R., Zakharyaschev, M.: The DL-Lite family and relations. J. Artif. Intell. Res. (JAIR) **36**, 1–69 (2009)
29. Donini, F.M., Lenzerini, M., Nardi, D., Schaerf, A.: Deduction in concept languages: from subsumption to instance checking. J. Log. Comput. (JLC) **4**(4), 423–452 (1994)

30. Lembo, D., Lenzerini, M., Rosati, R., Ruzzi, M., Savo, D.F.: Query rewriting for inconsistent DL-Lite ontologies. In: Rudolph, S., Gutierrez, C. (eds.) RR 2011. LNCS, vol. 6902, pp. 155–169. Springer, Heidelberg (2011). doi:10.1007/978-3-642-23580-1_12

31. Rosati, R., Ruzzi, M., Graziosi, M., Masotti, G.: Evaluation of techniques for inconsistency handling in OWL 2 QL ontologies. In: Cudré-Mauroux, P., et al. (eds.) ISWC 2012. LNCS, vol. 7650, pp. 337–349. Springer, Heidelberg (2012). doi:10.1007/978-3-642-35173-0_23

32. Lembo, D., Lenzerini, M., Rosati, R., Ruzzi, M., Savo, D.F.: Inconsistency-tolerant query answering in ontology-based data access. J. Web Semant. (JWS) **33**, 3–29 (2015)

33. Du, J., Qi, G., Shen, Y.-D.: Weight-based consistent query answering over inconsistent $\mathcal{SHIQ}$ knowledge bases. Knowl. Inf. Syst. **34**(2), 335–371 (2013)

34. Fuxman, A.D., Miller, R.J.: First-order query rewriting for inconsistent databases. In: Eiter, T., Libkin, L. (eds.) ICDT 2005. LNCS, vol. 3363, pp. 337–351. Springer, Heidelberg (2004). doi:10.1007/978-3-540-30570-5_23

35. Fuxman, A., Fazli, E., Miller, R.J.: Conquer: efficient management of inconsistent databases. In: Proceedings of the 31st ACM SIGMOD International Conference on Management of Data, pp. 155–166 (2005)

36. Chomicki, J., Marcinkowski, J., Staworko, S.: Computing consistent query answers using conflict hypergraphs. In: Proceedings of the 13th International Conference on Information and Knowledge Management (CIKM), pp. 417–426 (2004)

37. Chomicki, J., Marcinkowski, J., Staworko, S.: Hippo: a system for computing consistent answers to a class of SQL queries. In: Bertino, E., Christodoulakis, S., Plexousakis, D., Christophides, V., Koubarakis, M., Böhm, K., Ferrari, E. (eds.) EDBT 2004. LNCS, vol. 2992, pp. 841–844. Springer, Heidelberg (2004). doi:10.1007/978-3-540-24741-8_53

38. Marileo, M.C., Bertossi, L.E.: The consistency extractor system: Answer set programs for consistent query answering in databases. Data Knowl. Eng. **69**(6), 545–572 (2010)

39. Kolaitis, P.G., Pema, E., Tan, W.-C.: Efficient querying of inconsistent databases with binary integer programming. Proc. VLDB Endow. (PVLDB) **6**(6), 397–408 (2013)

40. Gottlob, G.: NP trees and Carnap's modal logic. J. ACM **42**(2), 421–457 (1995)

41. Wagner, K.W.: More complicated questions about maxima and minima, and some closures of NP. Theoret. Comput. Sci. **51**, 53–80 (1987)

42. Eiter, T., Gottlob, G.: The complexity class Θ_2^P: Recent results and applications in AI and modal logic. In: Chlebus, B.S., Czaja, L. (eds.) FCT 1997. LNCS, vol. 1279, pp. 1–18. Springer, Heidelberg (1997). doi:10.1007/BFb0036168

43. Kontchakov, R., Lutz, C., Toman, D., Wolter, F., Zakharyaschev, M.: The combined approach to query answering in DL-Lite. In: Proceedings of the 12th International Conference on the Principles of Knowledge Representation and Reasoning (KR) (2010)

44. Bienvenu, M., Bourgaux, C., Goasdoué, F.: Explaining inconsistency-tolerant query answering over description logic knowledge bases. In: Proceedings of the 30th AAAI Conference on Artificial Intelligence (AAAI) (2016)

45. Bienvenu, M., Bourgaux, C., Goasdoué, F.: Query-driven repairing of inconsistent dllite knowledge bases. In: Proceedings of the 25th International Joint Conference on Artificial Intelligence (IJCAI) (2016)

From Fuzzy to Annotated Semantic Web Languages

Umberto Straccia[1]([⊠]) and Fernando Bobillo[2]

[1] ISTI - CNR, Area Della Ricerca di Pisa, Pisa, Italy
umberto.straccia@isti.cnr.it
[2] Department of Computer Science and Systems Engineering,
Universidad de Zaragoza, Zaragoza, Spain
fbobillo@unizar.es

Abstract. The aim of this chapter is to present a detailed, self-contained and comprehensive account of the state of the art in representing and reasoning with fuzzy knowledge in Semantic Web Languages such as triple languages RDF/RDFS, conceptual languages of the OWL 2 family and rule languages. We further show how one may generalise them to so-called annotation domains, that cover also e.g. temporal and provenance extensions.

1 Introduction

Managing uncertainty and fuzziness is growing in importance in Semantic Web research as recognised by a large number of research efforts in this direction [1,2]. *Semantic Web Languages* (SWL) are the languages used to provide a formal description of concepts, terms, and relationships within a given domain, among which the *OWL 2 family* of languages [3], *triple languages* RDF & RDFS [4] and *rule languages* (such as RuleML [5], Datalog$^\pm$ [6] and RIF [7]) are major players.

While their syntactic specification is based on XML [8], their semantics is based on logical formalisms: briefly,

- RDFS is a logic having intensional semantics and the logical counterpart is ρdf [9];
- OWL 2 is a family of languages that relate to *Description Logics* (DLs) [10];
- rule languages relate roughly to the *Logic Programming* (LP) paradigm [11];
- both OWL 2 and rule languages have an extensional semantics.

Uncertainty Versus Fuzziness. One of the major difficulties, for those unfamiliar on the topic, is to understand the conceptual differences between uncertainty and fuzziness. Specifically, we recall that there has been a long-lasting misunderstanding in the literature of artificial intelligence and uncertainty modelling, regarding the role of probability/possibility theory and vague/fuzzy theory. A clarifying paper is [12]. We recall here the salient concepts.

© Springer International Publishing AG 2017
J.Z. Pan et al. (Eds.): Reasoning Web 2016, LNCS 9885, pp. 203–240, 2017.
DOI: 10.1007/978-3-319-49493-7_6

Uncertainty.

Under *uncertainty theory* fall all those approaches in which statements rather than being either true or false, are true or false to some *probability* or *possibility* (for example, "it will rain tomorrow"). That is, a statement is true or false in any world/interpretation, but we are "uncertain" about which world to consider as the right one, and thus we speak about e.g. a probability distribution or a possibility distribution over the worlds. For example, we cannot exactly establish whether it will rain tomorrow or not, due to our *incomplete* knowledge about our world, but we can estimate to which degree this is probable, possible, or necessary.

To be somewhat more formal, consider a propositional statement (formula) φ ("tomorrow it will rain") and a propositional interpretation (world) $\mathcal{I}$. We may see $\mathcal{I}$ as a function mapping propositional formulae into $\{0, 1\}$, i.e. $\mathcal{I}(\varphi) \in \{0, 1\}$. If $\mathcal{I}(\varphi) = 1$, denoted also as $\mathcal{I} \models \varphi$, then we say that the statement φ under $\mathcal{I}$ is true, false otherwise. Now, each interpretation $\mathcal{I}$ depicts some concrete world and, given n propositional letters, there are 2^n possible interpretations. In uncertainty theory, we do not know which interpretation $\mathcal{I}$ is the actual one and we say that we are *uncertain* about which world is the real one that will occur.

To deal with such a situation, one may construct a *probability distribution over the worlds*, that is a function Pr mapping interpretations in $[0, 1]$, i.e. $Pr(\mathcal{I}) \in [0, 1]$, with $\sum_{\mathcal{I}} Pr(\mathcal{I}) = 1$, where $Pr(\mathcal{I})$ indicates the probability that $\mathcal{I}$ is the actual world under which to interpret the propositional statement at hand. Then, the *probability* of a statement φ in Pr, denoted $Pr(\varphi)$, is the sum of all $Pr(\mathcal{I})$ such that $\mathcal{I} \models \varphi$, i.e.

$$Pr(\varphi) = \sum_{\mathcal{I} \models \varphi} Pr(\mathcal{I}).$$

Fuzziness.

On the other hand, under *fuzzy theory* fall all those approaches in which statements (for example, "heavy rain") are true to some *degree*, which is taken from a truth space (usually $[0, 1]$). That is, the convention prescribing that a proposition is either true or false is changed towards graded propositions. For instance, the compatibility of "heavy" in the phrase "heavy rain" is graded and the degree depends on the amount of rain is falling.[1] Often we may find rough definitions about rain types, such as:[2]

Rain. Falling drops of water larger than 0.5 mm in diameter. In forecasts, "rain" usually implies that the rain will fall steadily over a period of time;
Light rain. Rain falls at the rate of 2.6 mm or less an hour;
Moderate rain. Rain falls at the rate of 2.7 mm to 7.6 mm an hour;
Heavy rain. Rain falls at the rate of 7.7 mm an hour or more.

---

[1] More concretely, the intensity of precipitation is expressed in terms of a precipitation rate R: volume flux of precipitation through a horizontal surface, i.e. $m^3/m^2 s = ms^{-1}$. It is usually expressed in mm/h.

[2] http://usatoday30.usatoday.com/weather/wds8.htm.

It is evident that such definitions are quite harsh and resemble a bivalent (two-valued) logic: e.g. a precipitation rate of 7.7 mm/h is a heavy rain, while a precipitation rate of 7.6 mm/h is just a moderate rain. This is clearly unsatisfactory, as quite naturally the more rain is falling, the more the sentence "heavy rain" is true and, vice-versa, the less rain is falling the less the sentence is true.

In other words, this means essentially, that the sentence "heavy rain" is no longer either true or false as in the definition above, but is intrinsically graded.

A more fine grained way to define the various types of rains is illustrated in Fig. 1.

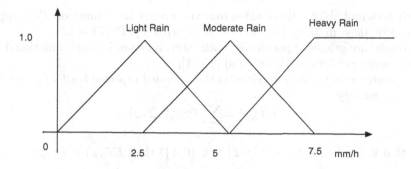

Fig. 1. Light, Moderate and Heavy Rain.

Light rain, moderate rain and heavy rain are called *Fuzzy Sets* in the literature [13] and are characterised by the fact that membership is a matter of degree. Of course, the definition of fuzzy sets is frequently context dependent and subjective: e.g. the definition of heavy rain is quite different from heavy person and the latter may be defined differently among human beings.

From a logical point of view, a propositional interpretation maps a statement φ to a truth degree in $[0, 1]$, i.e. $\mathcal{I}(\varphi) \in [0, 1]$. Essentially, we are unable to establish whether a statement is entirely true or false due to the involvement of *vague/fuzzy* concepts, such as "heavy".

Note that all fuzzy statements are truth-functional, that is, the degree of truth of every statement can be calculated from the degrees of truth of its constituents, while uncertain statements cannot always be a function of the uncertainties of their constituents [14]. For the sake of illustrative purpose, an example of truth functional interpretation of propositional statements is as follows:

$$\mathcal{I}(\varphi \wedge \psi) = min(\mathcal{I}(\varphi), \mathcal{I}(\psi))$$
$$\mathcal{I}(\varphi \vee \psi) = max(\mathcal{I}(\varphi), \mathcal{I}(\psi))$$
$$\mathcal{I}(\neg\varphi) \quad = 1 - \mathcal{I}(\varphi) \,.$$

In such a setting one may be interested in the so-called notions of *minimal (resp. maximal) degree of satisfaction* of a statement, i.e. $min_\mathcal{I}\mathcal{I}(\varphi)$ (resp. $max_\mathcal{I}\mathcal{I}(\varphi)$).

Uncertain fuzzy sentences.
Let us recap: in a probabilistic setting each statement is either true or false, but there is e.g. a probability distribution telling us how probable each interpretation is, i.e. $\mathcal{I}(\varphi) \in \{0, 1\}$ and $Pr(\mathcal{I}) \in [0, 1]$. In fuzzy theory instead, sentences are graded, i.e. we have $\mathcal{I}(\varphi) \in [0, 1]$.

A natural question is: can we have sentences combining the two orthogonal concepts? Yes, for instance, "there will be heavy rain tomorrow" is an uncertain fuzzy sentence. Essentially, there is uncertainty about the world we will have tomorrow, and there is fuzziness about the various types of rain we may have tomorrow.

From a logical point of view, we may model uncertain fuzzy sentences in the following way:

- we have a probability distribution over the worlds, i.e. a function Pr mapping interpretations in $[0, 1]$, i.e. $Pr(\mathcal{I}) \in [0, 1]$, with $\sum_{\mathcal{I}} Pr(\mathcal{I}) = 1$;
- sentences are graded. Specifically, each interpretation is truth functional and maps sentences into $[0, 1]$, i.e. $\mathcal{I}(\varphi) \in [0, 1]$;
- for a sentence φ, we are interested in the so-called *expected truth* of φ, denoted $ET(\varphi)$, namely

$$ET(\varphi) = \sum_{\mathcal{I}} Pr(\mathcal{I}) \cdot \mathcal{I}(\varphi).$$

Note that if $\mathcal{I}$ is bivalent (that is, $\mathcal{I}(\varphi) \in \{0, 1\}$) then $ET(\varphi) = Pr(\varphi)$.

Talk Overview.
We present here some salient aspects in representing and reasoning with fuzzy knowledge in Semantic Web Languages (SWLs) such as *triple languages* [4] (see, e.g. [15,16]), *conceptual languages* [3] (see, e.g. [17–19]) and *rule languages* (see, e.g. [1,20–25]). We refer the reader to [2] for an extensive presentation concerning fuzziness and semantic web languages. We then further show how one may generalise them to so-called annotation domains, that cover also e.g. temporal and provenance extensions (see, e.g. [26–28]).

2   Basics: From Fuzzy Sets to Mathematical Fuzzy Logic and Annotation Domains

2.1   Fuzzy Sets Basics

The aim of this section is to introduce the basic concepts of fuzzy set theory. To distinguish between fuzzy sets and classical (non fuzzy) sets, we refer to the latter as *crisp sets*. For an in-depth treatment we refer the reader to, e.g. [29,30].

From Crisp Sets to Fuzzy Sets.
To better highlight the conceptual shift from classical sets to fuzzy sets, we start with some basic definitions and well-known properties of classical sets. Let X be a *universal set* containing all possible elements of concern in each particular context. The *power set*, denoted 2^A, of a set $A \subset X$, is the set of subsets of A,

i.e., $2^A = \{B \mid B \subseteq A\}$. Often sets are defined by specifying a property satisfied by its members, in the form $A = \{x \mid P(x)\}$, where $P(x)$ is a statement of the form "x has property P" *that is either true or false* for any $x \in X$. Examples of universe X and subsets $A, B \in 2^X$ may be

$$X = \{x \mid x \text{ is a day}\}$$
$$A = \{x \mid x \text{ is a rainy day}\}$$
$$B = \{x \mid x \text{ is a day with precipitation rate } R \geq 7.5mm/h\}.$$

In the above case we have $B \subseteq A \subseteq X$.

The *membership function* of a set $A \subseteq X$, denoted χ_A, is a function mapping elements of X into $\{0, 1\}$, i.e. $\chi_A \colon X \to \{0, 1\}$, where $\chi_A(x) = 1$ iff $x \in A$. Note that for any sets $A, B \in 2^X$, we have that

$$A \subseteq B \text{ iff } \forall x \in X.\ \chi_A(x) \leq \chi_B(x). \tag{1}$$

The *complement* of a set A is denoted $\bar{A}$, i.e. $\bar{A} = X \setminus A$. Of course, $\forall x \in X.\ \chi_{\bar{A}}(x) = 1 - \chi_A(x)$. In a similar way, we may express set operations of intersection and union via the membership function as follows:

$$\forall x \in X.\ \chi_{A \cap B}(x) = min(\chi_A(x), \chi_B(x)) \tag{2}$$
$$\forall x \in X.\ \chi_{A \cup B}(x) = max(\chi_A(x), \chi_B(x)). \tag{3}$$

The *Cartesian product*, $A \times B$, of two sets $A, B \in 2^X$ is defined as $A \times B = \{\langle a, b \rangle \mid a \in A, b \in B\}$. A relation $R \subseteq X \times X$ is *reflexive* if for all $x \in X$ $\chi_R(x, x) = 1$, is *symmetric* if for all $x, y \in X$ $\chi_R(x, y) = \chi_R(y, x)$. The *inverse* of R is defined as function $\chi_{R^{-1}} \colon X \times X \to \{0, 1\}$ with membership function $\chi_{R^{-1}}(y, x) = \chi_R(x, y)$.

As defined so far, the membership function of a crisp set A assigns a value of either 1 or 0 to each individual of the universe set and, thus, discriminates between being a member or not being a member of A.

A *fuzzy set* [13] is characterised instead by a membership function $\chi_A \colon X \to [0, 1]$, or denoted simply $A \colon X \to [0, 1]$. With $\tilde{2}^X$ we denote the *fuzzy power set* over X, i.e. the set of all fuzzy sets over X. For instance, by referring to Fig. 1, the fuzzy set

$$C = \{x \mid x \text{ is a day with } heavy \text{ precipitation rate } R\}$$

is defined via the membership function

$$\chi_C(x) = \begin{cases} 1 & \text{if } R \geq 7.5 \\ (x - 5)/2.5 & \text{if } R \in [5, 7.5) \\ 0 & \text{otherwise}. \end{cases}$$

As pointed out previously, the definition of the membership function may depend on the context and may be subjective. Moreover, also the *shape* of such functions may be quite different. Luckily, the trapezoidal (Fig. 2(a)), the triangular

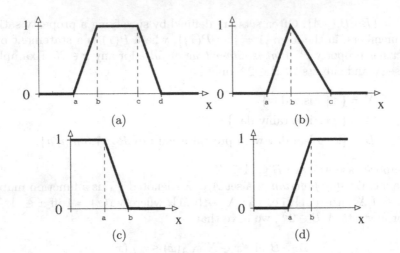

Fig. 2. (a) Trapezoidal function $trz(a, b, c, d)$; (b) Triangular function $tri(a, b, c)$; (c) L-function $ls(a, b)$; and (d) R-function $rs(a, b)$.

(Fig. 2(b)), the L-function (left-shoulder function, Fig. 2(c)), and the R-function (right-shoulder function, Fig. 2(d)) are simple, but most frequently used to specify membership degrees.

The usefulness of fuzzy sets depends critically on our capability to construct appropriate membership functions. The problem of constructing meaningful membership functions is a difficult one and we refer the interested reader to, e.g. [30, Chap. 10]. However, one easy and typically satisfactory method to define the membership functions (for a numerical domain) is to uniformly partition the range of, e.g. precipitation rates values (bounded by a minimum and maximum value), into 5 or 7 fuzzy sets using either trapezoidal functions (e.g. as illustrated in Fig. 3), or using triangular functions (as illustrated in Fig. 4). The latter one is the more used one, as it has less parameters.

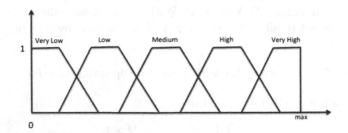

Fig. 3. Fuzzy sets construction using trapezoidal functions.

The standard fuzzy set operations are defined for any $x \in X$ as in Eqs. (2) and (3). Note also that the set inclusion defined as in Eq. (1) is indeed crisp in the sense that either $A \subseteq B$ or $A \nsubseteq B$.

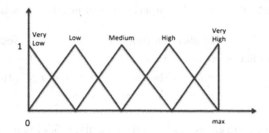

Fig. 4. Fuzzy sets construction using triangular functions.

Norm-Based Fuzzy Set Operations.
Standard fuzzy set operations are not the only ones that can be conceived to be suitable to generalise the classical Boolean operations. For each of the three types of operations there is a wide class of plausible fuzzy version. The most notable ones are characterised by the so-called class of *t-norms* $\otimes$ (called *triangular norms*), *t-conorms* $\oplus$ (also called *s-norm*), and *negation* $\ominus$ (see, e.g. [31]). An additional operator is used to define set inclusion (called *implication* $\Rightarrow$). Indeed, the *degree of subsumption* between two fuzzy sets A and B, denoted $A \sqsubseteq B$, is defined as $\inf_{x \in X} A(x) \Rightarrow B(x)$, where $\Rightarrow$ is an implication function.

An important aspect of such functions is that they satisfy some properties that one expects to hold (see Tables 1 and 2). Usually, the implication function $\Rightarrow$ is defined as *r-implication*, that is,

$$a \Rightarrow b = \sup \{c \mid a \otimes c \leq b\}.$$

Table 1. Properties for t-norms and s-norms.

Axiom name	T-norm	S-norm
Tautology/Contradiction	$a \otimes 0 = 0$	$a \oplus 1 = 1$
Identity	$a \otimes 1 = a$	$a \oplus 0 = a$
Commutativity	$a \otimes b = b \otimes a$	$a \oplus b = b \oplus a$
Associativity	$(a \otimes b) \otimes c = a \otimes (b \otimes c)$	$(a \oplus b) \oplus c = a \oplus (b \oplus c)$
Monotonicity	if $b \leq c$, then $a \otimes b \leq a \otimes c$	if $b \leq c$, then $a \oplus b \leq a \oplus c$

Of course, due to commutativity, $\otimes$ and $\oplus$ are monotone also in the first argument. We say that $\otimes$ is *indempotent* if $a \otimes a = a$, for any $a \in [0,1]$. For any $a \in [0,1]$, we say that a negation function $\ominus$ is *involutive* iff $\ominus \ominus a = a$. Salient negation functions are:

Standard or Łukasiewicz negation: $\ominus_l a = 1 - a$;
Gödelnegation: $\ominus_g a$ is 1 if $a = 0$, else is 0.

Table 2. Properties for implication and negation functions.

Axiom name	Implication function	Negation function
Tautology/contradiction	$0 \Rightarrow b = 1$, $a \Rightarrow 1 = 1$, $1 \Rightarrow 0 = 0$	$\ominus 0 = 1$, $\ominus 1 = 0$
Antitonicity	if $a \leq b$, then $a \Rightarrow c \geq b \Rightarrow c$	if $a \leq b$, then $\ominus a \geq \ominus b$
Monotonicity	if $b \leq c$, then $a \Rightarrow b \leq a \Rightarrow c$	

Of course, Łukasiewicznegation is involutive, while Gödel negation is not.

Salient t-norm functions are:

Gödel t-norm: $a \otimes_g b = min(a, b)$;
Bounded difference or Łukasiewicz t-norm: $a \otimes_l b = max(0, a + b - 1)$;
Algebraic product or product t-norm: $a \otimes_p b = a \cdot b$;
Drastic product: $a \otimes_d b = \begin{cases} 0 & \text{when } (a, b) \in [0, 1[\times [0, 1[\\ min(a, b) & \text{otherwise} \end{cases}$

Salient s-norm functions are:

Gödel s-norm: $a \oplus_g b = max(a, b)$;
Bounded sum or Łukasiewicz s-norm: $a \oplus_l b = min(1, a + b)$;
Algebraic sum or product s-norm: $a \oplus_p b = a + b - ab$;
Drastic sum: $a \oplus_d b = \begin{cases} 1 & \text{when } (a, b) \in]0, 1] \times]0, 1] \\ max(a, b) & \text{otherwise} \end{cases}$

We recall that the following important properties can be shown about t-norms and s-norms.

1. There is the following ordering among t-norms ($\otimes$ is any t-norm):

$$\otimes_d \leq \otimes \leq \otimes_g$$
$$\otimes_d \leq \otimes_l \leq \otimes_p \leq \otimes_g.$$

2. The only idempotent t-norm is $\otimes_g$.
3. The only t-norm satisfying $a \otimes a = 0$ for all $a \in [0, 1[$ is $\otimes_d$.
4. There is the following ordering among s-norms ($\oplus$ is any s-norm):

$$\oplus_g \leq \oplus \leq \oplus_d$$
$$\oplus_g \leq \oplus_p \leq \oplus_l \leq \oplus_d.$$

5. The only idempotent s-norm is $\oplus_g$.
6. The only s-norm satisfying $a \oplus a = 1$ for all $a \in]0, 1]$ is $\oplus_d$.

The *dual s-norm* of $\otimes$ is defined as

$$a \oplus b = 1 - (1 - a) \otimes (1 - b). \tag{4}$$

Some t-norms, s-norms, implication functions, and negation functions are shown in Table 3. One usually distinguishes three different sets of fuzzy set operations (called fuzzy logics), namely, Łukasiewicz, Gödel, and Product logic; the popular

Table 3. Combination functions of various fuzzy logics.

	Łukasiewicz logic	Gödel logic	Product logic	SFL
$a \otimes b$	$max(a + b - 1, 0)$	$min(a, b)$	$a \cdot b$	$min(a, b)$
$a \oplus b$	$min(a + b, 1)$	$max(a, b)$	$a + b - a \cdot b$	$max(a, b)$
$a \Rightarrow b$	$min(1 - a + b, 1)$	$\begin{cases} 1 & \text{if } a \leq b \\ b & \text{otherwise} \end{cases}$	$min(1, b/a)$	$max(1 - a, b)$
$\ominus a$	$1 - a$	$\begin{cases} 1 & \text{if } a = 0 \\ 0 & \text{otherwise} \end{cases}$	$\begin{cases} 1 & \text{if } a = 0 \\ 0 & \text{otherwise} \end{cases}$	$1 - a$

Table 4. Some additional properties of combination functions of various fuzzy logics.

Property	Łukasiewicz logic	Gödel logic	Product logic	SFL
$x \otimes \ominus x = 0$	+	−	−	−
$x \oplus \ominus x = 1$	+	−	−	−
$x \otimes x = x$	−	+	−	+
$x \oplus x = x$	−	+	−	+
$\ominus \ominus x = x$	+	−	−	+
$x \Rightarrow y = \ominus x \oplus y$	+	−	−	+
$\ominus (x \Rightarrow y) = x \otimes \ominus y$	+	−	−	+
$\ominus (x \otimes y) = \ominus x \oplus \ominus y$	+	+	+	+
$\ominus (x \oplus y) = \ominus x \otimes \ominus y$	+	+	+	+

Standard Fuzzy Logic (SFL) is a sublogic of Łukasiewicz logic as $min(a, b) = a \otimes_l (a \Rightarrow_l b)$ and $max(a, b) = 1 - min(1 - a, 1 - b)$. The importance of these three logics is due to the Mostert–Shields theorem [32] that states that any continuous t-norm can be obtained as an ordinal sum of these three (see also [33]).

The implication $x \Rightarrow y = max(1 - x, y)$ is called *Kleene-Dienes implication* in the fuzzy logic literature. Note that we have the following inferences: let $a \geq n$ and $a \Rightarrow b \geq m$. Then, under Kleene-Dienes implication, we infer that if $n > 1 - m$ then $b \geq m$. Under r-implication relative to a t-norm $\otimes$, we infer that $b \geq n \otimes m$.

The *composition* of two fuzzy relations $R_1 \colon X \times X \to [0, 1]$ and $R_2 \colon X \times X \to [0, 1]$ is defined as $(R_1 \circ R_2)(x, z) = \sup_{y \in X} R_1(x, y) \otimes R_2(y, z)$. A fuzzy relation R is *transitive* iff $R(x, z) \geqslant (R \circ R)(x, z)$.

Fuzzy Modifiers.
Fuzzy modifiers are an interesting feature of fuzzy set theory. Essentially, a fuzzy modifier, such as `very`, `more_or_less`, and `slightly`, apply to fuzzy sets to change their membership function.

Formally, a *fuzzy modifier* m represents a function
$f_m \colon [0, 1] \to [0, 1]$.

For example, we may define $f_{\mathtt{very}}(x) = x^2$ and $f_{\mathtt{slightly}}(x) = \sqrt{x}$. In this way, we may express the fuzzy set of very heavy rain by applying the modifier *very* to the fuzzy membership function of "heavy rain" i.e.

$$\chi_{\mathtt{very\,heavyrain}}(x) = f_{\mathit{very}}(\chi_{\mathtt{heavyrain}}(x)) = (\chi_{\mathtt{heavyrain}}(x))^2 = (rs(5,7.5)(x))^2.$$

A typical shape of modifiers is the so-called *linear modifiers*, as illustrated in Fig. 5. Note that such a modifier can be parameterized by means of one parameter c only, i.e. $lm(a,b) = lm(c)$, where $a = c/(c+1)$, $b = 1/(c+1)$.

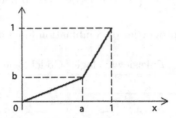

Fig. 5. Linear modifier $lm(a,b)$.

2.2   Mathematical Fuzzy Logic Basics

We recap here briefly that in *Mathematical Fuzzy Logic* [33], the convention prescribing that a statement is either true or false is changed and is a matter of degree measured on an ordered scale that is no longer $\{0,1\}$, but $[0,1]$. This degree is called *degree of truth* of the logical statement φ in the interpretation $\mathcal{I}$. *Fuzzy statements* have the form $\langle\varphi, r\rangle$, where $r \in [0,1]$ (see, e.g. [33,34]) and φ is a statement, which encodes that the degree of truth of φ is *greater or equal* r. A *fuzzy interpretation* $\mathcal{I}$ maps each basic statement p_i into $[0,1]$ and is then extended inductively to all statements:

$$\begin{aligned}
\mathcal{I}(\varphi \wedge \psi) &= \mathcal{I}(\varphi) \otimes \mathcal{I}(\psi) \\
\mathcal{I}(\varphi \vee \psi) &= \mathcal{I}(\varphi) \oplus \mathcal{I}(\psi) \\
\mathcal{I}(\varphi \to \psi) &= \mathcal{I}(\varphi) \Rightarrow \mathcal{I}(\psi) \\
\mathcal{I}(\varphi \leftrightarrow \psi) &= \mathcal{I}(\varphi \to \psi) \otimes \mathcal{I}(\psi \to \varphi) \\
\mathcal{I}(\neg\varphi) &= \ominus \mathcal{I}(\varphi) \\
\mathcal{I}(\exists x.\varphi) &= \sup_{a \in \Delta^{\mathcal{I}}} \mathcal{I}_x^a(\varphi) \\
\mathcal{I}(\forall x.\varphi) &= \inf_{a \in \Delta^{\mathcal{I}}} \mathcal{I}_x^a(\varphi),
\end{aligned} \qquad (5)$$

where $\Delta^{\mathcal{I}}$ is the domain of $\mathcal{I}$, and $\otimes$, $\oplus$, $\Rightarrow$, and $\ominus$ are the *t-norms, t-conorms, implication functions, a negation functions* we have seen in the previous section.[3]
One may also consider the following abbreviations:

$$\varphi \wedge_g \psi \stackrel{\mathtt{def}}{=} \varphi \wedge (\varphi \to \psi) \qquad (6)$$

$$\varphi \vee_g \psi \stackrel{\mathtt{def}}{=} ((\varphi \to \psi) \to \varphi) \wedge_g ((\psi \to \varphi) \to \psi) \qquad (7)$$

$$\neg_\otimes \varphi \stackrel{\mathtt{def}}{=} \varphi \to 0. \qquad (8)$$

---

[3] The function $\mathcal{I}_x^a$ is as $\mathcal{I}$ except that x is interpreted as a.

In case $\Rightarrow$ is the r-implication based on $\otimes$, then $\wedge_g$ (resp. $\vee_g$) is interpreted as Gödel t-norm (resp. s-norm), while $\neg_\otimes$ is interpreted as the negation function related to $\otimes$.

A fuzzy interpretation $\mathcal{I}$ *satisfies* a fuzzy statement $\langle\varphi, r\rangle$, or $\mathcal{I}$ is a *model* of $\langle\varphi, r\rangle$, denoted $\mathcal{I} \models \langle\varphi, r\rangle$, iff $\mathcal{I}(\varphi) \geq r$. We say that $\mathcal{I}$ is a *model* of φ if $\mathcal{I}(\varphi) = 1$. A *fuzzy knowledge base* (or simply knowledge base, if clear from context) is a set of fuzzy statements and an interpretation $\mathcal{I}$ *satisfies* (is a *model* of) a knowledge base, denoted $\mathcal{I} \models \mathcal{K}$, iff it satisfies each element in it.

We say $\langle\varphi, n\rangle$ is a *tight logical consequence* of a set of fuzzy statements $\mathcal{K}$ iff n is the infimum of $\mathcal{I}(\varphi)$ subject to all models $\mathcal{I}$ of $\mathcal{K}$. Notice that the latter is equivalent to $n = \sup\{r \mid \mathcal{K} \models \langle\varphi, r\rangle\}$. n is called the *best entailment degree* of φ w.r.t. $\mathcal{K}$ (denoted $bed(\mathcal{K}, \varphi)$), i.e.

$$bed(\mathcal{K}, \varphi) = \sup\{r \mid \mathcal{K} \models \langle\varphi, r\rangle\}. \tag{9}$$

On the other hand, the *best satisfiability degree* of φ w.r.t. $\mathcal{K}$ (denoted $bsd(\mathcal{K}, \varphi)$) is

$$bsd(\mathcal{K}, \varphi) = \sup_{\mathcal{I}}\{\mathcal{I}(\varphi) \mid \mathcal{I} \models \mathcal{K}\}. \tag{10}$$

Of course, the properties of Table 4 immediately translate into equivalence among formulae. For instance, the following equivalences hold (in brackets we indicate the logic for which the equivalences holds)

$$\neg\neg\varphi \equiv \varphi \quad (\mathcal{L})$$

$$\varphi \wedge \varphi \equiv \varphi \quad (G)$$

$$\neg(\varphi \wedge \neg\varphi) \equiv 1 \quad (\mathcal{L}, G, \Pi)$$

$$\varphi \vee \neg\varphi \equiv 1 \quad (\mathcal{L}).$$

Remark 1. Unlike the classical case, in general, we do not have that $\forall x.\varphi$ and $\neg\exists x.\neg\varphi$ are equivalent. They are equivalent for Łukasiewicz logic and SFL, but are neither equivalent for Gödel nor for Product logic. For instance, under Gödel negation, just consider an interpretation $\mathcal{I}$ with domain $\{a\}$ and $\mathcal{I}(p(a)) = u$, with $0 < u < 1$. Then $\mathcal{I}(\forall x.p(x)) = u$, while $\mathcal{I}(\neg\exists x.\neg p(x)) = 1$ and, thus, $\forall x.p(x) \not\equiv \neg\exists x.\neg p(x)$.

We refer the reader to [2] for an overview of reasoning algorithms for fuzzy propositional and First-Order Logics.

2.3 Conjunctive Queries

The classical case.
In case a KB is a classical knowledge base, a *conjunctive query* is a rule-like expression of the form

$$q(\mathbf{x}) \leftarrow \exists\mathbf{y}.\varphi(\mathbf{x}, \mathbf{y}) \tag{11}$$

where the rule body $\varphi(\mathbf{x}, \mathbf{y})$ is a conjunction[4] of predicates $P_i(\mathbf{z}_i)$ $(1 \leq i \leq n)$ and $\mathbf{z}_i$ is a vector of distinguished or non-distinguished variables.

For instance,

$$q(x, y) \leftarrow AdultPerson(x), Age(x, y)$$

is a conjunctive query, whose intended meaning is to retrieve all adult people and their age.

Given a vector $\mathbf{x} = \langle x_1, \ldots, x_k \rangle$ of variables, a *substitution* over $\mathbf{x}$ is a vector of individuals $\mathbf{t}$ replacing variables in $\mathbf{x}$ with individuals. Then, given a query $q(\mathbf{x}) \leftarrow \exists \mathbf{y}.\varphi(\mathbf{x}, \mathbf{y})$, and two substitutions $\mathbf{t}, \mathbf{t}'$ over $\mathbf{x}$ and $\mathbf{y}$, respectively, the *query instantiation* $\varphi(\mathbf{t}, \mathbf{t}')$ is derived from $\varphi(\mathbf{x}, \mathbf{y})$ by replacing $\mathbf{x}$ and $\mathbf{y}$ with $\mathbf{t}$ and $\mathbf{t}'$, respectively.

We adopt here the following notion of entailment. Given a knowledge base $\mathcal{K}$, a query $q(\mathbf{x}) \leftarrow \exists \mathbf{y}.\varphi(\mathbf{x}, \mathbf{y})$, and a vector $\mathbf{t}$ of individuals occurring in $\mathcal{K}$, we say that $q(\mathbf{t})$ is *entailed* by $\mathcal{K}$, denoted $\mathcal{K} \models q(\mathbf{t})$, if and only if there is a vector $\mathbf{t}'$ of individuals occurring in $\mathcal{K}$ such that in any two-valued model $\mathcal{I}$ of $\mathcal{K}$, $\mathcal{I}$ is a model of any atom in the query instantiation $\varphi(\mathbf{t}, \mathbf{t}')$.

If $\mathcal{K} \models q(\mathbf{t})$ then $\mathbf{t}$ is called a *answer* to q. We call these kinds of answers also *certain answers*. The *answer set* of q w.r.t. $\mathcal{K}$ is defined as

$$ans(\mathcal{K}, q) = \{\mathbf{t} \mid \mathcal{K} \models q(\mathbf{t})\}.$$

The fuzzy case.

Consider a new alphabet of *fuzzy variables* (denoted Λ). To start with, a *fuzzy query* is of the form

$$\langle q(\mathbf{x}), \Lambda \rangle \leftarrow \exists \mathbf{y} \exists \mathbf{\Lambda}'.\varphi(\mathbf{x}, \Lambda, \mathbf{y}, \mathbf{\Lambda}') \tag{12}$$

in which $\varphi(\mathbf{x}, \Lambda, \mathbf{y}, \mathbf{\Lambda}')$ is a conjunction (as for the crisp case, we use "," as conjunction symbol) of fuzzy predicates and built-in predicates, $\mathbf{x}$ and Λ are the distinguished variables, $\mathbf{y}$ and $\mathbf{\Lambda}'$ are the vectors of *non-distinguished variables* (existential quantified variables), and $\mathbf{x}$, Λ, $\mathbf{y}$ and $\mathbf{\Lambda}'$ are pairwise disjoint. Variable Λ and variables in $\mathbf{\Lambda}'$ can only appear in place of degrees of truth or built-in predicates. The query head contains at least one variable.

For instance, the query

$$\langle q(x), s \rangle \leftarrow \langle SportsCar(x), s_1 \rangle, hasPrice(x, y), s := s_1 \cdot ls(10000, 15000)(y)$$

has intended meaning to retrieve all cheap sports cars. Any answer x is scored according to the product of being cheap and a sports car, were cheap is encode as the fuzzy membership function $ls(10000, 15000)$.

From a semantics point of view, given a fuzzy KB $\mathcal{K}$, a query $\langle q(\mathbf{x}), \Lambda \rangle \leftarrow \exists \mathbf{y} \exists \mathbf{\Lambda}'.\varphi(\mathbf{x}, \Lambda, \mathbf{y}, \mathbf{\Lambda}')$, a vector $\mathbf{t}$ of individuals occurring in $\mathcal{K}$ and a truth degree λ in $[0, 1]$, we say that $\langle q(\mathbf{t}), \lambda \rangle$ is *entailed* by $\mathcal{K}$, denoted $\mathcal{K} \models \langle q(\mathbf{t}), \lambda \rangle$, if and only if there is a vector $\mathbf{t}'$ of individuals occurring $\mathcal{K}$ and a vector $\mathbf{\lambda}'$ of truth degrees in $[0, 1]$ such that for any model $\mathcal{I}$ of $\mathcal{K}$, $\mathcal{I}$ is a model of all fuzzy atoms

---

[4] We use the symbol "," to denote conjunction in the rule body.

occurring in $\varphi(\mathbf{t}, \lambda, \mathbf{t}', \lambda')$. If $\mathcal{K} \models \langle q(\mathbf{t}), \lambda \rangle$ then $\langle \mathbf{t}, \lambda \rangle$ is called an *answer* to q. The *answer set* of q w.r.t. $\mathcal{K}$ is

$$ans(\mathcal{K}, q) = \{ \langle \mathbf{t}, \lambda \rangle \mid \mathcal{K} \models \langle q(\mathbf{t}), \lambda \rangle, \lambda \neq 0 \text{ and}$$
$$\text{for any } \lambda' \neq \lambda \text{ such that } \mathcal{K} \models \langle q(\mathbf{t}), \lambda' \rangle, \lambda' \leq \lambda \text{ holds}\}.$$

That is, for any tuple $\mathbf{t}$, the truth degree λ is as large as possible.

Fuzzy queries with aggregation operators.
We may extend conjunctive queries to disjunctive queries and to queries including aggregation operators as well. Formally, let • be an aggregate function with
  • $\in \{\mathsf{SUM}, \mathsf{AVG}, \mathsf{MAX}, \mathsf{MIN}, \mathsf{COUNT}, \oplus, \otimes\}$
then a query with aggregates is of the form

$$\langle q(\mathbf{x}), \Lambda \rangle \leftarrow \exists \mathbf{y} \exists \mathbf{\Lambda}'. \varphi(\mathbf{x}, \mathbf{y}, \mathbf{\Lambda}'),$$
$$\mathsf{GroupedBy}(\mathbf{w}), \tag{13}$$
$$\Lambda := \bullet\, [f(\mathbf{z})],$$

where $\mathbf{w}$ are variables in $\mathbf{x}$ or $\mathbf{y}$ and each variable in $\mathbf{x}$ occurs in $\mathbf{w}$ and any variable in $\mathbf{z}$ occurs in $\mathbf{y}$ or $\mathbf{\Lambda}'$.

From a semantics point of view, we say that $\mathcal{I}$ *is a model of* (*satisfies*) $\langle q(\mathbf{t}), \lambda \rangle$, denoted $\mathcal{I} \models \langle q(\mathbf{t}), \lambda \rangle$ if and only if

$\lambda = \bullet[\lambda_1, \dots, \lambda_k]$ where $g = \{ \langle \mathbf{t}, \mathbf{t}'_1, \lambda'_1 \rangle, \dots, \langle \mathbf{t}, \mathbf{t}'_k, \lambda'_k \rangle \}$,
  is a group of k tuples with identical projection
  on the variables in $\mathbf{w}$, $\varphi(\mathbf{t}, \mathbf{t}'_r, \lambda'_r)$ is true in $\mathcal{I}$
  and $\lambda_r = f(\mathbf{t})$ where $\mathbf{t}$ is the projection of $\langle \mathbf{t}'_r, \lambda'_r \rangle$
  on the variables $\mathbf{z}$.

Now, the notion of $\mathcal{K} \models \langle q(\mathbf{t}), \lambda \rangle$ is as usual: any model of $\mathcal{K}$ is a model of $\langle q(\mathbf{t}), \lambda \rangle$.

The notion of answer and answer set of a disjunctive query is a straightforward extension of the ones for conjunctive queries.

Top-k Retrieval.
As now each answer to a query has a degree of truth (i.e. *score*), a basic inference problem that is of interest is the top-k retrieval problem, formulated as follows.

Given a fuzzy KB $\mathcal{K}$, and a query q, retrieve k answers $\langle \mathbf{t}, \lambda \rangle$ with maximal degree and rank them in decreasing order relative to the degree λ, denoted
  $ans_k(\mathcal{K}, q) = \mathsf{Top}_k \, ans(\mathcal{K}, q)$.

2.4   Annotation Domains

We have seen that fuzzy statements extend statements with an *annotation* $r \in [0, 1]$. Interestingly, we may further generalise this by allowing a statement being annotated with a value λ taken from a so-called *annotation domain* [16,26–28,39],[5] which allow to deal with several domains (such as, fuzzy, temporal,

---

[5] The readers familiar with the annotated logic programming framework [35], will notice the similarity of the approaches.

provenance) and their combination, in a uniform way. Formally, let us consider a non-empty set L. Elements in L are our annotation values. For example, in a fuzzy setting, $L = [0, 1]$, while in a typical temporal setting, L may be time points or time intervals. In the annotation framework, an interpretation will map statements to elements of the annotation domain. Now, an *annotation domain* is an idempotent, commutative semi-ring

$$D = \langle L, \oplus, \otimes, \bot, \top \rangle,$$

where $\oplus$ is $\top$-annihilating [39]. That is, for $\lambda, \lambda_i \in L$

1. $\oplus$ is idempotent, commutative, associative;
2. $\otimes$ is commutative and associative;
3. $\bot \oplus \lambda = \lambda$, $\top \otimes \lambda = \lambda$, $\bot \otimes \lambda = \bot$, and $\top \oplus \lambda = \top$;
4. $\otimes$ is distributive over $\oplus$, i.e. $\lambda_1 \otimes (\lambda_2 \oplus \lambda_3) = (\lambda_1 \otimes \lambda_2) \oplus (\lambda_1 \otimes \lambda_3)$;

It is well-known that there is a natural partial order on any idempotent semi-ring: an annotation domain $D = \langle L, \oplus, \otimes, \bot, \top \rangle$ induces a partial order $\preceq$ over L defined as:

$$\lambda_1 \preceq \lambda_2 \text{ if and only if } \lambda_1 \oplus \lambda_2 = \lambda_2.$$

The order $\preceq$ is used to express redundant/entailed/subsumed information. For instance, for temporal intervals, an annotated statement $\langle \varphi, [2000, 2006] \rangle$ entails $\langle \varphi, [2003, 2004] \rangle$, as $[2003, 2004] \subseteq [2000, 2006]$ (here, $\subseteq$ plays the role of $\preceq$).

Remark 2. $\oplus$ is used to combine information about the same statement. For instance, in temporal logic, from $\langle \varphi, [2000, 2006] \rangle$ and $\langle \varphi, [2003, 2008] \rangle$, we infer $\langle \varphi, [2000, 2008] \rangle$, as $[2000, 2008] = [2000, 2006] \cup [2003, 2008]$; here, $\cup$ plays the role of $\oplus$. In the fuzzy context, from $\langle \varphi, 0.7 \rangle$ and $\langle \varphi, 0.6 \rangle$, we infer $\langle \varphi, 0.7 \rangle$, as $0.7 = max(0.7, 0.6)$ (here, max plays the role of $\oplus$).

Remark 3. $\otimes$ is used to model the "conjunction" of information. In fact, a $\otimes$ is a generalisation of boolean conjunction to the many-valued case. In fact, $\otimes$ satisfies also that

1. $\otimes$ is bounded: i.e. $\lambda_1 \otimes \lambda_2 \preceq \lambda_1$.
2. $\otimes$ is $\preceq$-monotone, i.e. for $\lambda_1 \preceq \lambda_2$, $\lambda \otimes \lambda_1 \preceq \lambda \otimes \lambda_2$

For instance, on interval-valued temporal logic, from $\langle \varphi, [2000, 2006] \rangle$ and $\langle \varphi \to \psi, [2003, 2008] \rangle$, we may infer $\langle \psi, [2003, 2006] \rangle$, as $[2003, 2006] = [2000, 2006] \cap [2003, 2008]$; here, $\cap$ plays the role of $\otimes$. In the fuzzy context, one may chose any t-norm [31,33], e.g. product, and, thus, from $\langle \varphi, 0.7 \rangle$ and $\langle \varphi \to \psi, 0.6 \rangle$, we will infer $\langle \psi, 0.42 \rangle$, as $0.42 = 0.7 \cdot 0.6$ (here, $\cdot$ plays the role of $\otimes$).

Remark 4. Observe that the distributivity condition is used to guarantee that e.g. we obtain the same annotation $\lambda \otimes (\lambda_2 \oplus \lambda_3) = (\lambda_1 \otimes \lambda_2) \oplus (\lambda_1 \otimes \lambda_3)$ of ψ that can be inferred from $\langle \varphi, \lambda_1 \rangle$, $\langle \varphi \to \psi, \lambda_2 \rangle$ and $\langle \varphi \to \psi, \lambda_3 \rangle$.

Note that, conceptually, in order to build an annotation domain, one has to:

1. determine the set of annotation values L (typically a countable set[6]), identify the top and bottom elements;
2. define a suitable operations $\otimes$ and $\oplus$ that acts as "conjunction" and "disjunction" function, to support the intended inferences.

Eventually, *annotated queries* are as fuzzy queries in which annotation variables and terms are used in place of fuzzy variables and values $r \in [0,1]$ instead. We refer the reader to [28] for more details about annotation domains.

3 Fuzzy Logic and Semantic Web Languages

We have seen in the previous section how to "fuzzyfy" a classical language such as propositional logic and FOL, namely fuzzy statements are of the form $\langle \varphi, r \rangle$, where φ is a statement and $r \in [0,1]$.

The natural extension to SWLs consists then in replacing φ with appropriate expressions belonging to the logical counterparts of SWLs, namely ρdf, DLs and LPs, as we will illustrate next.

3.1 Fuzzy RDFS

The basic ingredients of *RDF* are *triples* of the form (s, p, o), such as $(umberto, likes, tomato)$, stating that *subject s* has *property p* with *value o*. In *RDF Schema* (RDFS), which is an extension of RDF, additionally some special keywords may be used as properties to further improve the expressivity of the language. For instance we may also express that the class of 'tomatoes are a subclass of the class of vegetables', $(tomato, \text{sc}, vegetables)$, while Zurich is an instance of the class of cities, $(zurich, \text{type}, city)$.

Form a computational point of view, one computes the so-called *closure* (denoted $cl(\mathcal{K})$) of a set of triples $\mathcal{K}$. That is, one infers all possible triples using inference rules [9, 36, 37], such as

$$\frac{(A, \text{sc}, B), (X, \text{type}, A)}{(X, \text{type}, B)}$$

"if A subclass of B and X instance of A then infer that X is instance of B",

and then store all inferred triples into a relational database to be used then for querying. We recall also that there also several ways to store the closure $cl(\mathcal{K})$ in a database (see [38, 40]). Essentially, either we may store all the triples in table with three columns *subject, predicate, object*, or we use a table for each predicate, where each table has two columns *subject, object*. The latter approach seems to be better for query answering purposes.

---

[6] Note that one may use XML decimals in $[0, 1]$ in place of real numbers for the fuzzy domain.

In *Fuzzy RDFS* (see [2,15] and references therein), triples are annotated with a degree of truth in $[0,1]$. For instance, "Rome is a big city to degree 0.8" can be represented with $\langle(Rome, \mathsf{type}, BigCity), 0.8\rangle$. More formally, *fuzzy triples* are expressions of the form $\langle\tau, r\rangle$, where τ is a RDFS triple (the truth value r may be omitted and, in that case, the value $r = 1$ is assumed).

The interesting point is that from a computational point of view the inference rules parallel those for "crisp" RDFS: indeed, the rules are of the form

$$\frac{\langle\tau_1, r_1\rangle, \ \ldots, \ \langle\tau_k, r_k\rangle, \{\tau_1, \ldots, \tau_k\} \vdash_{\mathsf{RDFS}} \tau}{\langle\tau, \bigotimes_i r_i\rangle} \tag{14}$$

Essentially, this rule says that if a classical RDFS triple τ can be inferred by applying a classical RDFS inference rule to triples $\tau_1, \ldots, \tau_k$ (denoted $\{\tau_1, \ldots, \tau_k\} \vdash_{\mathsf{RDFS}} \tau$), then the truth degree of τ will be $\bigotimes_i r_i$.

As a consequence, the rule system is quite easy to implement for current inference systems. Specifically, as for the crisp case, one may compute the closure $cl(\mathcal{K})$ of a set of fuzzy triples $\mathcal{K}$, store them in a relational database and thereafter query the database.

Concerning conjunctive queries, they are essentially the same as in Sect. 2.3, where predicates are replaced with triples. For instance, the query

$$\langle q(x), s\rangle \leftarrow \langle(x, \mathsf{type}, SportsCar), s_1\rangle, (x, hasPrice, y), s = s_1 \cdot cheap(y) \tag{15}$$

where e.g. $cheap(y) = ls(10000, 15000)(y)$, has intended meaning to retrieve all cheap sports car. Then, any answer is scored according to the product of being cheap and a sports car.

Annotation Domains and RDFS

The generalisation to annotation domains is conceptual easy, as now one may replace truth degrees with annotation terms taken from an appropriate domain. For further details see [28].

3.2 Fuzzy DLs

Description Logics (DLs) [10] are the logical counterpart of the family of OWL languages. So, to illustrate the basic concepts of fuzzy OWL, it suffices to show the fuzzy DL case (see [2,17,41], for a survey). We recap that the basic ingredients are the descriptions of classes, properties, and their instances, such as

- $a{:}C$, such as $a{:}\mathsf{Person} \sqcap \forall\mathsf{hasChild.Femal}$, meaning that individual a is an instance of concept/class C (here C is seen as a unary predicate);
- $(a, b){:}R$, such as $(\mathsf{tom}, \mathsf{mary}){:}\mathsf{hasChild}$, meaning that the pair of individuals $\langle a, b\rangle$ is an instance of the property/role R (here R is seen as a binary predicate);
- $C \sqsubseteq D$, such as $\mathsf{Person} \sqsubseteq \forall\mathsf{hasChild.Person}$, meaning that the class C is a subclass of class D;

So far, several *fuzzy* variants of DLs have been proposed: they can be classified according to

- the description logic resp. ontology language that they generalize [24,42–64];
- the allowed fuzzy constructs [60,65–89];
- the underlying fuzzy logic [90–97];
- their reasoning algorithms and computational complexity results [18,19,90, 98–134].

In general, fuzzy DLs allow expressions of the form $\langle a{:}C, r\rangle$, stating that a is an instance of concept/class C with degree at least r, i.e. the FOL formula $C(a)$ is true to degree at least r. Similarly, $\langle C_1 \sqsubseteq C_2, r\rangle$ states a vague subsumption relationships. Informally, $\langle C_1 \sqsubseteq C_2, r\rangle$ dictates that the FOL formula $\forall x.C_1(x) \rightarrow C_2(x)$ is true to degree at least r. Essentially, *fuzzy DLs* are then obtained by interpreting the statements as fuzzy FOL formulae and attaching a weight n to DL statements, thus, defining so *fuzzy DL statements*.

Example 1. Consider the following background knowledge about cars:

$$Car \sqsubseteq \exists HasPrice.Price$$
$$Sedan \sqsubseteq Car$$
$$Van \sqsubseteq Car$$
$$CheapPrice \sqsubseteq Price$$
$$ModeratePrice \sqsubseteq Price$$
$$ExpensivePrice \sqsubseteq Price$$
$$\langle CheapPrice \sqsubseteq ModeratePrice, 0.7\rangle$$
$$\langle ModeratePrice \sqsubseteq ExpensivePrice, 0.4\rangle$$
$$CheapCar = Car \sqcap \exists HasPrice.CheapPrice$$
$$ModerateCar = Car \sqcap \exists HasPrice.ModeratePrice$$
$$ExpensiveCar = Car \sqcap \exists HasPrice.ExpensivePrice$$

Essentially, the vague concepts here are *CheapPrice, ModeratePrice*, and *ExpensivePrice* and the graded GCIs declare to which extent there is a relationship among them.

The facts about two specific cars a and b are encoded with:

$$\langle a{:}Sedan \sqcap \exists HasPrice.CheapPrice, 0.7\rangle$$
$$\langle b{:}Van \sqcap \exists HasPrice.ModeratePrice, 0.8\rangle.$$

So, a is a sedan having a cheap price, while b is a van with a moderate price.

Under Gödel semantics it can be shown that

$$\mathcal{K} \models \langle a{:}ModerateCar, 0.7\rangle$$
$$\mathcal{K} \models \langle b{:}ExpensiveCar, 0.4\rangle.$$

From a decision procedure point of view, a popular approach consists of a set of inference rules that generate a set of In-equations (that depend on the t-norm and fuzzy concept constructors) that have to be solved by an operational research solver (see, e.g. [60,92]). An informal rule example is as follows:

"If individual a is instance of the class intersection $C_1 \sqcap C_2$ to degree greater or equal to $x_{a{:}C_1 \sqcap C_2}$,[7] then a is instance of C_i ($i = 1, 2$) to degree

---

[7] For a fuzzy DL formula φ we consider a variable x_φ with intended meaning: the degree of truth of φ is greater or equal to x_φ.

greater or equal to $x_{a:C_i}$, where additionally the following in-equation holds: $x_{a:C_1 \sqcap C_2} \leq x_{a:C_1} \otimes x_{a:C_2}$."

Concerning conjunctive queries, they are essentially the same as in Sect. 2.3, where predicates are replaced with unay and binary predicates. For instance, the fuzzy DL analogue of the RDFS query (15) is

$$\langle q(x), s \rangle \leftarrow \langle SportsCar(x), s_1 \rangle, HasPrice(x, y), s := s_1 \cdot cheap(y). \qquad (16)$$

Applications.
Fuzzy set theory and fuzzy logic [13] have proved to be suitable formalisms to handle fuzzy knowledge. Not surprisingly, *fuzzy ontologies* already emerge as useful in several applications, such as information retrieval [135–141], recommendation systems [142–145], image interpretation [146–152], the Semantic Web and the Internet [153–155], ambient intelligence [156–159], ontology merging [160,161], matchmaking [21,162–169], decision making [170], summarization [171], robotics [172,173], machine learning [174–182] and many others [88,183–193].

Representing Fuzzy OWL Ontologies in OWL.
OWL [194] and its successor OWL 2 [3,195] are standard W3C languages for defining and instantiating Web ontologies whose logical counterpart are classical DLs. So far, several fuzzy extensions of DLs exists and some fuzzy DL reasoners have been implemented, such as FUZZYDL [65,196], DELOREAN [42], FIRE [197,198], SOFTFACTS [199], *GURDL* [200], *GERDS* [201], *YADLR* [202], *FRESG* [203] and DLMEDIA [139,204]. Not surprisingly, each reasoner uses its own fuzzy DL language for representing fuzzy ontologies and, thus, there is a need for a standard way to represent such information. A first possibility would be to adopt as a standard one of the fuzzy extensions of the languages OWL and OWL 2 that have been proposed, such as [58,205,206]. However, as it is not expected that a fuzzy OWL extension will become a W3C proposed standard in the near future, [69,207,208] identifies the syntactic differences that a fuzzy ontology language has to cope with, and proposes to use OWL 2 *itself* to represent fuzzy ontologies [209].

Annotation Domains and OWL
The generalisation to annotation domains is conceptual easy, as now one may replace truth degrees with annotation terms taken from an appropriate domain (see, e.g. [97,114,116]).

3.3   Fuzzy Rule Languages

The foundation of the core part of rule languages is *Datalog* [210], i.e. a Logic Programming Language (LP) [11]. In LP, the management of imperfect information has attracted the attention of many researchers and numerous frameworks have been proposed. Addressing all of them is almost impossible, due to both the large number of works published in this field (early works date back to early

80-ties [211]) and the different approaches proposed (see, e.g. [1]). Below a list of references.[8]

Fuzzy set theory: [211–239]
Multi-valued logic: [20–25, 35, 50, 52, 53, 62, 96, 164–167, 240–303]

Basically [11], a Datalog program $\mathcal{P}$ is made out by a set of rules and a set of facts. *Facts* are ground *atoms* of the form $P(\mathbf{c})$. On the other hand rules are similar as conjunctive queries and are of the form
$$A(\mathbf{x}) \leftarrow \exists \mathbf{y}.\varphi(\mathbf{x}, \mathbf{y}),$$
where $\varphi(\mathbf{x}, \mathbf{y})$ is a conjunction of n-ary predicates. A *query* is a rule and the *answer set* of a query q w.r.t. a set $\mathcal{K}$ of facts and rules is the set of tuples $\mathbf{t}$ such that there exists $\mathbf{t}'$ such that the instantiation $\varphi(\mathbf{t}, \mathbf{t}')$ of the query body is true in *minimal model* of $\mathcal{K}$, which is guaranteed to exists.

In the *fuzzy* case, rules and facts are as for the crisp case, except that now a predicate is annotated. An example of fuzzy rule defining good hotels may be the following:

$$\langle GoodHotel(x), s \rangle \leftarrow Hotel(x), \langle Cheap(x), s_1 \rangle, \langle CloseToVenue(x), s_2 \rangle,$$
$$\langle Comfortable(x), s_3 \rangle, s := 0.3 \cdot s_1 + 0.5 \cdot s_2 + 0.2 \cdot s_3 \quad (17)$$

A *fuzzy query* is a fuzzy rule and, informally, the *fuzzy answer set* is the ordered set of weighted tuples $\langle \mathbf{t}, s \rangle$ such that all the fuzzy atoms in the rule body are true in the minimal model and s is the result of the scoring function f applied to its arguments. The existence of a minimal is guaranteed if the scoring functions in the query and in the rule bodies are *monotone* [1].

We conclude by saying that most works deal with logic programs without negation and some may provide some technique to answer queries in a top-down manner, as e.g. [23, 35, 235, 242, 269]. Deciding whether a wighted tuple $\langle \mathbf{t}, s \rangle$ is the answer set is undecidable in general, though is decidable if the truth space is finite and fixed a priory, as then the minimal model is finite.

Another rising problem is the problem to compute the top-k ranked answers to a query, without computing the score of all answers. This allows to answer queries such as "find the top-k closest hotels to the conference location". Solutions to this problem can be found in [25, 52, 299].

Annotation Domains and Rule Languages

The generalisation of fuzzy rule languages to the case in which an annotation $r \in [0, 1]$ is replaced with an annotation value λ taken from an annotation domain is straightforward and proceeds as for the other SWLs.

---

[8] The list of references is by no means intended to be all-inclusive. The author apologises both to the authors and with the readers for all the relevant works, which are not cited here.

4 Conclusions

In this chapter, we have provided a "crash course" through fuzzy DLs, by illustrating the basic concepts involved in. For a more in depth presentation, we refer the reader to [2].

References

1. Straccia, U.: Managing uncertainty and vagueness in description logics, logic programs and description logic programs. In: Baroglio, C., Bonatti, P.A., Małuszyński, J., Marchiori, M., Polleres, A., Schaffert, S. (eds.) Reasoning Web. LNCS, vol. 5224, pp. 54–103. Springer, Heidelberg (2008). doi:10.1007/978-3-540-85658-0_2
2. Straccia, U.: Foundations of Fuzzy Logic and Semantic Web Languages. CRC Studies in Informatics Series. Chapman & Hall, London (2013)
3. OWL 2 Web Ontology Language Document Overview. W3C (2009). http://www.w3.org/TR/2009/REC-owl2-overview-20091027/
4. Hayes, P.: RDF Semantics, W3C Recommendation, February 2004. http://www.w3.org/TR/rdf-mt
5. http://ruleml.org/index.html. The rule markup initiative
6. Calì, A., Gottlob, G., Lukasiewicz, T.: Datalog ±: a unified approach to ontologies and integrity constraints. In: Proceedings of the 12th International Conference on Database Theory, pp. 14–30. ACM, New York (2009). ISBN 978-1-60558-423-2. doi:10.1145/1514894.1514897
7. Rule Interchange Format (RIF). W3C (2011). http://www.w3.org/2001/sw/wiki/RIF
8. XML. W3C http://www.w3.org/XML/
9. Muñoz, S., Pérez, J., Gutierrez, C.: Minimal deductive systems for RDF. In: Franconi, E., Kifer, M., May, W. (eds.) ESWC 2007. LNCS, vol. 4519, pp. 53–67. Springer, Heidelberg (2007). doi:10.1007/978-3-540-72667-8_6
10. Baader, F., Calvanese, D., McGuinness, D., Nardi, D., Patel-Schneider, P.F. (eds.): The Description Logic Handbook: Theory, Implementation, and Applications. Cambridge University Press, Cambridge (2003)
11. Lloyd, J.W.: Foundations of Logic Programming. Springer, Heidelberg (1987)
12. Dubois, D., Prade, H.: Possibility theory, probability theory and multiple-valued logics: a clarification. Ann. Math. Artif. Intell. 32(1–4), 35–66 (2001). ISSN 1012-2443
13. Zadeh, L.A.: Fuzzy sets. Inf. Control 8(3), 338–353 (1965)
14. Dubois, D., Prade, H.: Can we enforce full compositionality in uncertainty calculi? In: Proceedings of the 12th National Conference on Artificial Intelligence (AAAI 1994), Seattle, Washington, pp. 149–154 (1994)
15. Straccia, U.: A minimal deductive system for general fuzzy RDF. In: Polleres, A., Swift, T. (eds.) RR 2009. LNCS, vol. 5837, pp. 166–181. Springer, Heidelberg (2009). doi:10.1007/978-3-642-05082-4_12
16. Straccia, U., Lopes, N., Lukacsy, G., Polleres, A.: A general framework for representing and reasoning with annotated semantic web data. In: Proceedings of the 24th AAAI Conference on Artificial Intelligence (AAAI 2010), pp. 1437–1442. AAAI Press (2010)

17. Lukasiewicz, T., Straccia, U.: Managing uncertainty and vagueness in description logics for thesemantic web. J. Web Semant. **6**, 291–308 (2008)
18. Straccia, U.: Reasoning within fuzzy description logics. J. Artif. Intell. Res. **14**, 137–166 (2001)
19. Straccia, U.: Answering vague queries in fuzzy DL-Lite. In: Proceedings of the 11th International Conference on Information Processing and Management of Uncertainty in Knowledge-Based Systems, (IPMU 2006), pp. 2238–2245. E.D.K., Paris (2006). ISBN 2-84254-112-X
20. Damasio, C.V., Pan, J.Z., Stoilos, G., Straccia, U.: Representing uncertainty rules in RuleMl. Fundam. Inform. **82**(3), 265–288 (2008)
21. Ragone, A., Straccia, U., Noia, T.D., Sciascio, E.D., Donini, F.M.: Fuzzy matchmaking in e-market places of peer entities using datalog. Fuzzy Sets Syst. **160**(2), 251–268 (2009)
22. Straccia, U.: Query answering in normal logic programs under uncertainty. In: Godo, L. (ed.) ECSQARU 2005. LNCS (LNAI), vol. 3571, pp. 687–700. Springer, Heidelberg (2005). doi:10.1007/11518655_58
23. Straccia, U.: Uncertainty management in logic programming: simple and effective top-down query answering. In: Khosla, R., Howlett, R.J., Jain, L.C. (eds.) KES 2005. LNCS (LNAI), vol. 3682, pp. 753–760. Springer, Heidelberg (2005). doi:10.1007/11552451_103
24. Straccia, U.: Fuzzy description logic programs. In: Proceedings of the 11th International Conference on Information Processing and Management of Uncertainty in Knowledge-BasedSystems, (IPMU 2006), pp. 1818–1825. E.D.K., Paris (2006). ISBN 2-84254-112-X
25. Straccia, U.: Towards top-k query answering in deductive databases. In: Proceedings of the 2006 IEEE International Conference on Systems, Man and Cybernetics (SMC 2006), pp. 4873–4879. IEEE (2006)
26. Lopes, N., Zimmermann, A., Hogan, A., Lukacsy, G., Polleres, A., Straccia, U., Decker, S.: RDF needs annotations. In: Proceedings of W3C Workshop – RDF Next Steps (2010).http://www.w3.org/2009/12/rdf-ws/
27. Lopes, N., Polleres, A., Straccia, U., Zimmermann, A.: AnQL: SPARQLing up annotated RDFS. In: Patel-Schneider, P.F., Pan, Y., Hitzler, P., Mika, P., Zhang, L., Pan, J.Z., Horrocks, I., Glimm, B. (eds.) ISWC 2010. LNCS, vol. 6496, pp. 518–533. Springer, Heidelberg (2010). doi:10.1007/978-3-642-17746-0_33
28. Zimmermann, A., Lopes, N., Polleres, A., Straccia, U.: A general framework for representing, reasoning and querying with annotated semantic web data. J. Web Semant. **11**, 72–95 (2012)
29. Dubois, D., Prade, H.: Fuzzy Sets and Systems. Academic Press, New York (1980)
30. Klir, G.J., Yuan, B.: Fuzzy Sets and Fuzzy Logic: Theory and Applications. Prentice-Hall Inc., Upper Saddle River (1995). ISBN 0-13-101171-5
31. Klement, E.P., Mesiar, R., Pap, E.: Triangular Norms. Trends in Logic – Studia Logica Library. Kluwer Academic Publishers, New York (2000)
32. Mostert, P.S., Shields, A.L.: On the structure of semigroups on a compact manifold with boundary. Ann. Math. **65**, 117–143 (1957)
33. Hájek, P.: Metamathematics of Fuzzy Logic. Kluwer, Dordrecht (1998)
34. Hähnle, R.: Advanced many-valued logics. In: Gabbay, D.M., Guenthner, F. (eds.) Handbook of Philosophical Logic, vol. 2, 2nd edn. Kluwer, Dordrecht (2001)
35. Kifer, M., Subrahmanian, V.: Theory of generalized annotated logic programming and itsapplications. J. Logic Program. **12**, 335–367 (1992)

36. Marin, D.: A formalization of RDF. Technical report TR/DCC-2006-8, Department of Computer Science, Universidad de Chile (2004). http://www.dcc.uchile.cl/cgutierr/ftp/draltan.pdf

37. RDF Semantics, W3C (2004). http://www.w3.org/TR/rdf-mt/

38. Abadi, D.J., Marcus, A., Madden, S., Hollenbach, K.: SW-store: a vertically partitioned DBMS for semantic web data management. VLDB J. **18**(2), 385–406 (2009)

39. Buneman, P., Kostylev, E.: Annotation algebras for RDFS. In: The Second International Workshop on the Role of Semantic Web in Provenance Management (SWPM 2010). CEUR Workshop Proceedings (2010)

40. Ianni, G., Krennwallner, T., Martello, A., Polleres, A.: A rule system for querying persistent RDFS data. In: The Semantic Web: Research and Applications, 6th European Semantic Web Conference (ESWC 2009), pp. 857–862 (2009)

41. Bobillo, F., Cerami, M., Esteva, F., García-Cerdaña, À., Peñaloza, R., Straccia, U.: Fuzzy description logics in the framework of mathematical fuzzylogic. In: Petr Cintula, C.N., Fermüller, C. (eds.) Handbook of Mathematical Fuzzy Logic, vol. 3. Studies in Logic Studies in Logic, Mathematical Logic and Foundations, vol. 58, pp. 1105–1181. College Publications, London (2015). Chapter 16, ISBN 978-1-84890-193-3

42. Bobillo, F., Delgado, M., Gómez-Romero, J.: DeLorean: a reasoner for fuzzy OWL 1.1. In: Proceedings of the 4th International Workshop on Uncertainty Reasoning for the Semantic Web (URSW 2008). CEUR Workshop Proceedings, vol. 423, 2008. ISSN 1613-0073

43. Bobillo, F., Straccia, U.: On qualified cardinality restrictions in fuzzy description logics under Lukasiewicz semantics. In: Magdalena, L., Ojeda-Aciego, M., Verdegay, J.L. (eds.), Proceedings of the 12th International Conference of Information Processing and Management of Uncertainty in Knowledge-Based Systems (IPMU 2008), pp. 1008–1015, June 2008

44. Bobillo, F., Straccia, U.: Extending datatype restrictions in fuzzy description logics. In: Proceedings of the 9th International Conference on Intelligent Systems Design and Applications (ISDA 2009), pp. 785–790. IEEE Computer Society (2009)

45. Bobillo, F., Straccia, U.: Fuzzy description logics with fuzzy truth values. In: Carvalho, J.P.B., Dubois, D., Kaymak, U., Sousa, J.M.C. (eds.), Proceedings of the 13th World Congress of the International Fuzzy Systems Association and 6th Conference of the European Society for Fuzzy Logic and Technology (IFSA-EUSFLAT 2009), pp. 189–194, July 2009. ISBN 978-989-95079-6-8

46. Bobillo, F., Straccia, U.: Supporting fuzzy rough sets in fuzzy description logics. In: Sossai, C., Chemello, G. (eds.) ECSQARU 2009. LNCS (LNAI), vol. 5590, pp. 676–687. Springer, Heidelberg (2009). doi:10.1007/978-3-642-02906-6_58

47. Dubois, D., Mengin, J., Prade, H.: Possibilistic uncertainty and fuzzy features in description logic. A preliminary discussion. In: Sanchez, E. (ed.) Capturing Intelligence: Fuzzy Logic and the Semantic Web. Elsevier, Amsterdam (2006)

48. Lukasiewicz, T.: Fuzzy description logic programs under the answer set semantics for the semantic web. In: Second International Conference on Rules and Rule Markup Languages for the Semantic Web (RuleML 2006), pp. 89–96. IEEE Computer Society (2006)

49. Lukasiewicz, T.: Fuzzy description logic programs under the answer set semantics forthe semantic web. Fundamenta Informaticae **82**(3), 289–310 (2008)

50. Lukasiewicz, T., Straccia, U.: Description logic programs under probabilistic uncertainty and fuzzy vagueness. In: Mellouli, K. (ed.) ECSQARU 2007. LNCS (LNAI), vol. 4724, pp. 187–198. Springer, Heidelberg (2007). doi:10.1007/978-3-540-75256-1_19

51. Lukasiewicz, T., Straccia, U.: Tightly integrated fuzzy description logic programs under the answer set semantics for the semantic web. In: Marchiori, M., Pan, J.Z., Marie, C.S. (eds.) RR 2007. LNCS, vol. 4524, pp. 289–298. Springer, Heidelberg (2007). doi:10.1007/978-3-540-72982-2_23

52. Lukasiewicz, T., Straccia, U.: Top-k retrieval in description logic programs under vagueness for the semantic web. In: Prade, H., Subrahmanian, V.S. (eds.) SUM 2007. LNCS (LNAI), vol. 4772, pp. 16–30. Springer, Heidelberg (2007). doi:10.1007/978-3-540-75410-7_2

53. Lukasiewicz, T., Straccia, U.: Tightly coupled fuzzy description logic programs under the answer set semantics for the semantic web. Int. J. Semant. Web, Inf. Syst. 4(3), 68–89 (2008)

54. Lukasiewicz, T., Straccia, U.: Description logic programs under probabilistic uncertainty and fuzzy vagueness. Int. J. Approx. Reason. 50(6), 837–853 (2009)

55. Sanchez, D., Tettamanzi, A.G.: Generalizing quantification in fuzzy description logics. In: Proceedings 8th Fuzzy Days in Dortmund (2004)

56. Sánchez, D., Tettamanzi, A.G.B.: Reasoning and quantification in fuzzy description logics. In: Bloch, I., Petrosino, A., Tettamanzi, A.G.B. (eds.) WILF 2005. LNCS (LNAI), vol. 3849, pp. 81–88. Springer, Heidelberg (2006). doi:10.1007/11676935_10

57. Sanchez, D., Tettamanzi, A.G.: Fuzzy quantification in fuzzy description logics. In: Sanchez, E. (ed.) Capturing Intelligence: Fuzzy Logic and the Semantic Web. Elsevier, Amsterdam (2006)

58. Stoilos, G., Stamou, G.: Extending fuzzy description logics for the semantic web. In: 3rd International Workshop of OWL: Experiences and Directions (2007). http://www.image.ece.ntua.gr/publications.php

59. Straccia, U.: A fuzzy description logic. In: Proceedings of the 15th National Conference on Artificial Intelligence (AAAI 1998), Madison, USA, pp. 594–599 (1998)

60. Straccia, U.: Description logics with fuzzy concrete domains. In: Bachus, F., Jaakkola, T. (eds.), 21st Conference on Uncertainty in Artificial Intelligence (UAI 2005), pp. 559–567. AUAI Press, Edinburgh (2005)

61. Straccia, U.: Fuzzy ALC with fuzzy concrete domains. In: Proceedings of the International Workshop on Description Logics (DL 2005), pp. 96–103. CEUR, Edinburgh (2005)

62. Straccia, U.: Fuzzy description logic programs. In: Bouchon-Meunier, C.M.B., Yager, R.R., Rifqi, M. (eds.) Uncertainty and Intelligent Information Systems, pp. 405–418. World Scientific, Singapore (2008). ISBN 978-981-279-234-1. Chap. 29

63. Venetis, T., Stoilos, G., Stamou, G., Kollias, S.: f-DLPs: extending description logic programs with fuzzy sets and fuzzy logic. In: IEEE International Conference on Fuzzy Systems (Fuzz-IEEE 2007) (2007). http://www.image.ece.ntua.gr/publications.php

64. Yen, J.: Generalizing term subsumption languages to fuzzy logic. In: Proceedings of the 12th International Joint Conference on Artificial Intelligence (IJCAI 1991), Sydney, Australia, pp. 472–477 (1991)

65. Bobillo, F., Straccia, U.: fuzzyDL: an expressive fuzzy description logic reasoner. In: 2008 International Conference on Fuzzy Systems (FUZZ 2008), pp. 923–930. IEEE Computer Society (2008)

66. Bobillo, F., Straccia, U.: Finite fuzzy description logics: a crisp representation for finite fuzzy $\mathcal{ALCH}$. In: Bobillo, F., Carvalho, R., da Costa, P.C.G., d'Amato, C., Fanizzi, N., Laskey, K.B., Laskey, K.J., Lukasiewicz, T., Martin, T., Nickles, M., Pool, M. (eds.) Proceedings of the 6th ISWC Workshop on Uncertainty Reasoning for the Semantic Web (URSW 2010). CEUR Workshop Proceedings, vol. 654, pp. 61–72, November 2010. ISSN 1613-0073

67. Bobillo, F., Straccia, U.: Aggregation operators and fuzzy OWL 2. In: Proceedings of the 20th IEEE International Conference on Fuzzy Systems (FUZZ-IEEE 2011), pp. 1727–1734. IEEE Press, June 2011

68. Bobillo, F., Straccia, U.: Fuzzy ontologies and fuzzy integrals. In: Proceedings of the 11th International Conference on Intelligent Systems Design and Applications (ISDA 2011), pp. 1311–1316. IEEE Press, November 2011

69. Bobillo, F., Straccia, U.: Fuzzy ontology representation using OWL 2. Int. J. Approx. Reason. **52**, 1073–1094 (2011)

70. Bobillo, F., Straccia, U.: Generalized fuzzy rough description logics. Inf. Sci. **189**, 43–62 (2012)

71. Bobillo, F., Straccia, U.: Aggregation operators for fuzzy ontologies. Appl. Soft Comput. **13**(9), 3816–3830 (2013). ISSN 1568-4946

72. Bobillo, F., Straccia, U.: General concept inclusion absorptions for fuzzy description logics: a first step. In: Proceedings of the 26th International Workshop on Description Logics (DL 2013). CEUR Workshop Proceedings, vol. 1014, pp. 513–525. CEUR-WS.org (2013). http://ceur-ws.org/Vol-1014/paper_3.pdf

73. Dinh-Khac, D., Hölldobler, S., Tran, D.-K.: The fuzzy linguistic description logic $\mathcal{ALC}_{FL}$. In: Proceedings of the 11th International Conference on Information Processing and Management of Uncertainty in Knowledge-Based Systems (IPMU 2006), pp. 2096–2103. E.D.K., Paris (2006). ISBN 2-84254-112-X

74. Hölldobler, S., Khang, T.D., Störr, H.-P.: A fuzzy description logic with hedges as concept modifiers. In: Phuong, N.H., Nguyen, H.T., Ho, N.C., Santiprabhob, P. (eds.) Proceedings InTech/VJFuzzy 2002, pp. 25–34. Institute of Information Technology, Vietnam Center for Natural Science and Technology, Science and Technics Publishing House, Hanoi (2002)

75. Hölldobler, S., Nga, N.H., Khang, T.D.: The fuzzy description logic $\mathcal{ALC}_{FH}$. In: Proceedings of the International Workshop on Description Logics (DL 2005) (2005)

76. Hölldobler, S., Störr, H.-P., Khang, T.D.: The fuzzy description logic $\mathcal{ALC}_{FH}$ with hedge algebras as concept modifiers. J. Adv. Comput. Intell. Intell. Inform. (JACIII) **7**(3), 294–305 (2003). doi:10.20965/jaciii.2003.p0294

77. Hölldobler, S., Störr, H.-P., Khang, T.D.: A fuzzy description logic with hedges and concept modifiers. In: Proceedings of the 10th International Conference on Information Processing and Management of Uncertainty in Knowledge-Based Systems (IPMU 2004) (2004)

78. Hölldobler, S., Störr, H.-P., Khang, T.D.: The subsumption problem of the fuzzy description logic $\mathcal{ALC}_{FH}$. In: Proceedings of the 10th International Conference on Information Processing and Management of Uncertainty in Knowledge-Based Systems (IPMU 2004) (2004)

79. Jiang, Y., Liu, H., Tang, Y., Chen, Q.: Semantic decision making using ontology-based soft sets. Math. Comput. Modell. **53**(5–6), 1140–1149 (2011)

80. Jiang, Y., Tang, Y., Chen, Q., Wang, J., Tang, S.: Extending soft sets with description logics. Comput. Math. Appl. **59**(6), 2087–2096 (2010). ISSN 0898-1221. http://dx.doi.org/10.1016/j.camwa.2009.12.014

81. Jiang, Y., Tang, Y., Wang, J., Deng, P., Tang, S.: Expressive fuzzy description logics over lattices. Knowl.-Based Syst. **23**, 150–161 (2010). ISSN 0950-7051. http://dx.doi.org/10.1016/j.knosys.2009.11.002
82. Jiang, Y., Tang, Y., Wang, J., Tang, S.: Reasoning within intuitionistic fuzzy rough description logics. Inf. Sci. **179**, 2362–2378 (2009)
83. Jiang, Y., Tang, Y., Wang, J., Tang, S.: Representation and reasoning of context-dependant knowledge in distributed fuzzy ontologies. Expert Syst. Appl. **37**(8), 6052–6060 (2010). ISSN 0957-4174. http://dx.doi.org/10.1016/j.eswa.2010.02.122
84. Jiang, Y., Wang, J., Deng, P., Tang, S.: Reasoning within expressive fuzzy rough description logics. Fuzzy Sets Syst. **160**(23), 3403–3424 (2009). doi:10.1016/j.fss.2009.01.004
85. Jiang, Y., Wang, J., Tang, S., Xiao, B.: Reasoning with rough description logics: an approximate concepts approach. Inf. Sci. **179**(5), 600–612 (2009). ISSN 0020-0255
86. Kang, B., Xu, D., Lu, J., Li, Y.: Reasoning for a fuzzy description logic with comparison expressions. In: Proceedings of the International Workshop on Description Logics (DL 2006). CEUR Workshop Proceedings (2006)
87. Mailis, T., Stoilos, G., Stamou, G.: Expressive reasoning with horn rules and fuzzy description logics. In: Marchiori, M., Pan, J.Z., Marie, C.S. (eds.) RR 2007. LNCS, vol. 4524, pp. 43–57. Springer, Heidelberg (2007). doi:10.1007/978-3-540-72982-2_4
88. Straccia, U.: Towards spatial reasoning in fuzzy description logics. In: 2009 IEEE International Conference on Fuzzy Systems (FUZZ-IEEE 2009), pp. 512–517. IEEE Computer Society (2009)
89. Tresp, C., Molitor, R.: A description logic for vague knowledge. In: Proceedings of the 13th European Conference on Artificial Intelligence (ECAI 1998), Brighton, England, August 1998
90. Bobillo, F., Delgado, M., Gómez-Romero, J., Straccia, U.: Fuzzy description logics under Gödel semantics. Int. J. Approx. Reason. **50**(3), 494–514 (2009)
91. Bobillo, F., Straccia, U.: A fuzzy description logic with product t-norm. In: Proceedings of the IEEE International Conference on Fuzzy Systems (Fuzz-IEEE 2007), pp. 652–657. IEEE Computer Society (2007)
92. Bobillo, F., Straccia, U.: Fuzzy description logics with general t-norms and datatypes. Fuzzy Sets Syst. **160**(23), 3382–3402 (2009)
93. Hájek, P.: Making fuzzy description logics more general. Fuzzy Sets Syst. **154**(1), 1–15 (2005)
94. Hájek, P.: What does mathematical fuzzy logic offer to description logic? In: Sanchez, E. (ed.) Fuzzy Logic and the Semantic Web, Capturing Intelligence, pp. 91–100. Elsevier, Amsterdam (2006). Chap. 5
95. Straccia, U.: Uncertainty in description logics: a lattice-based approach. In: Proceedings of the 10th International Conference on Information Processing and Management of Uncertainty in Knowledge-Based Systems (IPMU 2004), pp. 251 258 (2004)
96. Straccia, U.: Uncertainty and description logic programs over lattices. In: Sanchez, E. (ed.) Fuzzy Logic and the Semantic Web, Capturing Intelligence, pp. 115–133. Elsevier, Amsterdam (2006). Chap. 7
97. Straccia, U.: Description logics over lattices. Int. J. Uncertainty, Fuzziness Knowl.-Based Syst. **14**(1), 1–16 (2006)
98. Baader, F., Peñaloza, R.: Are fuzzy description logics with general concept inclusion axioms decidable? In: Proceedings of 2011 IEEE International Conference on Fuzzy Systems (Fuzz-IEEE 2011). IEEE Press (2011)

99. Baader, F., Peñaloza, R.: GCIs make reasoning in fuzzy DLs with the product t-norm undecidable. In: Proceedings of the 24th International Workshop on Description Logics (DL 2011). CEUR Electronic Workshop Proceedings (2011)

100. Bobillo, F., Bou, F., Straccia, U.: On the failure of the finite model property in some fuzzy description logics. Fuzzy Sets Syst. **172**(1), 1–12 (2011)

101. Bobillo, F., Delgado, M., Gómez-Romero, J.: A crisp representation for fuzzy $\mathcal{SHOIN}$ with fuzzy nominals and general concept inclusions. In: Proceedings of the 2nd Workshop on Uncertainty Reasoning for the Semantic Web (URSW 2006), November 2006

102. Bobillo, F., Delgado, M., Gómez-Romero, J.: A crisp representation for fuzzy $\mathcal{SHOIN}$ with fuzzy nominals and general concept inclusions. In: Costa, P.C.G., d'Amato, C., Fanizzi, N., Laskey, K.B., Laskey, K.J., Lukasiewicz, T., Nickles, M., Pool, M. (eds.) URSW 2005-2007. LNCS (LNAI), vol. 5327, pp. 174–188. Springer, Heidelberg (2008). doi:10.1007/978-3-540-89765-1_11

103. Bobillo, F., Delgado, M., Gómez-Romero, J.: Optimizing the crisp representation of the fuzzy description logic $\mathcal{SROIQ}$. In: Costa, P.C.G., d'Amato, C., Fanizzi, N., Laskey, K.B., Laskey, K.J., Lukasiewicz, T., Nickles, M., Pool, M. (eds.) URSW 2005-2007. LNCS (LNAI), vol. 5327, pp. 189–206. Springer, Heidelberg (2008). doi:10.1007/978-3-540-89765-1_12

104. Bobillo, F., Delgado, M., Gómez-Romero, J., Straccia, U.: Joining Gödel and Zadeh fuzzy logics in fuzzy description logics. Int. J. Uncertainty, Fuzziness Knowl.-Based Syst. **20**, 475–508 (2012)

105. Bobillo, F., Straccia, U.: Towards a crisp representation of fuzzy description logics under Łukasiewicz semantics. In: An, A., Matwin, S., Raś, Z.W., Ślęzak, D. (eds.) ISMIS 2008. LNCS (LNAI), vol. 4994, pp. 309–318. Springer, Heidelberg (2008). doi:10.1007/978-3-540-68123-6_34

106. Bobillo, F., Straccia, U.: Reasoning with the finitely many-valued Lukasiewicz fuzzy description logic $\mathcal{SROIQ}$. Inf. Sci. **181**, 758–778 (2011)

107. Bobillo, F., Straccia, U.: Finite fuzzy description logics and crisp representations. In: Bobillo, F., et al. (eds.) UniDL/URSW 2008-2010. LNCS (LNAI), vol. 7123, pp. 99–118. Springer, Heidelberg (2013). doi:10.1007/978-3-642-35975-0_6

108. Bobillo, F., Straccia, U.: A MILP-based decision procedure for the (fuzzy) description logic $\mathcal{ALCB}$. In: Proceedings of the 27th International Workshop on Description Logics (DL 2014), vol. 1193, pp. 378–390. CEUR Workshop Proceedings, ISSN 1613-0073, July 2014

109. Bobillo, F., Straccia, U.: On partitioning-based optimisations in expressive fuzzy description logics. In: Proceedings of the 24th IEEE International Conference on Fuzzy Systems (FUZZ-IEEE 2015). IEEE Press, August 2015. doi:10.1109/FUZZ-IEEE.2015.7337838

110. Bobillo, F., Straccia, U.: Optimising fuzzy description logic reasoners with general concept inclusions absorption. Fuzzy Sets Syst. **292**, 98–129 (2016). http://dx.doi.org/10.1016/j.fss.2014.10.029

111. Bonatti, P.A., Tettamanzi, A.G.B.: Some complexity results on fuzzy description logics. In: Gesú, V., Masulli, F., Petrosino, A. (eds.) WILF 2003. LNCS (LNAI), vol. 2955, pp. 19–24. Springer, Heidelberg (2006). doi:10.1007/10983652_3

112. Borgwardt, S., Distel, F., Peñaloza, R.: How fuzzy is my fuzzy description logic? In: Gramlich, B., Miller, D., Sattler, U. (eds.) IJCAR 2012. LNCS (LNAI), vol. 7364, pp. 82–96. Springer, Heidelberg (2012). doi:10.1007/978-3-642-31365-3_9

113. Borgwardt, S., Distel, F., Peñaloza, R.: Non-Gödel negation makes unwitnessed consistency undecidable. In: Proceedings of the 2012 International Workshop on Description Logics (DL 2012), vol. 846. CEUR-WS.org (2012)

114. Borgwardt, S., Peñaloza, R.: Description logics over lattices with multi-valued ontologies. In: Proceedings of the Twenty-Second International Joint Conference on Artificial Intelligence (IJCAI 2011), pp. 768–773 (2011)

115. Borgwardt, S., Peñaloza, R.: Finite lattices do not make reasoning in $\mathcal{ALCI}$ harder. In: Proceedings of the 7th International Workshop on Uncertainty Reasoning for the Semantic Web (URSW 2011), vol. 778, pp. 51–62. CEUR-WS.org (2011)

116. Borgwardt, S., Peñaloza, R.: Fuzzy ontologies over lattices with t-norms. In: Proceedings of the 24th International Workshop on Description Logics (DL 2011). CEUR Electronic Workshop Proceedings (2011)

117. Borgwardt, S., Peñaloza, R.: A tableau algorithm for fuzzy description logics over residuated De Morgan lattices. In: Krötzsch, M., Straccia, U. (eds.) RR 2012. LNCS, vol. 7497, pp. 9–24. Springer, Heidelberg (2012). doi:10.1007/978-3-642-33203-6_3

118. Borgwardt, S., Peñaloza, R.: Undecidability of fuzzy description logics. In: Proceedings of the 13th International Conference on Principles of Knowledge Representation and Reasoning (KR 2012), pp. 232–242. AAAI Press, Rome (2012)

119. Bou, F., Cerami, M., Esteva, F.: Finite-valued Lukasiewicz modal logic is PSPACE-complete. In: Proceedings of the 22nd International Joint Conference on Artificial Intelligence (IJCAI 2011), pp. 774–779 (2011)

120. Cerami, M., Esteva, F., Bou, F.: Decidability of a description logic over infinite-valued product logic. In: Proceedings of the Twelfth International Conference on Principles of Knowledge Representation and Reasoning (KR 2010). AAAI Press (2010)

121. Cerami, M., Straccia, U.: On the undecidability of fuzzy description logics with GCIs with lukasiewicz t-norm. Technical report, Computing Research Repository (2011). Available as CoRR technical report at http://arxiv.org/abs/1107.4212

122. Cerami, M., Straccia, U.: Undecidability of KB satisfiability for l-$\mathcal{ALC}$ with GCIs, July 2011. Unpublished manuscript

123. Fernando Bobillo, U.S.: Reducing the size of the optimization problems in fuzzy ontology reasoning. In: Proceedings of the 11th International Workshop on Uncertainty Reasoning for the Semantic Web (URSW 2015). CEUR Workshop Proceedings, vol. 1479, pp. 54–59. CEUR-WS.org (2015). http://ceur-ws.org/Vol-1479/paper6.pdf

124. Pan, J.Z., Stamou, G., Stoilos, G., Thomas, E.: Expressive querying over fuzzy DL-Lite ontologies. In: Twentieth International Workshop on Description Logics (2007). http://www.image.ece.ntua.gr/publications.php

125. Stoilos, G., Stamou, G., Pan, J., Tzouvaras, V., Horrocks, I.: The fuzzy description logic f-SHIN. In: International Workshop on Uncertainty Reasoning for the Semantic Web (2005). http://www.image.ece.ntua.gr/publications.php

126. Stoilos, G., Stamou, G.B., Pan, J.Z., Tzouvaras, V., Horrocks, I.: Reasoning with very expressive fuzzy description logics. J. Artif. Intell. Res. **30**, 273–320 (2007)

127. Stoilos, G., Straccia, U., Stamou, G., Pan, J.Z.: General concept inclusions in fuzzy description logics. In: Proceedings of the 17th Eureopean Conference on Artificial Intelligence (ECAI 2006), pp. 457–461. IOS Press (2006)

128. Straccia, U.: Transforming fuzzy description logics into classical description logics. In: Alferes, J.J., Leite, J. (eds.) JELIA 2004. LNCS (LNAI), vol. 3229, pp. 385–399. Springer, Heidelberg (2004). doi:10.1007/978-3-540-30227-8_33

129. Straccia, U.: Towards Top-k query answering in description logics: the case of DL-lite. In: Fisher, M., Hoek, W., Konev, B., Lisitsa, A. (eds.) JELIA 2006. LNCS (LNAI), vol. 4160, pp. 439–451. Springer, Heidelberg (2006). doi:10.1007/11853886_36

130. Straccia, U.: Reasoning in l-$\mathcal{SHIF}$: an expressive fuzzy description logic under lukasiewicz semantics. Technical report TR-2007-10-18, Istituto di Scienza e Tecnologie dell'Informazione, Consiglio Nazionale delle Ricerche, Pisa, Italy (2007)

131. Straccia, U., Bobillo, F.: Mixed integer programming, general concept inclusions and fuzzy description logics. In: Proceedings of the 5th Conference of the European Society for Fuzzy Logic and Technology (EUSFLAT 2007), vol. 2, pp. 213–220. University of Ostrava, Ostrava (2007)

132. Straccia, U., Bobillo, F.: Mixed integer programming, general concept inclusions and fuzzy description logics. Mathware Soft Comput. **14**(3), 247–259 (2007)

133. Li, Y., Xu, B., Lu, J., Kang, D.: Discrete tableau algorithms for $\mathcal{FSHI}$. In: Proceeedings of the International Workshop on Description Logics (DL 2006). CEUR (2006). http://ceur-ws.org/Vol-189/submission_14.pdf

134. Zhou, Z., Qi, G., Liu, C., Hitzler, P., Mutharaju, R.: Reasoning with fuzzy-$\mathcal{EL}^+$ ontologies using map reduce. In: 20th European Conference on Artificial Intelligence (ECAI 2012), pp. 933–934. IOS Press (2012)

135. Andreasen, T., Bulskov, H.: Conceptual querying through ontologies. Fuzzy Sets Syst. **160**(15), 2159–2172 (2009). ISSN 0165-0114. doi:10.1016/j.fss.2009.02.019

136. Calegari, S., Sanchez, E.: Object-fuzzy concept network: an enrichment of ontologies in semantic information retrieval. J. Am. Soc. Inf. Sci. Technol. **59**(13), 2171–2185 (2008). ISSN 1532-2882. doi:10.1002/asi.v59:13

137. Liu, C., Liu, D., Wang, S.: Fuzzy geospatial information modeling in geospatial semantic retrieval. Adv. Math. Comput. Methods **2**(4), 47–53 (2012)

138. Straccia, U., Visco, G.: DL-Media: an ontology mediated multimedia information retrieval system. In: Proceedings of the International Workshop on Description Logics (DL 2007), vol. 250. CEUR, Insbruck (2007). http://ceur-ws.org

139. Straccia, U., Visco, G.: DLMedia: an ontology mediated multimedia information retrieval system. In: Proceedings of the Fourth International Workshop on Uncertainty Reasoning for the Semantic Web, Karlsruhe, Germany, 26 October (URSW 2008). CEUR Workshop Proceedings, vol. 423. CEUR-WS.org (2008). http://sunsite.informatik.rwth-aachen.de/Publications/CEUR-WS/Vol-423/paper4.pdf

140. Wallace, M.: Ontologies and soft computing in flexible querying. Control Cybern. **38**(2), 481–507 (2009)

141. Zhang, L., Yu, Y., Zhou, J., Lin, C., Yang, Y.: An enhanced model for searching in semantic portals. In: WWW 2005: Proceedings of the 14th International Conference on World Wide Web, pp. 453–462. ACM Press, New York (2005). ISBN 1-59593-046-9. http://doi.acm.org/10.1145/1060745.1060812

142. Carlsson, C., Brunelli, M., Mezei, J.: Decision making with a fuzzy ontology. Soft Comput. **16**(7), 1143–1152 (2012). ISSN 1432-7643. doi:10.1007/s00500-011-0789-x

143. Lee, C.-S., Wang, M.H., Hagras, H.: A type-2 fuzzy ontology and its application to personal diabetic-diet recommendation. IEEE Trans. Fuzzy Syst. **18**(2), 374–395 (2010)

144. Pérez, I.J., Wikström, R., Mezei, J., Carlsson, C., Herrera-Viedma, E.: A new consensus model for group decision making using fuzzy ontology. Soft Comput. **17**(9), 1617–1627 (2013)

145. Yaguinuma, C.A., Santos, M.T.P., Camargo, H.A., Reformat, M.: A FML-based hybrid reasoner combining fuzzy ontology and mamdani inference. In: Proceedings of the 22nd IEEE International Conference on Fuzzy Systems (FUZZ-IEEE 2013) (2013)

146. Dasiopoulou, S., Kompatsiaris, I.: Trends and issues in description logics frameworks for image interpretation. In: Konstantopoulos, S., Perantonis, S., Karkaletsis, V., Spyropoulos, C.D., Vouros, G. (eds.) SETN 2010. LNCS (LNAI), vol. 6040, pp. 61–70. Springer, Heidelberg (2010). doi:10.1007/978-3-642-12842-4_10

147. Dasiopoulou, S., Kompatsiaris, I., Strintzis, M.G.: Applying fuzzy DLS in the extraction of image semantics. J. Data Semant. 14, 105–132 (2009)

148. Dasiopoulou, S., Kompatsiaris, I., Strintzis, M.G.: Investigating fuzzy DLS-based reasoning in semantic image analysis. Multimedia Tools Appl. 49(1), 167–194 (2010). ISSN 1380-7501. doi:10.1007/s11042-009-0393-6

149. Meghini, C., Sebastiani, F., Straccia, U.: A model of multimedia information retrieval. J. ACM 48(5), 909–970 (2001)

150. Stoilos, G., Stamou, G., Tzouvaras, V., Pan, J.Z., Horrock, I.: A fuzzy description logic for multimedia knowledge representation. In: Proceedings of the International Workshop on Multimedia and the Semantic Web (2005)

151. Straccia, U.: Foundations of a logic based approach to multimedia document retrieval. Ph.D. thesis, Department of Computer Science, University of Dortmund, Dortmund, Germany, June 1999

152. Straccia, U.: A framework for the retrieval of multimedia objects based on four-valued fuzzy description logics. In: Crestani, F., Pasi, G. (eds.) Soft Computing in Information Retrieval: Techniques and Applications. SFSC, vol. 50, pp. 332–357. Springer, Heidelberg (2000)

153. Costa, P.C.G., Laskey, K.B., Lukasiewicz, T.: Uncertainty representation and reasoning in the semantic web. In: Semantic Web Engineering in the Knowledge Society, pp. 315–340. IGI Global (2008)

154. Quan, T.T., Hui, S.C., Fong, A.C.M., Cao, T.H.: Automatic fuzzy ontology generation for semantic web. IEEE Trans. Knowl. Data Eng. 18(6), 842–856 (2006)

155. Sanchez, E. (ed.): Fuzzy Logic and the Semantic Web. Capturing Intelligence, vol. 1. Elsevier Science, Amsterdam (2006)

156. Díaz-Rodríguez, N., León-Cadahía, O., Pegalajar-Cuéllar, M., Lilius, J., Delgado, M.: Handling real-world context-awareness, uncertainty and vagueness in real-time human activity tracking and recognition with a fuzzy ontology-based hybrid method. Sensors 14(10), 18131–18171 (2014)

157. Díaz-Rodríguez, N., Pegalajar-Cuéllar, M., Lilius, J., Delgado, M.: A fuzzy ontology for semantic modelling and recognition of human behaviour. Knowl.-Based Syst. 66, 46–60 (2014)

158. Liu, C., Liu, D., Wang, S.: Situation modeling and identifying under uncertainty. In: Proceedings of the 2nd Pacific-Asia Conference on Circuits, Communications and System (PACCS 2010), pp. 296–299 (2010)

159. Rodríguez, N.D., Cuéllar, M.P., Lilius, J., Calvo-Flores, M.D.: A survey on ontologies for human behavior recognition. ACM Comput. Surveys 46(4), 43:1–43:33 (2014). ISSN 0360-0300. doi:10.1145/2523819

160. Chen, R.-C., Bau, C.T., Yeh, C.-J.: Merging domain ontologies based on the WordNet system and fuzzy formal concept analysis techniques. Appl. Soft Comput. 11(2), 1908–1923 (2011)

161. Todorov, K., Hudelot, C., Popescu, A., Geibel, P.: Fuzzy ontology alignment using background knowledge. Int. J. Uncertainty Fuzziness Knowl.-Based Syst. **22**(1), 75–112 (2014)
162. Agarwal, S., Lamparter, S.: SMART: a semantic matchmaking portal for electronic markets. In: CEC 2005: Proceedings of the Seventh IEEE International Conference on E-Commerce Technology (CEC 2005), pp. 405–408. IEEE Computer Society, Washington, DC (2005). ISBN 0-7695-2277-7. http://dx.doi.org/10.1109/ICECT.2005.84
163. Colucci, S., Noia, T.D., Ragone, A., Ruta, M., Straccia, U., Tinelli, E.: Informative top-k retrieval for advanced skill management. In: de Virgilio, R., Giunchiglia, F., Tanca, L. (eds.) Semantic Web Information Management, pp. 449–476. Springer, Heidelberg (2010). doi:10.1007/978-3-642-04329-1_19. Chap. 19
164. Ragone, A., Straccia, U., Bobillo, F., Noia, T., Sciascio, E.: Fuzzy bilateral matchmaking in e-marketplaces. In: Lovrek, I., Howlett, R.J., Jain, L.C. (eds.) KES 2008. LNCS (LNAI), vol. 5179, pp. 293–301. Springer, Heidelberg (2008). doi:10.1007/978-3-540-85567-5_37
165. Ragone, A., Straccia, U., Noia, T.D., Sciascio, E.D., Donini, F.M.: Extending datalog for matchmaking in P2P E-marketplaces. In: Ceci, M., Malerba, D., Tanca, L. (eds.) 15th Italian Symposium on Advanced Database Systems (SEBD 2007), pp. 463–470 (2007). ISBN 978-88-902981-0-3
166. Ragone, A., Straccia, U., Noia, T., Sciascio, E., Donini, F.M.: Vague knowledge bases for matchmaking in P2P E-marketplaces. In: Franconi, E., Kifer, M., May, W. (eds.) ESWC 2007. LNCS, vol. 4519, pp. 414–428. Springer, Heidelberg (2007). doi:10.1007/978-3-540-72667-8_30
167. Ragone, A., Straccia, U., Noia, T., Sciascio, E., Donini, F.M.: Towards a fuzzy logic for automated multi-issue negotiation. In: Hartmann, S., Kern-Isberner, G. (eds.) FoIKS 2008. LNCS, vol. 4932, pp. 381–396. Springer, Heidelberg (2008). doi:10.1007/978-3-540-77684-0_25
168. Straccia, U., Tinelli, E., Colucci, S., Noia, T., Sciascio, E.: Semantic-based top-k retrieval for competence management. In: Rauch, J., Raś, Z.W., Berka, P., Elomaa, T. (eds.) ISMIS 2009. LNCS (LNAI), vol. 5722, pp. 473–482. Springer, Heidelberg (2009). doi:10.1007/978-3-642-04125-9_50
169. Straccia, U., Tinelli, E., Noia, T.D., Sciascio, E.D., Colucci, S.: Top-k retrieval for automated human resource management. In: Proceedings of the 17th Italian Symposium on Advanced Database Systems (SEBD 2009), pp. 161–168 (2009)
170. Straccia, U.: Multi criteria decision making in fuzzy description logics: a first step. In: Velásquez, J.D., Ríos, S.A., Howlett, R.J., Jain, L.C. (eds.) KES 2009. LNCS (LNAI), vol. 5711, pp. 78–86. Springer, Heidelberg (2009). doi:10.1007/978-3-642-04595-0_10
171. Lee, C.-S., Jian, Z.-W., Huang, L.-K.: A fuzzy ontology and its application to news summarization. IEEE Trans. Syst. Man Cybern. Part B **35**(5), 859–880 (2005)
172. Eich, M., Hartanto, R., Kasperski, S., Natarajan, S., Wollenberg, J.: Towards coordinated multirobot missions for lunar sample collection in an unknown environment. J. Field Robot. **31**(1), 35–74 (2014)
173. Eich, T.: An application of fuzzy DL-based semantic perception to soil container classification. In: IEEE International Conference on Technologies for Practical Robot Applications (TePRA 2013), pp. 1–6. IEEE Press (2013)
174. Lisi, F.A., Straccia, U.: A logic-based computational method for the automated induction of fuzzy ontology axioms. Fundamenta Informaticae **124**(4), 503–519 (2013)

175. Lisi, F.A., Straccia, U.: A system for learning GCI axioms in fuzzy description logics. In: Proceedings of the 26th International Workshop on Description Logics (DL 2013). CEUR Workshop Proceedings, vol. 1014, pp. 760–778. CEUR-WS.org (2013). http://ceur-ws.org/Vol-1014/paper_42.pdf

176. Lisi, F.A., Straccia, U.: Can ILP deal with incomplete and vague structured knowledge? In: Muggleton, S.H., Watanabe, H. (eds.) Latest Advances in Inductive Logic Programming, chapter 21, pp. 199–206. World Scientific (2014). doi:10.1142/9781783265091_0021

177. Lisi, F.A., Straccia, U.: Learning in description logics with fuzzy concrete domains. Fundamenta Informaticae 140(3–4), 373–391 (2015). ISSN 1875-8681. doi:10.3233/FI-2015-1259

178. Lisi, F.A., Straccia, U.: An inductive logic programming approach to learning inclusion axiomsin fuzzy description logics. In: 26th Italian Conference on Computational Logic (CILC 2011). CEUR Electronic Workshop Proceedings, vol. 810, pp. 57–71 (2011). http://ceur-ws.org/Vol-810/paper-l04.pdf

179. Lisi, F.A., Straccia, U.: Towards learning fuzzy DL inclusion axioms. In: Fanelli, A.M., Pedrycz, W., Petrosino, A. (eds.) WILF 2011. LNCS (LNAI), vol. 6857, pp. 58–66. Springer, Heidelberg (2011). doi:10.1007/978-3-642-23713-3_8

180. Lisi, F.A., Straccia, U.: Dealing with incompleteness and vagueness in inductive logic programming. In: 28th Italian Conference on Computational Logic (CILC 2013). CEUR Electronic Workshop Proceedings, vol. 1068, pp. 179–193 (2013). ISSN 1613-0073. http://ceur-ws.org/Vol-1068/paper-l12.pdf

181. Lisi, F.A., Straccia, U.: A FOIL-like method for learning under incompleteness and vagueness. In: Zaverucha, G., Santos Costa, V., Paes, A. (eds.) ILP 2013. LNCS (LNAI), vol. 8812, pp. 123–139. Springer, Heidelberg (2014). doi:10.1007/978-3-662-44923-3_9

182. Straccia, U., Mucci, M.: pFOIL-DL: learning (fuzzy) $\mathcal{EL}$ concept descriptions from crisp owl data using a probabilistic ensemble estimation. In: Proceedings of the 30th Annual ACM Symposium on Applied Computing (SAC 2015), pp. 345–352. ACM, Salamanca (2015)

183. Balaj, R., Groza, A.: Detecting influenza epidemics based on real-time semantic analysis of Twitter streams. In: Proceedings of the 3rd International Conference on Modelling and Development of Intelligent Systems (MDIS 2013), pp. 30–39 (2013)

184. d'Aquin, M., Lieber, J., Napoli, A.: Towards a semantic portal for oncology using a description logic with fuzzy concrete domains. In: Sanchez, E. (ed.) Fuzzy Logic and the Semantic Web. Capturing Intelligence, pp. 379–393. Elsevier (2006)

185. Fernández, C.: Understanding image sequences: the role of ontologies in cognitive vision systems. Ph.D. thesis, Universitat Autònoma de Barcelona, Spain (2010)

186. Iglesias, J., Lehmann, J.: Towards integrating fuzzy logic capabilities into an ontology-based inductive logic programming framework. In: Proceedings of the 11th International Conference on Intelligent Systems Design and Applications (ISDA 2011), pp. 1323–1328 (2011)

187. Konstantopoulos, S., Karkaletsis, V., Bilidas, D.: An intelligent authoring environment for abstract semantic representations of cultural object descriptions. In: Proceedings of the EACL 2009 Workshop on Language Technology and Resources for Cultural Heritage, Social Sciences, Humanities, and Education (LaTeCH SHELT&R 2009), pp. 10–17 (2009)

188. Letia, I.A., Groza, A.: Modelling imprecise arguments in description logic. Adv. Electr. Comput. Eng. 9(3), 94–99 (2009)

189. Liu, O., Tian, Q., Ma, J.: A fuzzy description logic approach to model management in R&D project selection. In: Proceedings of the 8th Pacific Asia Conference on Information Systems (PACIS 2004) (2004)

190. Martínez-Cruz, C., van der Heide, A., Sánchez, D., Triviño, G.: An approximation to the computational theory of perceptions using ontologies. Expert Syst. Appl. **39**(10), 9494–9503 (2012). doi:10.1016/j.eswa.2012.02.107

191. Quan, T.T., Hui, S.C., Fong, A.C.M.: Automatic fuzzy ontology generation for semantic help-desk support. IEEE Trans. Ind. Inf. **2**(3), 155–164 (2006)

192. Rodger, J.A.: A fuzzy linguistic ontology payoff method for aerospace real optionsvaluation. Expert Syst. Appl. **40**(8) (2013)

193. Slavíček, V.: An ontology-driven fuzzy workflow system. In: Emde Boas, P., Groen, F.C.A., Italiano, G.F., Nawrocki, J., Sack, H. (eds.) SOFSEM 2013. LNCS, vol. 7741, pp. 515–527. Springer, Heidelberg (2013). doi:10.1007/978-3-642-35843-2_44

194. OWL Web Ontology Language Overview. http://www.w3.org/TR/owl-features/. W3C (2004)

195. Cuenca-Grau, B., Horrocks, I., Motik, B., Parsia, B., Patel-Schneider, P., Sattler, U.: OWL 2: the next step for OWL. J. Web Semant. **6**(4), 309–322 (2008)

196. Bobillo, F., Straccia, U.: The fuzzy ontology reasoner fuzzy DL. Knowl.-Based Syst. **95**, 12–34 (2016). doi:10.1016/j.knosys.2015.11.017. http://www.sciencedirect.com/science/article/pii/S0950705115004621

197. Fire. http://www.image.ece.ntua.gr/nsimou/FiRE/

198. Stoilos, G., Simou, N., Stamou, G., Kollias, S.: Uncertainty and the semantic web. IEEE Intell. Syst. **21**(5), 84–87 (2006)

199. Straccia, U.: Softfacts: a top-k retrieval engine for ontology mediated access to relational databases. In: Proceedings of the 2010 IEEE International Conference on Systems, Man and Cybernetics (SMC 2010), pp. 4115–4122. IEEE Press (2010)

200. Haarslev, V., Pai, H.-I., Shiri, N.: Optimizing tableau reasoning in ALC extended with uncertainty. In: Proceedings of the 2007 International Workshop on Description Logics (DL 2007) (2007)

201. Habiballa, H.: Resolution strategies for fuzzy description logic. In: Proceedings of the 5th Conference of the European Society for Fuzzy Logic and Technology (EUSFLAT 2007), vol. 2, pp. 27–36 (2007)

202. Konstantopoulos, S., Apostolikas, G.: Fuzzy-DL reasoning over unknown fuzzy degrees. In: Meersman, R., Tari, Z., Herrero, P. (eds.) OTM 2007. LNCS, vol. 4806, pp. 1312–1318. Springer, Heidelberg (2007). doi:10.1007/978-3-540-76890-6_59. ISBN 3-540-76889-0, 978-3-540-76889-0

203. Wang, H., Ma, Z.M., Yin, J.: FRESG: a kind of fuzzy description logic reasoner. In: Bhowmick, S.S., Küng, J., Wagner, R. (eds.) DEXA 2009. LNCS, vol. 5690, pp. 443–450. Springer, Heidelberg (2009). doi:10.1007/978-3-642-03573-9_38

204. Straccia, U.: An ontology mediated multimedia information retrieval system. In: Proceedings of the the the 40th International Symposium on Multiple-Valued Logic (ISMVL 2010), pp. 319–324. IEEE Computer Society (2010)

205. Gao, M., Liu, C.: Extending OWL by fuzzy description logic. In: Proceedings of the 17th IEEE International Conference on Tools with Artificial Intelligence (ICTAI 2005), pp. 562–567. IEEE Computer Society, Washington, DC (2005). ISBN 0-7695-2488-5. http://dl.acm.org/citation.cfm?id=1105924.1106115

206. Stoilos, G., Stamou, G., Pan, J.Z.: Fuzzy extensions of OWL: logical properties and reduction to fuzzy description logics. Int. J. Approx. Reason. **51**(6), 656–679 (2010). ISSN 0888-613X. doi:10.1016/j.ijar.2010.01.005

207. Bobillo, F., Straccia, U.: An OWL ontology for fuzzy OWL 2. In: Rauch, J., Raś, Z.W., Berka, P., Elomaa, T. (eds.) ISMIS 2009. LNCS (LNAI), vol. 5722, pp. 151–160. Springer, Heidelberg (2009). doi:10.1007/978-3-642-04125-9_18

208. Bobillo, F., Straccia, U.: Representing fuzzy ontologies in OWL 2. In: Proceedings of the 19th IEEE International Conference on Fuzzy Systems (FUZZ-IEEE 2010), pp. 2695–2700. IEEE Press, July 2010

209. Fuzzy OWL 2 Web Ontology Language. http://www.straccia.info/software/FuzzyOWL/. ISTI - CNR (2011)

210. Ullman, J.D.: Principles of Database and Knowledge Base Systems, vol. 1, 2. Computer Science Press, Potomac (1989)

211. Shapiro, E.Y.: Logic programs with uncertainties: a tool for implementing rule-based systems. In: Proceedings of the 8th International Joint Conference on Artificial Intelligence (IJCAI 1983), pp. 529–532 (1983)

212. Baldwin, J.F., Martin, T.P., Pilsworth, B.W.: Fril - Fuzzy and Evidential Reasoning in Artificial Intelligence. Research Studies Press Ltd., Baldock (1995)

213. Baldwin, J.F., Martin, T.P., Pilsworth, B.W.: Applications of fuzzy computation: knowledge based systems: knowledge representation. In: Ruspini, E.H., Bonnissone, P., Pedrycz, W. (eds.) Handbook of Fuzzy Computing. IOP Publishing, Bristol (1998)

214. Bueno, F., Cabeza, D., Carro, M., Hermenegildo, M., López-García, P., Puebla, G.: The Ciao prolog system. Reference manual. Technical report CLIPS3/97.1. School of Computer Science, Technical University of Madrid (UPM) (1997). http://www.cliplab.org/Software/Ciao/

215. Cao, T.H.: Annotated fuzzy logic programs. Fuzzy Sets Syst. **113**(2), 277–298 (2000)

216. Chortaras, A., Stamou, G.B., Stafylopatis, A.: Adaptation of weighted fuzzy programs. In: 16th International Conference on Artificial Neural Networks - ICANN 2006, Part II, pp. 45–54 (2006)

217. Chortaras, A., Stamou, G., Stafylopatis, A.: Integrated query answering with weighted fuzzy rules. In: Mellouli, K. (ed.) ECSQARU 2007. LNCS (LNAI), vol. 4724, pp. 767–778. Springer, Heidelberg (2007). doi:10.1007/978-3-540-75256-1_67

218. Chortaras, A., Stamou, G.B., Stafylopatis, A.: Top-down computation of the semantics of weighted fuzzy logic programs. In: First International Conference on Web Reasoning and Rule Systems (RR 2007), pp. 364–366 (2007)

219. Ebrahim, R.: Fuzzy logic programming. Fuzzy Sets Syst. **117**(2), 215–230 (2001)

220. Guller, D.: Procedural semantics for fuzzy disjunctive programs. In: Baaz, M., Voronkov, A. (eds.) LPAR 2002. LNCS (LNAI), vol. 2514, pp. 247–261. Springer, Heidelberg (2002). doi:10.1007/3-540-36078-6_17. ISBN 3-540-00010-0

221. Guller, D.: Semantics for fuzzy disjunctive programs with weak similarity. In: Abraham, A., Köppen, M. (eds.) Hybrid Information Systems. AINSC, vol. 14, pp. 285–299. Physica-Verlag, Heidelberg (2002). doi:10.1007/978-3-7908-1782-9_21. ISBN 3-7908-1480-6

222. Hinde, C.: Fuzzy prolog. Int. J. Man-Mach. Stud. **24**, 569–595 (1986)

223. Ishizuka, M., Kanai, N.: Prolog-ELF: incorporating fuzzy logic. In: Proceedings of the 9th International Joint Conference on Artificial Intelligence (IJCAI 1985), Los Angeles, CA, pp. 701–703 (1985)

224. Klawonn, F., Kruse, R.: A Lukasiewicz logic based Prolog. Mathware Soft Comput. **1**(1), 5–29 (1994). https://citeseer.ist.psu.edu/klawonn94lukasiewicz.html

225. Magrez, P., Smets, P.: Fuzzy modus ponens: a new model suitable for applications inknowledge-based systems. Int. J. Intell. Syst. **4**, 181–200 (1989)

226. Martin, T.P., Baldwin, J.F., Pilsworth, B.W.: The implementation of FProlog -a fuzzy prolog interpreter. Fuzzy Sets Syst. **23**(1), 119–129 (1987). ISSN 0165-0114. http://dx.doi.org/10.1016/0165-0114(87)90104-7

227. Mukaidono, M.: Foundations of fuzzy logic programming. In: Advances in Fuzzy Systems - Application and Theory, vol. 1. World Scientific, Singapore (1996)

228. Mukaidono, M., Shen, Z., Ding, L.: Fundamentals of fuzzy prolog. Int. J. Approx. Reason. **3**(2), 179–193 (1989). ISSN 0888-613X. http://dx.doi.org/10.1016/0888-613X(89)90005-4

229. Paulik, L.: Best possible answer is computable for fuzzy SLD-resolution. In: Haják, P. (ed.) Gödel 1996: Logical Foundations of Mathematics, Computer Science, and Physics. LNL, vol. 6, pp. 257–266. Springer, Heidelberg (1996)

230. Rhodes, P.C., Menani, S.M.: Towards a fuzzy logic programming system: a clausal form fuzzy logic. Knowl.-Based Syst. **8**(4), 174–182 (1995)

231. Sessa, M.I.: Approximate reasoning by similarity-based SLD resolution. Theoret. Comput. Sci. **275**, 389–426 (2002)

232. Shen, Z., Ding, L., Mukaidono, M.: Fuzzy Computing. In: Theoretical Framework of Fuzzy Prolog Machine, pp. 89–100. Elsevier Science Publishers B.V. (1988). Chap. A

233. Subramanian, V.: On the semantics of quantitative logic programs. In: Proceedings of the 4th IEEE Symposium on Logic Programming, pp. 173–182. Computer Society Press (1987)

234. van Emden, M.: Quantitative deduction and its fixpoint theory. J. Log. Program. **4**(1), 37–53 (1986)

235. Vojtás, P.: Fuzzy logic programming. Fuzzy Sets Syst. **124**, 361–370 (2001)

236. Vojtás, P., Paulík, L.: Soundness and completeness of non-classical extended SLD-resolution. In: Dyckhoff, R., Herre, H., Schroeder-Heister, P. (eds.) ELP 1996. LNCS, vol. 1050, pp. 289–301. Springer, Heidelberg (1996). doi:10.1007/3-540-60983-0_20

237. Vojtás, P., Vomlelová, M.: Transformation of deductive and inductive tasks between models of logic programming with imperfect information. In: Proceedings of the 10th International Conference on Information Processing and Management of Uncertainty in Knowledge-Based Systems (IPMU 2004), pp. 839–846 (2004)

238. Wagner, G.: Negation in fuzzy and possibilistic logic programs. In: Martin, T., Arcelli, F. (eds.) Logic Programming and Soft Computing. Research Studies Press (1998)

239. Yasui, H., Hamada, Y., Mukaidono, M.: Fuzzy prolog based on Lukasiewicz implication and bounded product. IEEE Trans. Fuzzy Syst. **2**, 949–954 (1995)

240. Calmet, J., Lu, J.J., Rodriguez, M., Schü, J.: Signed formula logic programming: Operational semantics and applications (extended abstract). In: Raś, Z.W., Michalewicz, M. (eds.) ISMIS 1996. LNCS, vol. 1079, pp. 202–211. Springer, Heidelberg (1996). doi:10.1007/3-540-61286-6_145

241. Damásio, C.V., Medina, J., Ojeda-Aciego, M.: Sorted multi-adjoint logic programs: termination results and applications. In: Alferes, J.J., Leite, J. (eds.) JELIA 2004. LNCS (LNAI), vol. 3229, pp. 252–265. Springer, Heidelberg (2004). doi:10.1007/978-3-540-30227-8_23

242. Damásio, C.V., Medina, J., Ojeda-Aciego, M.: A tabulation proof procedure for residuated logic programming. In: Proceedings of the 6th European Conference on Artificial Intelligence (ECAI 2004) (2004)

243. Damásio, C.V., Medina, J., Ojeda-Aciego, M.: Termination results for sorted multi-adjoint logic programs. In: Proceedings of the 10th International Conference on Information Processing and Management of Uncertainty in Knowledge-Based Systems (IPMU 2004), pp. 1879–1886 (2004)
244. Damásio, C.V., Pan, J.Z., Stoilos, G., Straccia, U.: An approach to representing uncertainty rules in RuleML. In: Second International Conference on Rules and Rule Markup Languages for the Semantic Web (RuleML 2006), pp. 97–106. IEEE (2006)
245. Damásio, C.V., Pereira, L.M.: A survey of paraconsistent semantics for logic programs. In: Gabbay, D., Smets, P. (eds.) Handbook of Defeasible Reasoning and Uncertainty Management Systems, pp. 241–320. Kluwer, Alphen aan den Rijn (1998)
246. Damásio, C.V., Pereira, L.M.: Antitonic logic programs. In: Eiter, T., Faber, W., Truszczyński, M. (eds.) LPNMR 2001. LNCS (LNAI), vol. 2173, pp. 379–393. Springer, Heidelberg (2001). doi:10.1007/3-540-45402-0_28
247. Damásio, C.V., Pereira, L.M.: Monotonic and residuated logic programs. In: Benferhat, S., Besnard, P. (eds.) ECSQARU 2001. LNCS (LNAI), vol. 2143, pp. 748–759. Springer, Heidelberg (2001). doi:10.1007/3-540-44652-4_66. ISBN 3-540-42464-4
248. Damásio, C.V., Pereira, L.M.: Sorted monotonic logic programs and their embeddings. In: Proceedings of the 10th International Conference on Information Processing and Management of Uncertainty in Knowledge-Based Systems (IPMU 2004), pp. 807–814 (2004)
249. Damásio, C., Medina, J., Ojeda-Aciego, M.: A tabulation procedure for first-order residuated logic programs. In: Proceedings of the 11th International Conference on Information Processing and Management of Uncertainty in Knowledge-Based Systems (IPMU 2006) (2006)
250. Damásio, C., Medina, J., Ojeda-Aciego, M.: A tabulation procedure for first-order residuated logic programs. In: Proceedings of the IEEE World Congress on Computational Intelligence (section Fuzzy Systems) (WCCI 2006), pp. 9576–9583 (2006)
251. Damásio, C., Medina, J., Ojeda-Aciego, M.: Termination of logic programs with imperfect information: applications and query procedure. J. Appl. Log. **7**(5), 435–458 (2007)
252. Denecker, M., Marek, V., Truszczyński, M.: Approximations, stable operators, well-founded fixpoints and applications in nonmonotonic reasoning. In: Minker, J. (ed.) Logic-Based Artifical Intelligence, pp. 127–144. Kluwer Academic Publishers, Alphen aan den Rijn (2000)
253. Denecker, M., Pelov, N., Bruynooghe, M.: Ultimate well-founded and stable semantics for logic programs with aggregates. In: Codognet, P. (ed.) ICLP 2001. LNCS, vol. 2237, pp. 212–226. Springer, Heidelberg (2001). doi:10.1007/3-540-45635-X_22. ISBN 3-540-42935-2
254. Denecker, M., Marek, V.W., Truszczyński, M.: Uniform semantic treatment of default and autoepistemic logics. In: Cohn, A., Giunchiglia, F., Selman, B. (eds.) Proceedings of the 7th International Conference on Principles of Knowledge Representation and Reasoning, pp. 74–84. Morgan Kaufman, Burlington (2000)
255. Denecker, M., Marek, V.W., Truszczyński, M.: Ultimate approximations. Technical report CW 320. Katholieke Iniversiteit Leuven, September 2001

256. Denecker, M., Marek, V.W., Truszczyński, M.: Ultimate approximations in non-monotonic knowledge representation systems. In: Fensel, D., Giunchiglia, F., McGuinness, D., Williams, M. (eds.) Principles of Knowledge Representation and Reasoning: Proceedings of the 8th International Conference, pp. 177–188. Morgan Kaufmann, Burlington (2002)

257. Fitting, M.C.: The family of stable models. J. Log. Programm. **17**, 197–225 (1993)

258. Fitting, M.C.: Fixpoint semantics for logic programming - a survey. Theoret. Comput. Sci. **21**(3), 25–51 (2002)

259. Fitting, M.: A Kripke-Kleene-semantics for general logic programs. J. Log. Program. **2**, 295–312 (1985)

260. Fitting, M.: Pseudo-Boolean valued Prolog. Stud. Logica **XLVII**(2), 85–91 (1987)

261. Fitting, M.: Bilattices and the semantics of logic programming. J. Log. Program. **11**, 91–116 (1991)

262. Hähnle, R.: Uniform notation of tableaux rules for multiple-valued logics. In: Proceedings of the International Symposium on Multiple-Valued Logic, pp. 238–245. IEEE Press, Los Alamitos (1991)

263. Khamsi, M., Misane, D.: Disjunctive signed logic programs. Fundamenta Informaticae **32**, 349–357 (1996)

264. Khamsi, M., Misane, D.: Fixed point theorems in logic programming. Ann. Math. Artif. Intell. **21**, 231–243 (1997)

265. Kifer, M., Li, A.: On the semantics of rule-based expert systems with uncertainty. In: Gyssens, M., Paredaens, J., Gucht, D. (eds.) ICDT 1988. LNCS, vol. 326, pp. 102–117. Springer, Heidelberg (1983). doi:10.1007/3-540-50171-1_6

266. Kulmann, P., Sandri, S.: An annotaded logic theorem prover for an extended possibilistic logic. Fuzzy Sets Syst. **144**, 67–91 (2004)

267. Lakshmanan, L.V.S.: An epistemic foundation for logic programming with uncertainty. In: Thiagarajan, P.S. (ed.) FSTTCS 1994. LNCS, vol. 880, pp. 89–100. Springer, Heidelberg (1994). doi:10.1007/3-540-58715-2_116

268. Lakshmanan, L.V., Sadri, F.: Uncertain deductive databases: a hybrid approach. Inf. Syst. **22**(8), 483–508 (1997)

269. Lakshmanan, L.V., Shiri, N.: A parametric approach to deductive databases with uncertainty. IEEE Trans. Knowl. Data Eng. **13**(4), 554–570 (2001)

270. Loyer, Y., Straccia, U.: Uncertainty and partial non-uniform assumptions in parametric deductive databases. In: Flesca, S., Greco, S., Ianni, G., Leone, N. (eds.) JELIA 2002. LNCS (LNAI), vol. 2424, pp. 271–282. Springer, Heidelberg (2002). doi:10.1007/3-540-45757-7_23

271. Loyer, Y., Straccia, U.: The well-founded semantics in normal logic programs with uncertainty. In: Hu, Z., Rodríguez-Artalejo, M. (eds.) FLOPS 2002. LNCS, vol. 2441, pp. 152–166. Springer, Heidelberg (2002). doi:10.1007/3-540-45788-7_9

272. Loyer, Y., Straccia, U.: The approximate well-founded semantics for logic programs with uncertainty. In: Rovan, B., Vojtáš, P. (eds.) MFCS 2003. LNCS, vol. 2747, pp. 541–550. Springer, Heidelberg (2003). doi:10.1007/978-3-540-45138-9_48

273. Loyer, Y., Straccia, U.: Default knowledge in logic programs with uncertainty. In: Palamidessi, C. (ed.) ICLP 2003. LNCS, vol. 2916, pp. 466–480. Springer, Heidelberg (2003). doi:10.1007/978-3-540-24599-5_32

274. Loyer, Y., Straccia, U.: Epistemic foundation of the well-founded semantics over bilattices. In: Fiala, J., Koubek, V., Kratochvíl, J. (eds.) MFCS 2004. LNCS, vol. 3153, pp. 513–524. Springer, Heidelberg (2004). doi:10.1007/978-3-540-28629-5_39

275. Loyer, Y., Straccia, U.: Any-world assumptions in logic programming. Theoret. Comput. Sci. **342**(2–3), 351–381 (2005)

276. Loyer, Y., Straccia, U.: Epistemic foundation of stable model semantics. J. Theory Pract. Log. Program. **6**, 355–393 (2006)

277. Lu, J.J.: Logic programming with signs and annotations. J. Log. Comput. **6**(6), 755–778 (1996)

278. Lu, J.J., Calmet, J., Schü, J.: Computing multiple-valued logic programs. Mathware Soft Comput. **2**(4), 129–153 (1997)

279. Lukasiewicz, T., Straccia, U.: Tightly integrated fuzzy description logic programs under the answer semantics for the semantic web. Infsys Research report 1843-07-03. Institut FüR Informations Systeme Arbeitsbereich Wissensbasierte Systeme, Technische Universität Wien (2007)

280. Lukasiewicz, T., Straccia, U.: Tightly integrated fuzzy description logic programs under the answer semantics for the semantic web. In: Sheth, M.L.A. (ed.) Progressive Concepts for Semantic Web Evolution: Applications and Developments, pp. 237–256. IGI Global (2010). Chap. 11

281. Madrid, N., Straccia, U.: On top-k retrieval for a family of non-monotonic ranking functions. In: Larsen, H.L., Martin-Bautista, M.J., Vila, M.A., Andreasen, T., Christiansen, H. (eds.) FQAS 2013. LNCS (LNAI), vol. 8132, pp. 507–518. Springer, Heidelberg (2013). doi:10.1007/978-3-642-40769-7_44

282. Majkic, Z.: Coalgebraic semantics for logic programs. In: 18th Workshop on (Constraint) Logic Programming (WCLP 2005), Ulm, Germany (2004)

283. Majkic, Z.: Many-valued intuitionistic implication and inference closure in abilattice-based logic. In: 35th International Symposium on Multiple-Valued Logic (ISMVL 2005), pp. 214–220 (2005)

284. Majkic, Z.: Truth and knowledge fixpoint semantics for many-valued logic programming. In: 19th Workshop on (Constraint) Logic Programming (WCLP 2005), pp. 76–87, Ulm, Germany (2005)

285. Marek, V.W., Truszczyński, M.: Logic programming with costs. Technical report, University of Kentucky (2000). ftp://al.cs.engr.uky.edu/cs/manuscripts/lp-costs.ps

286. Mateis, C.: Extending disjunctive logic programming by T-norms. In: Gelfond, M., Leone, N., Pfeifer, G. (eds.) LPNMR 1999. LNCS (LNAI), vol. 1730, pp. 290–304. Springer, Heidelberg (1999). doi:10.1007/3-540-46767-X_21

287. Mateis, C.: Quantitative disjunctive logic programming: semantics and computation. AI Commun. **13**, 225–248 (2000)

288. Medina, J., Ojeda-Aciego, M.: Multi-adjoint logic programming. In: Proceedings of the 10th International Conference on Information Processing and Management of Uncertainty in Knowledge-Based Systems (IPMU 2004), pp. 823–830 (2004)

289. Medina, J., Ojeda-Aciego, M., Vojtaš, P.: Multi-adjoint logic programming with continous semantics. In: Eiter, T., Faber, W., Truszczyński, M. (eds.) LPNMR 2001. LNCS (LNAI), vol. 2173, pp. 351–364. Springer, Heidelberg (2001). doi:10.1007/3-540-45402-0_26

290. Medina, J., Ojeda-Aciego, M., Vojtáš, P.: A procedural semantics for multi-adjoint logic programming. In: Brazdil, P., Jorge, A. (eds.) EPIA 2001. LNCS (LNAI), vol. 2258, pp. 290–297. Springer, Heidelberg (2001). doi:10.1007/3-540-45329-6_29

291. Medina, J., Ojeda-Aciego, M., Vojtás, P.: Similarity-based unification: a multi-adjoint approach. Fuzzy Sets Syst. **1**(146), 43–62 (2004)

292. Rounds, W.C., Zhang, G.-Q.: Clausal logic and logic programming in algebraic domains. Inf. Comput. **171**, 183–200 (2001). https://citeseer.ist.psu.edu/276602.html

293. Schroeder, M., Schweimeier, R.: Fuzzy argumentation and extended logic programming. In: Proceedings of ECSQARU Workshop Adventures in Argumentation (2001)

294. Schroeder, M., Schweimeier, R.: Arguments and misunderstandings: fuzzy unification for negotiating agents. In: Proceedings of the ICLP Workshop CLIMA 2002. Elsevier (2002)

295. Schweimeier, R., Schroeder, M.: Fuzzy unification and argumentation for well-founded semantics. In: Emde Boas, P., Pokorný, J., Bieliková, M., Štuller, J. (eds.) SOFSEM 2004. LNCS, vol. 2932, pp. 102–121. Springer, Heidelberg (2004). doi:10.1007/978-3-540-24618-3_9

296. Straccia, U.: Annotated answer set programming. In: Proceedings of the 11th International Conference on Information Processing and Management of Uncertainty in Knowledge-Based Systems (IPMU 2006), pp. 1212–1219. E.D.K., Paris (2006). ISBN 2-84254-112-X

297. Straccia, U.: Query answering under the any-world assumption for normal logic programs. In: Proceedings of the 10th International Conference on Principles of Knowledge Representation (KR 2006), pp. 329–339. AAAI Press (2006)

298. Straccia, U.: A top-down query answering procedure for normal logic programs under the any-world assumption. In: Mellouli, K. (ed.) ECSQARU 2007. LNCS (LNAI), vol. 4724, pp. 115–127. Springer, Heidelberg (2007). doi:10.1007/978-3-540-75256-1_13

299. Straccia, U.: Towards vague query answering in logic programming for logic-based information retrieval. In: Melin, P., Castillo, O., Aguilar, L.T., Kacprzyk, J., Pedrycz, W. (eds.) IFSA 2007. LNCS (LNAI), vol. 4529, pp. 125–134. Springer, Heidelberg (2007). doi:10.1007/978-3-540-72950-1_13

300. Straccia, U.: On the top-k retrieval problem for ontology-based access to databases. In: Pivert, O., Zadrożny, S. (eds.) Flexible Approaches in Data, Information and Knowledge Management. SCI, vol. 497, pp. 95–114. Springer, Heidelberg (2014). doi:10.1007/978-3-319-00954-4_5. ISBN 978-3-319-00953-7

301. Straccia, U., Madrid, N.: A top-k query answering procedure for fuzzy logic programming. Fuzzy Sets Syst. **205**, 1–29 (2012)

302. Straccia, U., Ojeda-Aciego, M., Damásio, C.V.: On fixed-points of multi-valued functions on complete lattices and their application to generalized logic programs. SIAM J. Comput. **8**(5), 1881–1911 (2009)

303. Turner, H.: Signed logic programs. In: Bruynooghe, M. (ed.) Proceedings of the 1994 International Symposium on Logic Programming, pp. 61–75. The MIT Press (1994). https://citeseer.ist.psu.edu/turner94signed.html

Applying Machine Reasoning and Learning in Real World Applications

Freddy Lecue[1,2(✉)]

[1] Accenture Technology Labs, Dublin, Ireland
freddy.lecue@inria.fr, freddy.lecue@accenture.com
[2] Inria, Rocquencourt, France

Abstract. Knowledge discovery, as an area focusing upon methodologies for extracting knowledge through deduction (a priori) or from data (a posteriori), has been largely studied in Database and Artificial Intelligence. Deductive reasoning such as logic reasoning gains logically knowledge from pre-established (certain) knowledge statements, while inductive inference such as data mining or learning discovers knowledge by generalising from initial information. While deductive reasoning and inductive learning are conceptually addressing knowledge discovery problems from different perspectives, they are inference techniques that nicely complement each other in real-world applications. In this chapter we will present how techniques from machine learning and reasoning can be reconciled and integrated to address large scale problems in the context of (i) transportation in cities of Bologna, Dublin, Miami, Rio and (ii) spend optimisation in finance.

1 Introduction and Context

The *Semantic Web* [1], where the semantics of information is indicated using machine-processable languages such as the *Web Ontology Language* (OWL) [2], is considered to provide many advantages over the "syntactic" version of the current World-Wide-Web. OWL, for example, is underpinned by Description Logics (DL) [3] and ontologies [4] (a formal conceptualization of a particular domain). This allows automatic processing of information assets tagged with OWL, focusing on their semantics rather than on the way they are shown on the Web.

As claimed by the World Wide Web Consortium (W3C)[1], the semantic Web - also known as the Linked Data Web or Web 3.0 - is a Web of knowledge. Such Web could benefit any methodology which aims at tackling any issue related to variety of data i.e., one of the four dimensions (volume, velocity, variety and veracity) of the so-called "Big Data". Indeed the semantic Web exposes common vocabularies, representation layers, querying and reasoning techniques to process the Web of data at scale. In particular many machine reasoning mechanisms have been studied for automating complex tasks on top of large amount of heterogenous data with expressive descriptions e.g., ontology classification [5],

---

[1] http://www.w3.org/standards/semanticweb/data.

J.Z. Pan et al. (Eds.): Reasoning Web 2016, LNCS 9885, pp. 241–257, 2017.
DOI: 10.1007/978-3-319-49493-7_7

inconsistency justification [6] among others. Although many other open research challenges have also emerged from some of the dimensional features of the Web of data e.g., volume, velocity and variety, future reasoning system can benefits from such unique characteristics of data. From faster to better decision, these are examples of advantages that future reasoning techniques can deliver on top of the Web of data.

From a more data-driven perspective, machine learning [7] has been emerged as an area focusing upon methodologies for extracting knowledge a posteriori i.e., from data. The underlying inference mechanisms, also referenced as inductive inference such as data mining or learning discovers knowledge by generalising from initial information. Machine learning algorithms are being used in many different domains, and are becoming increasingly ubiquitous with more and more applications in places. From anomaly detection in the context of fraud analytics such as money laundering [8] or suspicious trading in a stock market [9], to classification of news article from millions of sources [10]. In particular many machine learning mechanisms have been studied for retrieving systematic patterns and models on top of large amount of data. Machine learning is getting its momentum, partly because of the "Big Data" context and new powerful infrastructures which are now available.

In this work we aim at showing how learning and reasoning techniques can be combined. For instance using reasoning for interpreting and understanding how models are extracted from data using machine learning, or explaining how and why abnormal patterns can be extracted from data. All are examples of hybrid reasoning where learning and reasoning are applied seamlessly. In particular we present some real-world applications, combining the strength of machine reasoning and learning to scale in industry settings. We focused on (i) smart transportation in cities, and studied the problems of prediction and diagnosis of road traffic condition and congestion using semantic Web technologies, and (ii) spend optimisation in finance, and studied the problem of explaining abnormal expenses in large companies.

2   Road Transportation in Cities

As the number of vehicles on the road steadily increases and the expansion of roadways is remained static, congestion in cities and urban areas became one of the major transportation issues in most industrial countries [11]. Urban traffic costs 5.5 billion hours of travel delay and 2.9 billion gallons of wasted fuel in the USA alone, all at the price of $121 billion per year. More important, it is getting worse, year by year i.e., the costs of the costs of extra time and wasted fuel has quintupled over the past 30 years. It also used to (i) stress and frustrate motorists, encouraging road rage and reducing health of motorists [12], and more dramatically, (ii) interfere with the passage of emergency vehicles traveling to destinations where they are urgently needed. of them. All are examples of negative effects of congestion in cities. Four ways can be considered to reduce congestion [13]; one is to improve the infrastructure e.g., by increasing the road capacity, but this requires enormous expenditure which is often not viable. Promoting public transport in large cities is another way but it is not always convenient. Another solution, which has been studied in this work is to:

Determine the future states of roads segments, which will support transportation departments and their managers to proactively manage the traffic before congestion is reached e.g., changing traffic light strategy.

Another approach, which we also focused on, is to:

Diagnose road traffic congestion as it enables city managers to have a solid understanding of the traffic issues that lead to an unexpected situation, whom can then take appropriate corrective actions (e.g., rerouting or changing traffic light strategy in case of an accident or a broken traffic light) and better plan the city traffic.

The task of road traffic diagnosis as the identification of clear and descriptive explanations of their reasons is not straightforward, especially in quasi real-time situations. Understanding potential causes is important for informing interested parties, for instance, car drivers and public authorities, in quasi real time. This is also crucial for ensuring that public authorities will take optimal decisions and appropriate actions in time, especially in case of emergency. Towards this challenge we investigated an innovative system combining machine learning and reasoning towards the problems of prediction and diagnosis. Semantics has been identified as a crucial representation layer for reasoning, and thus a major breakthrough to unlock the various industrial problems of integration, diagnosis and prediction.

Most notably, the integration and exploitation of such a system have be conducted in major cities i.e., Bologna (Italy), Dublin (Ireland), Miami (USA) and Rio (Brazil) through the IBM DALI (Data Annotation, Linkage and Integration) [22] and IBM STAR-CITY[2] [24] (Semantic Traffic Analytics and Reasoning for CITY[3] - Fig. 1) projects.

2.1 Semantics-Driven Prediction of Road Traffic Conditions

We addressed the problem of predicting road traffic conditions such as congestion by automatically integrating and contextualising numerous sensors. Such sensors expose heterogenous, exogenous, raw data streams such as weather information, road works and city events. Table 1 synthesizes the main important details of the data sets we have considered for this reasoning task.

Figure 2 describes the architecture for generating OWL EL representations from raw CSV, tweets, XML, PDF data in Table 1, all accessed through different mechanisms.

---

[2] http://researcher.watson.ibm.com/researcher/view_group.php?id=5101.

[3] http://208.43.99.116:9080/simplicity/demo.jsp.

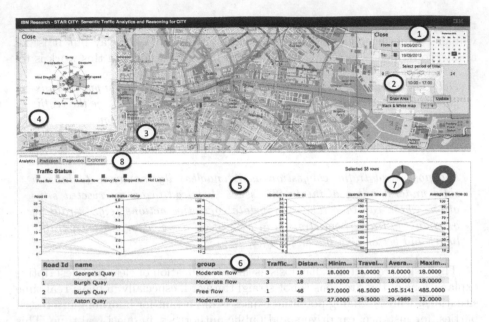

Fig. 1. Interface of STAR-CITY. ①, ②: temporal (Date and Time) context (subject to user selection), ③: spatial (Map Area) context (subject to user selection), ④: Weather context for the spatio-temporal analysis, ⑤: Travel time status vs. historical min., max., average travel time (segments are select-able), ⑥: Detailed version of travel time information (records are select-able with automated update on the parallel chart and map), ⑦: Spatio-temporal proportion of travel time status. ⑧: Tab-based selection of analytics and reasoning: prediction, diagnosis, exploration.

All representations have the same core static background knowledge to capture time (W3C Time Ontology[4]), space (W3C Geo Ontology[5] for encoding location) but differ only in some domain-related vocabularies e.g., traffic flow type, weather phenomenon, road condition, event type. These ontologies have been mainly used for enriching raw data, facilitating its integration, comparison, and matching. The DBpedia vocabulary has been used for cross-referencing entities (not described here). The spatial representation is either directly parsed from the raw data sources or retrieved through DB2 Spatial extender when text-based descriptions such as street name are retrieved. We did not make use of the Semantic Sensor Network ontology[6] as it is mainly designed for reasoning over sensor-related descriptions rather than its data and associated phenomenons. We served real-time ontology streams by using IBM InfoSphere Streams, where different mapping techniques are used depending on the data format. The main benefits of packaging our approach using stream processing are:

---

[4] http://www.w3.org/TR/owl-time/.

[5] http://www.w3.org/2003/01/geo/.

[6] http://www.w3.org/2005/Incubator/ssn/.

Table 1. (Raw) Data sources for STAR-CITY in Dublin, Bologna, Miami and Rio.

Source Type	Data Source	Description	City			
			Dublin (Ireland)	Bologna (Italy)	Miami (USA)	Rio (Brazil)
Traffic Anomaly	Journey travel times across the city	Traffic Department's TRIPS system[a]	CSV format (47 routes, 732 sensors) 0.1 GB per day[b]	✗ (not available)		
	Dublin Bus Dynamics	Vehicle activity (GPS location, line number, delay, stop flag)	✗ (not used)	SIRI: XML format[c] (596 buses, 80KB per update 11GB per day[d])	CSV format (893 buses, 225 KB per update 43 GB per day[e])	CSV format (1,349 buses, 181 KB per update 14 GB per day[f])
Traffic Diagnosis	Social-Media Related Feeds	Reputable sources of road traffic conditions in Dublin City	"Tweet" format - Accessed through Twitter streaming API[g]			
			Approx. 150 tweets per day[h] (approx. 0.001 GB)	✗ (not available)	Approx. 500 tweets per day[i] (approx. 0.003 GB)	✗ (not available)
	Road Works and Maintenance		PDF format (approx. 0.003 GB per day[j])	XML format (approx. 0.001 GB per day[k])	HTML format (approx. 0.001 GB per day[l])	✗ (not available)
	Social events e.g., music event, political event	Planned events with small attendance	XML format - Accessed once a day through Eventbrite"APIs			
			Approx. 85 events per day (0.001 GB)	Approx. 35 events per day (0.001 GB)	Approx. 285 events per day (0.005 GB)	Approx. 232 events per day (0.01 GB)
		Planned events with large attendance	XML format - Accessed once a day through Eventful"APIs			
			Approx. 180 events per day (0.05 GB)	Approx. 110 events per day (0.04 GB)	Approx. 425 events per day (0.1 GB)	Approx. 310 events per day (0.08 GB)
	Bus Passenger Loading / Unloading (information related to number of passenger getting in / out)		✗ (not available)	✗ (not available)	CSV format (approx. 0.8 GB per day[e])	CSV format (approx. 0.1 GB per day[e])

[a] Travel-time Reporting Integrated Performance System - http://www.advantechdesign.com.au/trips
[b] http://dublinked.ie/datastore/datasets/dataset-215.php (live)
[c] Service Interface for Real Time Information - http://siri.org.uk
[d] http://82.187.83.50/GoogleServlet/ElaboratedDataPublication (live)
[e] Private Data - No Open data
[f] http://data.rio.rj.gov.br/dataset/gps-de-onibus/resource/cfeb367c-c1c3-4fa7-b742-65c2c99d8d90 (live)
[g] https://sitestream.twitter.com/1.1/site.json?follow=ID
[h] https://twitter.com/LiveDrive - https://twitter.com/aaroadwatch - https://twitter.com/GardaTraffic
[i] https://twitter.com/fl511_southeast
[j] http://www.dublincity.ie/RoadsandTraffic/ScheduledDisruptions/Documents/TrafficNews.pdf
[k] http://82.187.83.50/TMC_DATEX/
[l] http://www.fl511.com/events.aspx
[m] https://www.eventbrite.com/api - http://api.eventful.com

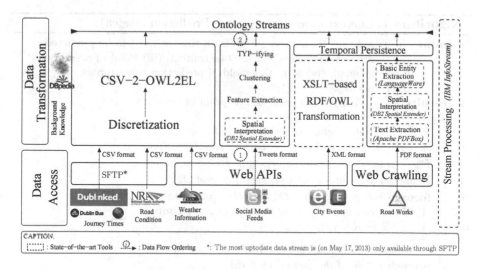

Fig. 2. Semantic stream enrichment.

1. Easy synchronisation of streams (with different frequency updates) and their OWL2 EL transformation, all streams come with different frequency updates and capturing ontology stream snapshots required synchronisation.

2. Flexible and scalable composition of stream operations (e.g., transformation, aggregation, filtering) by adjusting its processing units.
3. Identification of patterns and rules over different time windows, (iv) possible extension to higher throughput sensors.

All above points are all natively supported by stream processing engines. The underlying data is used for contextualizing traffic conditions. In particular we aimed at correlating and discovering association of knowledge across time and space to improve accuracy and consistency of traffic congestion prediction. Our approach [14, 25, 26] for prediction of road traffic conditions, and exploits semantic Web technologies and adapts recent research work in semantic predictive reasoning [15] as a way to annotate and interpret semantics of stream data. In details Algorithm 1 sketches the approach of predicting travel time data stream $\mathcal{O}_m^n$ at point of time $j \in [m, n]$ using weather information data stream $\mathcal{P}_m^n$.

The approach mainly consists of:

- auto-correlation of data on a time basis (lines 5, 6 of Algorithm 1) for retrieving all similar past weather conditions,
- association rules mining (lines 7–10 of Algorithm 1) for inferring correlation and rules between past weather condition and travel time e.g., rule (1) which explicits how weather condition may impact traffic status on road r_1,

---

Algorithm 1. Context-aware Travel Time Prediction (sketch)

---

1 **Input**: (i) Travel time data stream $\mathcal{O}_m^n$ (from time m to n) to be predicted, (ii) Weather data stream $\mathcal{P}_m^n$ used as context, (iii) Point of prediction time $j \in [m, n]$, (iv) Min. threshold of prediction rule support m_s, confidence m_c.

2 **Result**: $\mathcal{P}_m^n(j)$: Travel time predicted at point of time j.

3 **begin**

4    $\mathcal{R} \leftarrow \emptyset$; % *Initialization of prediction rules set.*

5   % *Auto-correlation of contextual weather information.*

6    $\tilde{\mathcal{P}}_m^n \leftarrow$ retrieve all similar contexts of $\mathcal{P}_m^n(j)$ in $[m, n]$;

7   % *Stream association rules between $\tilde{\mathcal{P}}_m^n$ and $\mathcal{O}_m^n$.*

8   **foreach** *rule* $\rho \in \tilde{\mathcal{P}}_m^n(k) \times \mathcal{O}_m^n(k)$, $\forall k \in [m, n]$ **do**

9   **if** $support(\rho) > m_s \wedge confidence(\rho) > m_c$ **then**

10          $\mathcal{R} \leftarrow \mathcal{R} \cup \{\rho\}$;

11   % *Semantic evaluation of rule $\rho \in \mathcal{R}$ at point of time j.*

12   **foreach** $\rho \in \mathcal{R}$ *of the form* $\mathcal{G} \twoheadrightarrow h$ **do**

13   **if** h *in semantically consistent with* $\mathcal{O}_m^n(j)$ **then**

14   Apply rule ρ at point of time j of $\mathcal{O}_m^n$;

15   **return** $\mathcal{O}_m^n(j)$;

$$HeavyTrafficFlow(s) \leftarrow Road(r_1) \wedge Road(r_2) \wedge isAdjacentTo(r_1, r_2) \wedge$$
$$hasTravelTimeStatus(r_1, s) \wedge$$
$$hasWeatherPhenomenon(r_1, w) \wedge$$
$$OptimunHumidity(w) \wedge$$
$$hasTrafficPhenomenon(r_2, a) \wedge$$
$$RoadTrafficAccident(a) \tag{1}$$

- validation of prediction results in $\mathcal{O}_m^n(j)$ by analyzing its semantic consistency (line 13 of Algorithm 1) i.e., checking that no conflict of knowledge is occurring (e.g., free flow prediction while all connected roads are all congested).

All predictive rules are extracted through association mining of semantic descriptions (across streams). Then they are filtered based on their occurrence (i.e., support) and confidence in line 9 of Algorithm 1. The rule Eq. 1, extended with temporal dimension, is one example of such semantic rules. In the context of social media, all tweets are semantically described (if possible) through incidents, accident, obstruction, closures by analysing their content. Their semantics is then associated with knowledge from other sources e.g., weather condition, traffic condition, types of events to infer weighted recurring rules.

Figure 3 illustrates how predictions are handled in STAR-CITY[7]. The future status of road segments (of the selected boundary box) and their proportion are

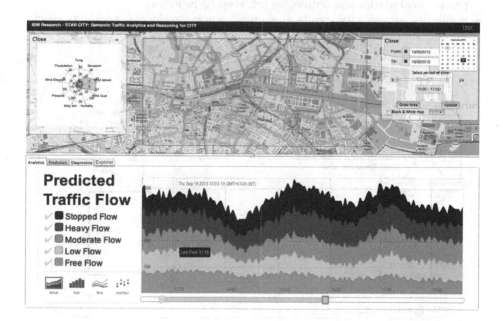

Fig. 3. Traffic congestion severity prediction user interface. (Color figure online)

[7] Prediction part of the live IBM STAR-CITY system (http://208.43.99.116:9080/simplicity/demo.jsp).

reported. In particular it reports a 180-minutes ahead prediction of the severity of traffic congestion in Dublin city. The bottom part reports the proportion of free (green), stopped (brown) flow roads of the selected area (top part). Different visualizations are offered i.e., area, bar, line, scatter to provide better comparison experiences.

2.2 Semantics-Driven Diagnosis of Road Traffic Conditions

STAR-CITY provides reasoning mechanisms to deliver insight on historical and real-time (road) traffic conditions, all supporting efficient urban planning. In particular inference techniques to diagnose and predict road traffic conditions in real-time using heterogenous streaming data from various types of sensors in the aforementioned cities.

Our work on prediction has been complemented with traffic diagnosis to explain traffic congestion in quasi-real-time. In both works data is lifted at semantic level (cf. Fig. 2) but the diagnosis and predictive approaches are different techniques. Diagnosis is based on semantic matching of events and a probabilistic model while prediction is based on stream auto-correlation, association mining.

An overview of our diagnosis approach [24, 27–30] is illustrated in Fig. 4. In particular:

- All relevant data is transformed and uplifted to a semantic level using appropriate vocabularies and ontologies (cf. step ① in Fig. 4).
- Similarity among city events (i.e., road works, accidents, incidents, social events) are pre-computed for faster processing (cf. step ② in Fig. 4).

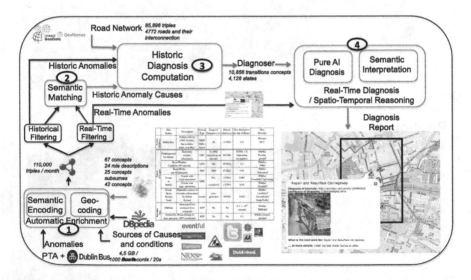

Fig. 4. Architecture of diagnosis approach for traffic congestion. ①: Data transformation and semantic uplift, ②: entity (semantic) similarity evaluation, ③: off-line diagnosis, ④ on-line diagnosis.

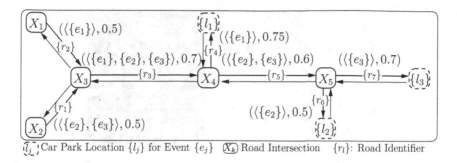

$\lceil l_j \rceil$ Car Park Location $\{l_j\}$ for Event $\{e_j\}$ (X_k) Road Intersection $\{r_l\}$: Road Identifier

Fig. 5. A traffic congestion diagnoser (a road is labelled if its traffic congestion could be explained).

- Then our approach compiled off-line all historic diagnosis information into a deterministic finite state machine, following the structure of a city road network (using linkgeodata.org), here Bologna, Dublin, Miami and Rio (cf. step ③ in Fig. 4). The network is used (i) to properly connect roads and (ii) for exploiting congestion propagation. Figure 5 depicts a sample of a road network as a graph where nodes are junctions and edges are roads. Each edge is annotated with events occurring on the road with its probability to have congestion in the past. Events could be atomic or association of events depending in the number of events occurring.
The state machine is augmented with respect to all semantic-augmented city events and weather conditions where a subset of them are connected to past congestion and the probability with which they have caused it.
- The real-time diagnosis of traffic condition is performed by analyzing the historical versions of the state machines to retrieve similar contexts i.e., congestion, potential explanation (i.e., city events), weather information. The similarity is estimated by comparing semantic descriptions of the context (cf. step ④ in Fig. 4) through matchmaking functions introduced by [16,17].

From an end-user perspective Fig. 6 illustrates the number of traffic congestion (red-highlighted number) which has been retrieved given the time window.

Fig. 6. Traffic congestion anomaly. The red-highlighted numbers refer to the number of traffic congestion in the spatio-temporal context of Fig. 1. (Color figure online)

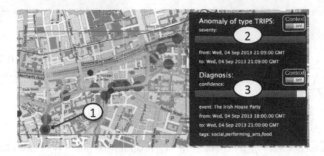

Fig. 7. Diagnosis interpretation. An example of traffic congestion (red points) where one i.e., ① is diagnosed by 4 (blue points) explanations. ②: Detailed descriptions (severity, date, time) of congestion, ③: Detailed description (city event name, date, time, type) of diagnosis. (Color figure online)

Our system also presents the spatial representation of traffic conditions and their diagnosis (red and blue icons for respectively congestion and diagnosis in Fig. 7).

Finally our system exposes a spider chart (① and ③) together with pie charts (② and ④) to compile and compare the impact of city events on the traffic conditions in Fig. 8. ① (respectively ②) represents all types (respectively subtypes) of city events occurring in the selected area of the city while ③ (respectively ④) captures all types (respectively subtypes) of city events which negatively impact (i.e., diagnose) the traffic conditions. For instance most of congestion is due to social events ④ i.e., 72% and more particularly music events ④ although 34% of city events are road works in the spatio-temporal context of Fig. 8. All results in Figs. 6, 7, 8 can be interpreted by city managers to understand how traffic condition is impacted by any type of city event.

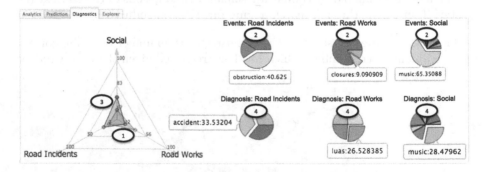

Fig. 8. STAR-CITY diagnosis: impact of city events on traffic conditions. ①: Proportion of city events which occurs in the spatio-temporal context, ②: Clustering and proportion of city events by subtypes, ③: Proportion of diagnosis (i.e., types of city events which cause a traffic congestion), ④ clustering and proportion of diagnosis by subtypes.

2.3   Spend Optimisation in Finance

In the context of finance, the challenge is the following: identifying abnormal accommodation expenses and providing their explanation together with contributing factors. In particular the *Intelligent Finance* Accenture system (Fig. 9) is addressing the problem of *spend optimisation* i.e., *how determining when to save and spend*. $546 Billion worldwide has been estimated to be lost because of non spend optimisation. The system, designed to be used by large companies with enormous variety of expenses, is detecting abnormal expenses using machine learning techniques and providing explanation of abnormal expenses (flight, accommodation, entertainment). The system is mainly used by (i) travel and expenses business owner to better manage spend optimisation, (ii) expenses auditor for tracking abnormal expenses, and (iii) internal travel system administrator for defining expenses policy based on the reasoning results.

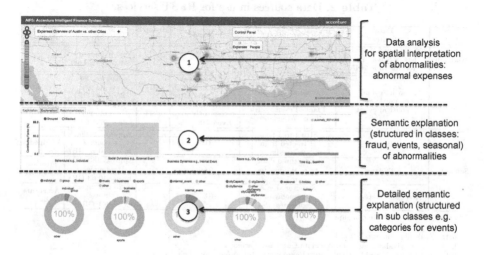

Fig. 9. Accenture intelligent finance application. ① refers to a spatial representation of abnormalities. ② refers to an histogram representation of one level of diagnosis (not exposed in this resource). ③ refers to a pie chart representation of the level of diagnosis exposed by our resource cf. semantic categories in accommodation/∗/diagnosis.

Explanatory reasoning has been achieved by applying core principles of [27] and revisiting the application domain in [28]. In a nutshell, the reasoning task of explaining anomalies (abnormal high price of accommodation) consists in interpreting the impact of external events (e.g., social events, seasonal effect). The interpretation is achieved by measuring the semantic similarity (through ontology matching) of any 4-uple of the form $< city, accommodation, event, impact >$ where city, accommodation, event and impact are defined by respectively 89, 41, 24 and 11 semantic properties using vocabularies of DBpedia and wikidata.

Table 2 describes all data and vocabularies exploited by our reasoning service.

• **Background Data in Use**: We define background data as data consumed silently by the reasoning service with no user specification in the service input parameters. Only relevant background data (i.e., data which could be mapped to DBpedia, wikidata for semantic comparison cf. [27,28]) is transformed in RDF for reasoning purpose. The size in column *"Size per day"* is the maximal amount of data collected i.e., for diagnosing abnormal price of accommodation in all cities.

• **Output Data**: An average number of 61 (resp. 4,509) RDF triples, with an average size of 3.1 (resp. 165.8) KB is retrieved from the diagnosis (resp. abnormality) service. All results of the diagnosis services are maintained in different internal RDF stores (through various size of graphs) for scalability purpose. The current size of the knowledge base is $31,897,325$ RDF triples, with a monthly growth of 41%.

Table 2. Data sources in use for REST services.

Source Type	Data Source	Description	Format	Historic (Year)	Size per day (GBytes)	Data Provider
Anomaly	300,000+ unique travellers in 500+ cities recorded for 2015	Min. and max. number of respectively 2,521 and 24,800 expenses per city[a]	CSV	2015	.93 (complete) .41 (aggregated)	Private
Diagnosis	Social events e.g., music event, political event	Planned events with small attendance	JSON format Accessed through Eventbrite APIs[a]	2011	Approx. 94 events per day (0.49 GB)	Eventbrite
		Planned events with large attendance	JSON format Accessed through Eventful APIs[b]	2011	Approx. 198 events per day (0.39 GB)	Eventful
Semantics	DBpedia	Structured facts extracted from wikipedia	RDF[c]	-	Approx. 33,000+ resources in use (0.23 GB)	Wikipedia
	wikidata	Structured data from Wikimedia projects	RDF[d]	-	Approx. 189,000+ resources in use (0.63 GB)	Freebase Google inc.
	Accenture Categories	Structured is-A taxonomy of event categories	RDF[e]	-	25 resources in use (0.001 GB)	Accenture inc.

[a] https://www.eventbrite.com/api
[b] http://api.eventful.com
[c] http://wiki.dbpedia.org/Datasets
[d] https://www.wikidata.org/wiki/Wikidata:Database_download
[e] http://54.194.213.178:8111/ExplanatoryReasoning/ontology/categories.n3

3   Experimentation

3.1   Semantics-Driven Prediction of Road Traffic Conditions

Our experimental results emphasized the advantage of our approach [14] which uses semantic Web technologies for predicting knowledge in streams. Figure 10 presents the accuracy of our approach against some pure statistical approaches: [18] (noted [DR97]) using inductive logic, [19] (noted [SA95]) using basic taxonomies and [15] using ABox axioms-related rules (noted [LP13b]).

Since state-of-the-art approaches fail to encode text-based streams in pure value-based time series, they simply fail to interpret their semantics. On the contrary, our approach interpreted their semantics to enrich the prediction model,

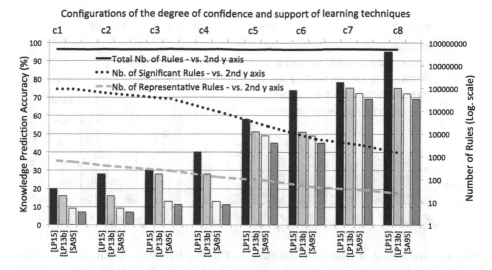

Fig. 10. Prediction accuracy: Our approach vs. State-of-the-art approaches. 1st x axis: approaches on 8 test cases. 1st y axis: accuracy of predicted knowledge. 2nd x axis: 8 types of (increasing) support/confidence configurations. 2nd y axis: search space of candidate rules.

ensuring better accuracy. Our results also pointed out the scalability limitation of our approach. Indeed the scalability of our predictive reasoning approach in the context of traffic has been shown to be negatively impacted by the number of data sources and their size. Their number are critical as they drive heterogeneity in rules, which could improve accuracy, but not scalability. It would be worst with more expressive DLs due to binding and containment checks. Therefore it would be interesting to adapt the approach in distributed settings.

3.2   Semantics-Driven Diagnosis of Road Traffic Conditions

We experimented the scalability and correctness of our diagnosis reasoning by evaluating its computation time and accuracy in different cities and contexts. Figure 11 reports the scalability of our diagnosis reasoning and its core components (i.e., data transformation, OWL/RDF loading in Jena TDB, anomaly detection) by comparing their computation time in different cities and contexts. Similarly to data transformation and OWL/RDF loading, the anomaly detection and diagnosis reasoning have been performed over one day of traffic.

Figure 12 reports the impact of historical information and size and number of data sets on accuracy of diagnosis results in Dublin, Bologna, Miami and Rio. The accuracy has been evaluated by comparing our explanation results against those estimated by transportation experts (used as ground truth) in their respective cities. A basis of one complete day of experimentation has been used i.e., 2,800, 240, 1,190 and 3,100 traffic anomalies for respectively Dublin, Bologna, Miami and Rio.

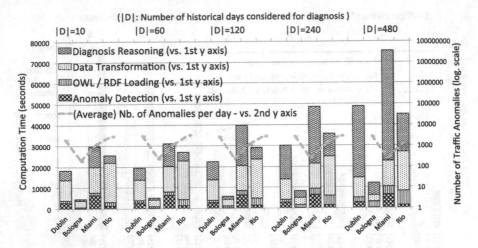

Fig. 11. Scalability of diagnosis reasoning in Dublin, Bologna, Miami and Rio. The experiments have been conducted on a server of 6 Intel(R) Xeon(R) X5650, 3.46 GHz cores, and 6 GB RAM.

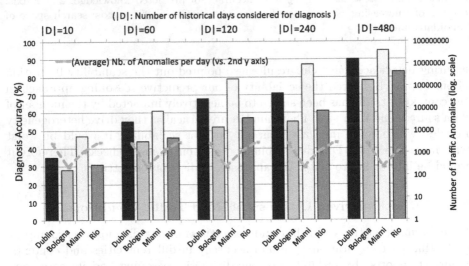

Fig. 12. Accuracy of diagnosis reasoning in Dublin, Bologna, Miami and Rio.

4   Lessons Learnt

From the implementation of such systems we learnt that the more data sets the more overhead on transformation, loading, and reasoning. For instance our app-roach performs better in contexts with less data sets (e.g., Bologna for the trans-portation asset, and Austin for finance asset) than in contexts with more data sets (e.g., Miami for the transportation asset, and Chicago for the finance asset), although the latter results remain scalable. Interestingly the more historical information the more computation time, specifically for diagnosis reasoning. This was caused by the intensive event similarity search over historical events performed by the diagnosis. It has also been demonstrated that the more data sources the more accurate the diagnosis results. Reducing the number of historical events decreases accuracy of diagnosis. The more similar historical events the higher the probability to catch accurate diagnosis. For instance the accuracy of diagnosis results is improved by a factor of 1.5 by multiplying the number of historical days by a factor 8 (from 60 to 480 days).

Similarly to the application of prediction, the application of diagnosis to traffic has shown some limitations in scalability.

5 Conclusion

In this chapter, we presented some real-world applications, combining the strength of machine reasoning and learning to scale in industry settings. We focused on (i) smart transportation in cities, and studied the problems of prediction and diagnosis of road traffic condition and congestion using semantic Web technologies, and (ii) spend optimisation in finance, and studied the problems of identifying and explaining abnormal expenses in large companies. Importantly the systems have been shown to work efficiently with real, live and heterogeneous stream data. The positive results have been highly emphasised to attract a few other engagements over years.

As future directions we expect to study the trade-off between (i) expressivity of the representation layer of data, (ii) scalability of the model and (iii) accuracy of diagnosis. This is particularly valid for the traffic and finance domains. Indeed it is expected to obtain different levels of expressivity for different cities since specialised data could expose different levels of representation. The definition of a common framework for comparing systems from different contexts and cities would be required for evaluation purpose.

Congratulations for having gone though all the chapters. We hope you enjoyed this journey. It might be an idea to read Chap. 1 once more, which is meant to be a bit more general than the other chapters and to connect these chapters altogether. These connections might give you further insights after you finish reading the remaining chapters.

In case you would like to read more on related topics, here are some suggestions:

- If you want to know more about how to apply knowledge graphs in enterprise, you could read 'Exploiting Linked Data and Knowledge Graphs for Large Organisations' [20].
- If you want to learn something new about how knowledge graphs and ontologies can be used in software engineering in general, you should try 'Ontology-Driven Software Development' [21].
- If you want to know more about Description Logics, you could read 'Description Logic Handbook' [23], which contains a collection of good quality research papers written before OWL was standardised.

• Last but not least, if you want to know more about Reasoning Web, you should check the previous and future lecture notes of the Reasoning Web Summer School.

References

1. Berners-Lee, T., Hendler, J., Lassila, O.: The semantic Web. Sci. Am. **284**(5), 34–43 (2001)
2. Smith, M.K.: OWL Web Ontology Language Guide (2004). http://www.w3c.org/TR/owl-guide/
3. Baader, F., Nutt, W.: Basic description logics. In: The Description Logic Handbook: Theory Implementation and Applications. Cambridge University Press (2003). ISBN 0521781760
4. Gruber, T.R.: A translation approach to portable ontology specifications. Knowl. Acquis. **5**(2), 199–220 (1993)
5. Seidenberg, J., Rector, A.: Web ontology segmentation: analysis, classification and use. In: Proceedings of the 15th International Conference on World Wide Web, pp. 13–22. ACM (2006)
6. Kalyanpur, A., Parsia, B., Horridge, M., Sirin, E.: Finding All Justifications of OWL DL Entailments. Springer, Heidelberg (2007)
7. Bishop, C.M.: Pattern Recognition and Machine Learning. Springer, New York (2006)
8. Kingdon, J.: AI fights money laundering. IEEE Intell. Syst. **19**(3), 87–89 (2004)
9. Yu, L., Chen, H., Wang, S., Lai, K.K.: Evolving least squares support vector machines for stock market trend mining. IEEE Trans. Evol. Comput. **13**(1), 87–102 (2009)
10. Joachims, T.: Text categorization with Support Vector Machines: learning with many relevant features. In: Nédellec, C., Rouveirol, C. (eds.) ECML 1998. LNCS, vol. 1398, pp. 137–142. Springer, Heidelberg (1998). doi:10.1007/BFb0026683
11. Schrank, D., Eisele, B.: 2012 urban mobility report (2012). http://goo.gl/Ke2xU
12. Koslowsky, M., Krausz, M.: On the relationship between commuting, stress symptoms, and attitudinal measures: a LISREL application. J. Appl. Behav. Sci. **29**(4), 485–492 (1993)
13. Bando, M., Hasebe, K., Nakayama, A., Shibata, A., Sugiyama, Y.: Dynamical model of traffic congestion and numerical simulation. Phys. Rev. E **51**, 1035–1042 (1995)
14. Lecue, F., Pan, J.Z.: Consistent knowledge discovery from evolving ontologies. In: AAAI (2015)
15. Lécué, F., Pan, J.Z.: Predicting knowledge in an ontology stream. In: IJCAI (2013)
16. Li, L., Horrocks, I.: A software framework for matchmaking based on semantic web technology. In: WWW, pp. 331–339 (2003)
17. Paolucci, M., Kawamura, T., Payne, T.R., Sycara, K.: Semantic matching of web services capabilities. In: Horrocks, I., Hendler, J. (eds.) ISWC 2002. LNCS, vol. 2342, pp. 333–347. Springer, Heidelberg (2002). doi:10.1007/3-540-48005-6_26
18. Dehaspe, L., Raedt, L.: Mining association rules in multiple relations. In: Lavrač, N., Džeroski, S. (eds.) ILP 1997. LNCS, vol. 1297, pp. 125–132. Springer, Heidelberg (1997). doi:10.1007/3540635149_40
19. Srikant, R., Agrawal, R.: Mining generalized association rules. In: VLDB, pp. 407–419 (1995)

20. Pan, J.Z., Vetere, G., Gomez-Perez, J.M., Wu, H.: Exploiting Linked Data and Knowledge Graphs for Large Organisations. Springer, Heidelberg (2016)
21. Pan, J., Staab, S., Amann, U., Ebert, J., Zhao, Y.E.: Ontology-Driven Software Development. Springer, Heidelberg (2012)
22. Kotoulas, S., Lopez, V., Lloyd, R., Sbodio, M.L., Lécué, F., Stephenson, M., Daly, E.M., Bicer, V., Gkoulalas-Divanis, A., Lorenzo, G.D., Schumann, A., Aonghusa, P.M.: SPUD - semantic processing of urban data. J. Web Sem. **24**, 11–17 (2014). doi:10.1016/j.websem.2013.12.003
23. Baader, F., Calvanese, D., McGuinness, D.L., Nardi, D., Patel-Schneider, P.F. (eds.): The Description Logic Handbook: Theory, Implementation, and Applications. Cambridge University Press, Cambridge (2003). ISBN 0-521-78176-0
24. Lécué, F., Tallevi-Diotallevi, S., Hayes, J., Tucker, R., Bicer, V., Sbodio, M.L., Tommasi, P.: Smart traffic analytics in the semantic web with STAR-CITY: scenarios, system and lessons learned in Dublin city. J. Web Sem. **27**, 26–33 (2014). doi:10.1016/j.websem.2014.07.002
25. Lécué, F., Tucker, R., Bicer, V., Tommasi, P., Tallevi-Diotallevi, S., Sbodio, M.: Predicting Severity of Road Traffic Congestion Using Semantic Web Technologies. In: Presutti, V., d'Amato, C., Gandon, F., d'Aquin, M., Staab, S., Tordai, A. (eds.) ESWC 2014. LNCS, vol. 8465, pp. 611–627. Springer, Heidelberg (2014). doi:10.1007/978-3-319-07443-6_41
26. Lécué, F., Pan, J.Z.: Predicting knowledge in an ontology stream. In: IJCAI 2013, Proceedings of the 23rd International Joint Conference on Artificial Intelligence, Beijing, China, 3–9 August 2013 (2013). http://www.aaai.org/ocs/index.php/IJCAI/IJCAI13/paper/view/6608
27. Lécué, F.: Diagnosing changes in an ontology stream: a DL reasoning approach. In: Proceedings of the Twenty-Sixth AAAI Conference on Artificial Intelligence, Toronto, Ontario, Canada, 22–26 July 2012 (2012). http://www.aaai.org/ocs/index.php/AAAI/AAAI12/paper/view/4988
28. Lécué, F., Schumann, A., Sbodio, M.L.: Applying semantic web technologies for diagnosing road traffic congestions. In: Cudré-Mauroux, P., et al. (eds.) ISWC 2012. LNCS, vol. 7650, pp. 114–130. Springer, Heidelberg (2012). doi:10.1007/978-3-642-35173-0_8
29. Lécué, F.: Towards scalable exploration of diagnoses in an ontology stream. In: Proceedings of the Twenty-Eighth AAAI Conference on Artificial Intelligence, Québec City, Québec, Canada, 27–31 July 2014, pp. 87–93 (2014). http://www.aaai.org/ocs/index.php/AAAI/AAAI14/paper/view/8194
30. Lécué, F., et al.: Semantic traffic diagnosis with STAR-CITY: architecture and lessons learned from deployment in Dublin, Bologna, Miami and Rio. In: Mika, P., et al. (eds.) ISWC 2014. LNCS, vol. 8797, pp. 292–307. Springer, Heidelberg (2014). doi:10.1007/978-3-319-11915-1_19

Author Index

Printed in the United States
By Bookmasters